Glioma Signaling

Advances in Experimental Medicine and Biology

For further volumes:
http://www.springer.com/series/5584

Jolanta Barańska
Editor

Glioma Signaling

Springer

Editor
Jolanta Barańska
Biochemistry
Nencki Institute of Experimental
Biology
Warsaw, Poland

ISSN 0065-2598
ISBN 978-94-024-0627-6 ISBN 978-94-007-4719-7 (eBook)
DOI 10.1007/978-94-007-4719-7
Springer Dordrecht Heidelberg New York London

Softcover re-print of the Hardcover 1st edition 2013

Printed on acid-free paper

Springer is part of Springer Science+Business Media (www.springer.com)

Preface

The book *Glioma Signaling* reflects the need to bring together data covering the wide spectrum of molecular mechanisms responsible for glioma physiology and pathobiology. It concentrates on regulation of glioma growth, progression and invasion, areas where studies have reached the stage of critical evaluation.

The book is divided in two parts. The first half (Chaps. 1, 2, 3, 4, 5 and 6) is focused on the mechanism of nucleotide receptor signaling. It is introduced by Chap. 1, in which Geoffrey Burnstock presents nucleotide signaling in the brain. Burnstock was the first who, in the 1970s, discovered the role of ATP as a transmitter in the nervous system and termed a large family of the responding receptors "purinergic". Since it was later shown that not only adenine but also uridine extracellular nucleotides act on the same class of receptors, they are presently termed "nucleotide receptors" although the previous designation is still used. Chapter 1 introduces the reader to the nucleotide signaling in the brain, both in neurons and glial cells, its involvement in co-transmission, neuron-glia interaction and pathobiology, including gliomas.

Chapters 2 and 3 are focused on P1 and P2 nucleotide receptor signaling in glioma cells, respectively. In Chap. 2, adenosine signaling by A_1, A_{2A}, A_{2B} and A_3 subtypes of P1 receptor and pathways controlling survival and death of astrocytoma cells are presented. The A_3 receptor, highly expressed in a variety of cancer cell lines and characterized by strong pro-apoptotic effects, is suggested as a possible new target for chemotherapy of brain tumors. Chapter 3 dwells on the expression and functionality of P2 receptors activated by extracellular ATP, ADP, UTP and UDP, and describes the molecular mechanisms of $P2Y_1$ and $P2Y_{12}$ cross-talk, proposed to be involved in cell survival, proliferation and up-regulated growth, in glioma C6 cells.

The next three chapters concentrate on special aspects of nucleotide receptor activity in glioma cells, such as generation of calcium signals (Chap. 4), cross-talk with ecto-enzyme activity (Chap. 5) and modulation of actin cytoskeleton dynamics (Chap. 6). Moreover, Chap. 4 compares the nature of calcium signals in glioma cells with that in astrocytes and microglia, whereas Chap. 6 presents the peculiar role of the $P2Y_2$ receptor in the cross-talk between pathways that regulate C6 cell cytoskeleton. Chapter 5 shows the effect of ATP and AMP degradation by ecto-nucleotidases on tumor growth. In comparison to astrocytes, gliomas have low ATPase and high AMPase activity. Such alterations in extracellular ATP metabolism, associated with P1 and P2 nucleotide receptor activity, are suggested to play an important role in glioma progression.

In the second half of the book (Chaps. 7, 8, 9, 10 and 11) authors' attention is concentrated on signaling induced by growth factors, cytokines and cannabinoids and their role in glioma pathobiology. Chapters 7 and 8 provide new data concerning signaling determinants and functions of receptor tyrosine kinases in glioma cell invasion, describe signaling elements stimulated by integrins (Chap. 7), and discuss in details the role of different growth factor receptors and multiple kinase inhibitors in glioma invasion (Chap. 8). Both chapters present recent advances in the understanding of the molecular mechanism controlling glioma invasion and suggest personalized anti-invasion therapies for genetically distinct subgroups of patients.

Chapter 9 deals with transforming growth factor beta (TGFβ), a multifunctional cytokine, activating membrane serine/threonine kinase receptors and leading to the activation of the cytoplasmic transcription factor, SMAD. TGFβ plays a special role in oncogenesis – being a strong proliferation inhibitor for endothelial cells and astrocytes, it acts as a tumor suppressor on many tumor cells, especially high-grade gliomas. STAT (signal transducer and activator of transcription) protein family is another group of cytoplasmic transcription factors, which function as downstream effectors of cytokine and growth factor receptor signaling. The mechanisms, complexity and dysfunction of STAT signaling in gliomas and its role in pathobiology are described in Chap. 10.

The book ends with a description of cannabinoid signaling at the molecular level. Cannabinoids are currently considered as anti-cancer agents. Chapter 11 describes anti-survival, pro-apoptotic and anti-proliferative effects of cannabinoids on glioma cells and suggests their high therapeutic potential.

Thus, *Glioma Signaling* summarizes a number of subjects of interest to both scientific and clinical investigators. Since glioma is a group of brain tumors very refractory to conventional treatments, such as surgery, radiotherapy and chemotherapy, the search for new directions in glioma therapy is still needed. Data presented in this book give some hope that investigations in the area of molecular biology and neurobiology, concerning mechanisms of intracellular signaling responsible for glioma pathogenesis, up-regulated growth, progression and invasion, will benefit glioma patients.

With grateful thanks to all authors for their cooperation and to Springer SBM for encouraging me to undertake the effort of preparing this book.

Jolanta Barańska
Editor

Contents

Contributors

Maria P. Abbracchio Laboratory of Molecular and Cellular Pharmacology of Purinergic Transmission, Department of Pharmacological and Biomolecular Sciences, Università degli Studi di Milano, Milan, Italy

Jolanta Barańska Nencki Institute of Experimental Biology, Polish Academy of Sciences, Warsaw, Poland

Ana Maria Oliveira Battastini Departamento de Bioquímica, ICBS, Universidade Federal do Rio Grande do Sul, Porto Alegre, RS, Brazil

Elizandra Braganhol Centro de Ciências Químicas, Farmacêuticas e de Alimentos (CCQFA), Universidade Federal de Pelotas (UFPel), Capão do Leão, RS, Brazil

Geoffrey Burnstock Autonomic Neuroscience Centre, University College Medical School, London, UK

Stefania Ceruti Laboratory of Molecular and Cellular Pharmacology of Purinergic Transmission, Department of Pharmacological and Biomolecular Sciences, Università degli Studi di Milano, Milan, Italy

Iwona Ciechomska Laboratory of Transcription Regulation, Department of Cell Biology, Nencki Institute of Experimental Biology, Polish Academy of Sciences, Warsaw, Poland

Aleksandra Ellert-Miklaszewska Laboratory of Transcription Regulation, Department of Cell Biology, Nencki Institute of Experimental Biology, Polish Academy of Sciences, Warsaw, Poland

Jun-ichiro Hamada Department of Neurosurgery, Division of Neuroscience, Graduate School of Medical Science, Kanazawa University, Kanazawa, Ishikawa, Japan

Yutaka Hayashi Department of Neurosurgery, Division of Neuroscience, Graduate School of Medical Science, Kanazawa University, Kanazawa, Ishikawa, Japan

Bozena Kaminska Laboratory of Transcription Regulation, Department of Cell Biology, Nencki Institute of Experimental Biology, Polish Academy of Sciences, Warsaw, Poland

Magdalena Kijewska Laboratory of Transcription Regulation, Department of Cell Biology,, Nencki Institute of Experimental Biology, Polish Academy of Sciences, Warsaw, Poland

Daisuke Kita Department of Neurosurgery, Division of Neuroscience, Graduate School of Medical Science, Kanazawa University, Kanazawa, Ishikawa, Japan

Wanda Kłopocka Nencki Institute of Experimental Biology, Polish Academy of Sciences, Warsaw, Poland

Marta Kocyk Laboratory of Transcription Regulation, Department of Cell Biology, Nencki Institute of Experimental Biology, Polish Academy of Sciences, Warsaw, Poland

Postgraduate School of Molecular Medicine, Warsaw Medical University, Warsaw, Poland

Jarosław Korczyński Nencki Institute of Experimental Biology, Polish Academy of Sciences, Warsaw, Poland

Aneta Kwiatkowska Center for Oncology and Cell Biology, Laboratory for Brain Tumor Biology, Feinstein Institute for Medical Research, Manhasset, NY, USA

Guido Lenz Departamento de Biofísica, IB e Centro de Biotecnologia, Universidade Federal do Rio Grande do Sul, Porto Alegre, RS, Brazil

Mitsutoshi Nakada Department of Neurosurgery, Division of Neuroscience, Graduate School of Medical Science, Kanazawa University, Kanazawa, Ishikawa, Japan

Paweł Pomorski Laboratory of Transcription Regulation, Department of Cell Biology, Nencki Institute of Experimental Biology, Polish Academy of Sciences, Warsaw, Poland

Ilya V. Pyko Department of Neurosurgery, Division of Neuroscience, Graduate School of Medical Science, Kanazawa University, Kanazawa, Ishikawa, Japan

Karolina Swiatek-Machado Nencki Institute of Experimental Biology, Polish Academy of Sciences, Warsaw, Poland

Marc Symons Center for Oncology and Cell Biology, Laboratory for Brain Tumor Biology, Feinstein Institute for Medical Research, Manhasset, NY, USA

Lei Teng Department of Neurosurgery, Division of Neuroscience, Graduate School of Medical Science, Kanazawa University, Kanazawa, Ishikawa, Japan

Department of Neurosurgery, The First Clinical College of Harbin Medical University, Harbin, Heilongjiang Province, People's Republic of China

Takuya Watanabe Department of Neurosurgery, Division of Neuroscience, Graduate School of Medical Science, Kanazawa University, Kanazawa, Ishikawa, Japan

Márcia Rosângela Wink Departamento de Ciências Básicas da Saúde, Universidade Federal de Ciências da Saúde de Porto Alegre, Porto Alegre, RS, Brazil

Dorota Wypych Nencki Institute of Experimental Biology, Polish Academy of Sciences, Warsaw, Poland

Laboratory of Molecular and Cellular Pharmacology of Purinergic Transmission, Department of Pharmacological and Biomolecular Sciences, Università degli Studi di Milano, Milan, Italy

Chapter 1
Introduction to Purinergic Signalling in the Brain

Geoffrey Burnstock

Abstract ATP is a cotransmitter with glutamate, noradrenaline, GABA, acetylcholine and dopamine in the brain. There is a widespread presence of both adenosine (P1) and P2 nucleotide receptors in the brain on both neurons and glial cells. Adenosine receptors play a major role in presynaptic neuromodulation, while P2X ionotropic receptors are involved in fast synaptic transmission and synaptic plasticity. P2Y G protein-coupled receptors are largely involved in presynaptic activities, as well as mediating long-term (trophic) signalling in cell proliferation, differentiation and death during development and regeneration. Both P1 and P2 receptors participate in neuron-glial interactions. Purinergic signalling is involved in control of cerebral vascular tone and remodelling and has been implicated in learning and memory, locomotor and feeding behaviour and sleep. There is increasing interest in the involvement of purinergic signalling in the pathophysiology of the CNS, including trauma, ischaemia, epilepsy, neurodegenerative diseases, neuropsychiatric and mood disorders, and cancer, including gliomas.

Keywords ATP • Adenosine • Purinoceptors • Cotransmission • Neuromodulation • Glia • Neuron-gial interactions • Trophic signalling • Memory • Sleep • Neurodegenerative diseases • Gliomas

Abbreviations

ACh	Acetylcholine
ADP	Adenosine diphosphate
ATP	Adenosine 5′-triphosphate
bFGF	Basic fibroblast growth factor
CNS	Central nervous system
CREB	cAMP response element-binding protein
E-NPPs	Ecto-nucleotide pyrophosphatase/phosphodiesterases
E-NTPDases	Ecto-nucleoside triphosphate diphosphohydrolase
GABA	γ-Amino butyric acid

G. Burnstock (✉)
Autonomic Neuroscience Centre, University College Medical School,
Royal Free Campus, Rowland Hill Street, NW3 2PF London, UK
e-mail: g.burnstock@ucl.ac.uk

J. Barańska (ed.), *Glioma Signaling*, Advances in Experimental Medicine and Biology 986,
DOI 10.1007/978-94-007-4719-7_1, © Springer Science+Business Media Dordrecht 2013

IL-6	Interleukin-6
NA	Noradrenaline
UDP	Uridine diphosphate
UTP	Uridine 5′-triphosphate
TMZ	Temozolomide

1.1 Introduction

The concept of purinergic neurotransmission was proposed in 1972 (Burnstock 1972), after it was shown that adenosine 5′-triphosphate (ATP) was a transmitter in non-adrenergic, non-cholinergic inhibitory nerves in the guinea-pig taenia coli. Subsequently ATP was identified as a co-transmitter in sympathetic and parasympathetic nerves (Burnstock 1976) and it is now recognised that ATP acts as either the sole transmitter or a co-transmitter in most nerves in both the peripheral nervous system and central nervous system (CNS) (see Burnstock 2007a). Since 1992, there has been an explosion of interest in purinergic transmission in the different regions of the brain and spinal cord (Burnstock 2007a; North and Verkhratsky 2006). Various purinergic receptor subtypes have been shown to be widely distributed throughout the CNS, being present in neurones and glia (see Abbracchio and Burnstock 1998). It is now well established that ATP acts both as a fast excitatory neurotransmitter or neuromodulator and has potent long-term (trophic) roles in cell proliferation, differentiation and death in development and regeneration, as well as in disease (Abbracchio and Burnstock 1998; Zimmermann 2006).

In 1976, purinergic receptors were first defined (Burnstock 1976) and 2 years later a basis for distinguishing two types of purinoceptor, identified as P1 and P2 (for adenosine and ATP/adenosine diphosphate [ADP], respectively) was proposed (Burnstock 1978). At about the same time, two subtypes of the P1 (adenosine) receptor were recognised (Londos et al. 1980; van Calker et al. 1979), but it was not until 1985 that a proposal suggesting a pharmacological basis for distinguishing two types of P2 receptor (P2X and P2Y) was made (Burnstock and Kennedy 1985). A year later, two further P2 purinoceptor subtypes were identified, namely, a P2T receptor selective for ADP on platelets and a P2Z receptor on macrophages (Gordon 1986). Further subtypes followed, perhaps the most important being the P2U receptor, which could recognize pyrimidines such as uridine 5′-triphosphate (UTP) and uridine diphosphate (UDP) as well as ATP (O'Connor et al. 1991). On the basis of studies of transduction mechanisms (Dubyak 1991), and the cloning of nucleotide receptors (Brake et al. 1994; Lustig et al. 1993; Valera et al. 1994; Webb et al. 1993), it was proposed that purinoceptors should belong to two major families: a P2X family of ligand-gated ion channel receptors and a P2Y family of G protein-coupled receptors (Abbracchio and Burnstock 1994). This nomenclature has been widely adopted and currently seven P2X subunits and eight P2Y receptor subtypes are recognised, including receptors that are sensitive to pyrimidines as well as purines (see Burnstock 2007b and Fig. 1.1).

Exocytotic neuronal vesicular release of ATP is well established (Pankratov et al. 2007) and there is also evidence for vesicular release of ATP from astrocytes, perhaps involving lysosomes (Zhang et al. 2007). Evidence has been provided for additional mechanisms of nucleotide release, including ATP-binding cassette transporters, connexin or pannexin hemichannels, plasmalemmal voltage-dependent anion channels, as well as $P2X_7$ receptors (Dubyak 2006; Scemes et al. 2007). After release, ATP and other nucleotides undergo rapid enzymatic degradation by ectonucleotidases, which is functionally important as ATP metabolites act as physiological ligands for various purinergic receptors (Zimmermann 2006). Ectonucleotidases include the E-NTPDases (ecto-nucleoside triphosphate diphosphohydrolases), E-NPPs (ecto-nucleotide pyrophosphatase/phosphodiesterases), alkaline phosphatases and ecto-5′-nucleotidase. Although generally adenosine is produced by ectoenzymatic

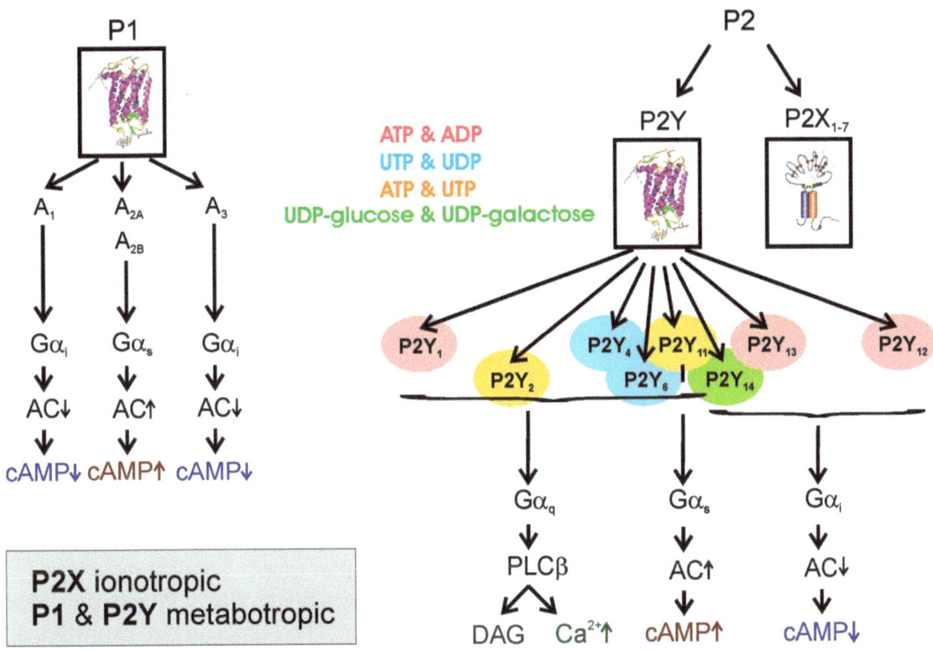

Fig. 1.1 Summary of receptor subtypes for purines and pyrimidines with emphasis on the G proteins involved in metabotropic P1 and P2Y receptors (Courtesy of Jolanta Baranska)

breakdown of ATP, there may be subpopulations of neurones and/or astrocytes that release adenosine directly (Wall and Dale 2007).

1.2 Purinergic Signalling in the CNS

ATP is present in high concentrations within the brain, varying from approximately 2 mM/kg in the cortex to 4 mM/kg in the putamen and hippocampus (Kogure and Alonso 1978). Much is now known about the breakdown of ATP released in the CNS (Kukulski et al. 2004). Cortex and hippocampus synaptic membranes exhibit higher activities of NTPDase1 and NTPDase2 than cerebellum and medulla oblongata, while ecto-5′nucleotidases and adenosine deaminase were found in most brain regions.

Adenosine actions in the CNS have been recognised for many years (see Dunwiddie 1985; Phillis and Wu 1981; Snyder 1985; Williams 1984). However, consideration of the role(s) of ATP in the CNS received less attention until more recently (see Abbracchio 1997; Bo and Burnstock 1994; Burnstock 1996, 2003; Gibb and Halliday 1996; Illes and Alexandre Ribeiro 2004; Illes and Zimmermann 1999; Inoue et al. 1996; Masino and Dunwiddie 2001; North and Verkhratsky 2006; Robertson et al. 2001). In particular, fast purinergic synaptic transmission has been clearly identified in the brain (Khakh 2001). It was first observed in the medial habenula (Edwards et al. 1992) and has now been described in a number of other areas of the CNS, including spinal cord (Bardoni et al. 1997), locus coeruleus (Nieber et al. 1997), hippocampus (Mori et al. 2001; Pankratov et al. 1999) and somatic-sensory cortex (Pankratov et al. 2002). Electron microscopic immunocytochemical studies support these functional experiments. Although adenosine, following ectoenzymatic breakdown of ATP, is the predominant, presynaptic modulator of transmitter release in the CNS (see Dunwiddie 1985), ATP itself

Table 1.1 Purinergic cotransmitters in the brain

Central nervous system	Cotransmitters
Cortex, caudate nucleus	ATP+ACh
Hypothalamus, locus ceruleus	ATP+NA
Hypothalamus, dorsal horn, retina	ATP+GABA
Mesolimbic system	ATP+DA
Hippocampus, dorsal horn	ATP+glutamate

can also act presynaptically (Cunha and Ribeiro 2000). A strong case is made for coordinated purinergic regulatory systems in the CNS controlling local network behaviours by regulating the balance between the effects of ATP, adenosine and ectonucleotidases on synaptic transmission (Kato et al. 2004; Matsuoka and Ohkubo 2004).

In situ hybridisation of P2 receptor subtype mRNA and immunohistochemistry of receptor subtype proteins have been carried out in recent years to show wide, but heterogeneous distribution in the CNS of both P2X receptors (Burnstock and Knight 2004; Kanjhan et al. 1999; Llewellyn-Smith and Burnstock 1998; Loesch and Burnstock 1998; Rubio and Soto 2001) and P2Y receptors (Burnstock 2003; Moore et al. 2000; Moran-Jimenez and Matute 2000). $P2X_2$, $P2X_4$ and $P2X_6$ receptors are widespread in the brain and often form heteromultimers. $P2X_1$ receptors are found in some regions such as cerebellum and $P2X_3$ receptors in the brain stem. $P2X_7$ receptors are probably largely pre-junctional. $P2Y_1$ receptors are also abundant and widespread in the brain. The hippocampus expresses all P2X receptor subtypes and $P2Y_1$, $P2Y_2$, $P2Y_4$, $P2Y_6$ and $P2Y_{12}$ receptors.

Evidence has been presented that nucleotides can act synergistically with growth factors to regulate trophic events (Neary et al. 1994; Rathbone et al. 1999). However, a recent paper has shown that ATP can also stimulate neurite outgrowth from neuroblastoma cells independent of nerve growth factor (Lakshmi and Joshi 2006).

1.2.1 Cotransmission

Evidence for purinergic cotransmission in the CNS has lagged behind that presented for purinergic cotransmission in the periphery (see Burnstock 2009). However, in the last few years a number of such studies have been reported (see Table 1.1).

Release of ATP from synaptosomal preparations and slices from discrete areas of the rat and guinea-pig brain including cortex, hypothalamus, medulla, and habenula, has been measured (Barberis and McIlwain 1976; Sperlagh et al. 1998; White 1977). In cortical synaptosomes, a proportion of the ATP appears to be coreleased with acetylcholine (ACh), and a smaller proportion with noradrenaline (NA) (Potter and White 1980). In preparations of affinity-purified cholinergic nerve terminals from the rat caudate nucleus, ATP and ACh are coreleased (Richardson and Brown 1987). There is evidence for corelease of ATP with catecholamines from neurons in the locus coeruleus (Poelchen et al. 2001) and hypothalamus (Buller et al. 1996; Sperlagh et al. 1998). Purinergic and adrenergic agonist synergism for vasopressin and oxytocin release from hypothalamic supraoptic neurons is consistent with ATP cotransmission in the hypothalamus (Kapoor and Sladek 2000). Corelease of ATP with γ-amino butyric acid (GABA) has been demonstrated in the rabbit retina (Perez and Bruun 1987) and in dorsal horn and lateral hypothalamic neurons (Jo and Role 2002). There is evidence for corelease of ATP with glutamate in the hippocampus (Mori et al. 2001) as well as widespread and pronounced modulatory effects of ATP on glutamatergic mechanisms (Illes et al. 2001). A recent study has shown that in central neuronal terminals, ATP is primarily stored and released from a distinct pool of vesicles and that the release of ATP is not synchronized either with the cotransmitters GABA or glutamate (Pankratov et al. 2006). Cooperativity between extracellular ATP and N-methyl-D-aspartate receptors in long-term potentiation induction in hippocampal CA1 neurons (Fujii et al. 2004) is consistent with ATP/glutamate cotransmission. Colocalisation of functional nicotinic and ionotropic nucleotide receptors has also been identified in isolated cholinergic synaptic terminals in midbrain (Diaz-Hernandez et al. 2002). Interactions between $P2X_2$ and both $\alpha_4\beta_4$ and $\alpha_4\beta_2$ nicotinic receptor channels have been shown in oocyte expression studies (Khakh et al. 2005). There is indirect evidence supporting the possibility that dopamine and ATP are cotransmitters in the CNS (Krugel et al. 2003). After cerebellar lesions in rats producing axotomy of mossy and climbing fibre systems, nitrergic and purinergic systems were activated with similar time courses on pre-cerebellar stations (Viscomi et al. 2004). This raises the possibility that, as in a subpopulation of neurons in the gut, nitric oxide and ATP are cotransmitters.

1.2.2 *Glial Cells*

Multiple P1 and P2 receptor subtypes are expressed by astrocytes, oligodendrocytes and microglia (see Burnstock and Knight 2004). The patterns of expression are complex and can change with physiological and developmental conditions. Adenosine stimulates glutamate release from astrocytes via A_{2A} receptors (Nishizaki 2004). A_3 receptors mediate chemokine CCL2 synthesis in cultured mouse astrocytes (Wittendorp et al. 2004). Astrocytes in the cortex and cerebellum express $P2Y_{13}$ as well as $P2Y_1$ and $P2X_2$ receptors (Carrasquero et al. 2005). NTPDase2 is the dominant ectonucleotidase expressed by rat astrocytes (Wink et al. 2006).

ATP participates in both short-term calcium signalling events and in long-term proliferation, differentiation and death of glia (Cotrina et al. 2000). Both adenosine and ATP induce astroglial cell proliferation and the formation of reactive astrocytes (Neary et al. 1996). ATP and basic fibroblast growth factor (bFGF) signals merge at the mitogen-activated protein kinase cascade, and this integration may underlie the synergistic interactions of ATP and bFGF in astrocytes. Activation of adenosine A_{2B} receptors in astroglioma cells has been shown to increase interleukin-6 (IL-6) mRNA and IL-6 protein synthesis. Blockade of A_{2A} receptors prevents bFGF-induced reactive astrogliosis in rat striated primary astrocytes (Brambilla et al. 2003). Extracellular nucleotide signalling has also been identified in adult neural stem cells (Mishra et al. 2006).

Release of ATP through connexin hemichannels in astrocytes has been reported (Stout et al. 2002), although vesicular release has also been described (Bowser and Khakh 2007; Coco et al. 2003; Montana et al. 2006; Pankratov et al. 2006). It has also been suggested that $P2X_7$ receptor pores may directly mediate efflux of cytosolic ATP, glutamate and GABA from glial cells in the CNS (Duan and Neary 2006). Calcium rises in rat cortical astrocytes are mediated by $P2Y_1$ and $P2X_7$ receptors, but additional P2 receptors ($P2X_2$, $P2X_4$ $P2X_5$, $P2Y_2$, $P2Y_4$ and $P2Y_{14}$) may also contribute (Fumagalli et al. 2003). Another study has shown that cultured astrocytes are able to release UTP either at rest or following hypoxia and that $P2Y_2$ receptor mRNA increased by 2-fold during glucose-oxygen deprivation (Ballerini et al. 2006). $P2Y_2$ and $P2Y_4$ receptors are strongly expressed in glial endfeet apposed to blood vessel walls (Paemeleire and Leybaert 2000).

1.2.3 *Neuron-Glial Interactions*

Purinergic signalling is emerging as a major means of integrating functional activity between neurons, glial and vascular cells in the CNS. These interactions mediate effects of neural activity, in development and in association with neurodegeneration, myelination, inflammation and cancer (see Abbracchio and Burnstock 1998; Fields and Burnstock 2006). New findings from purinergic research began to

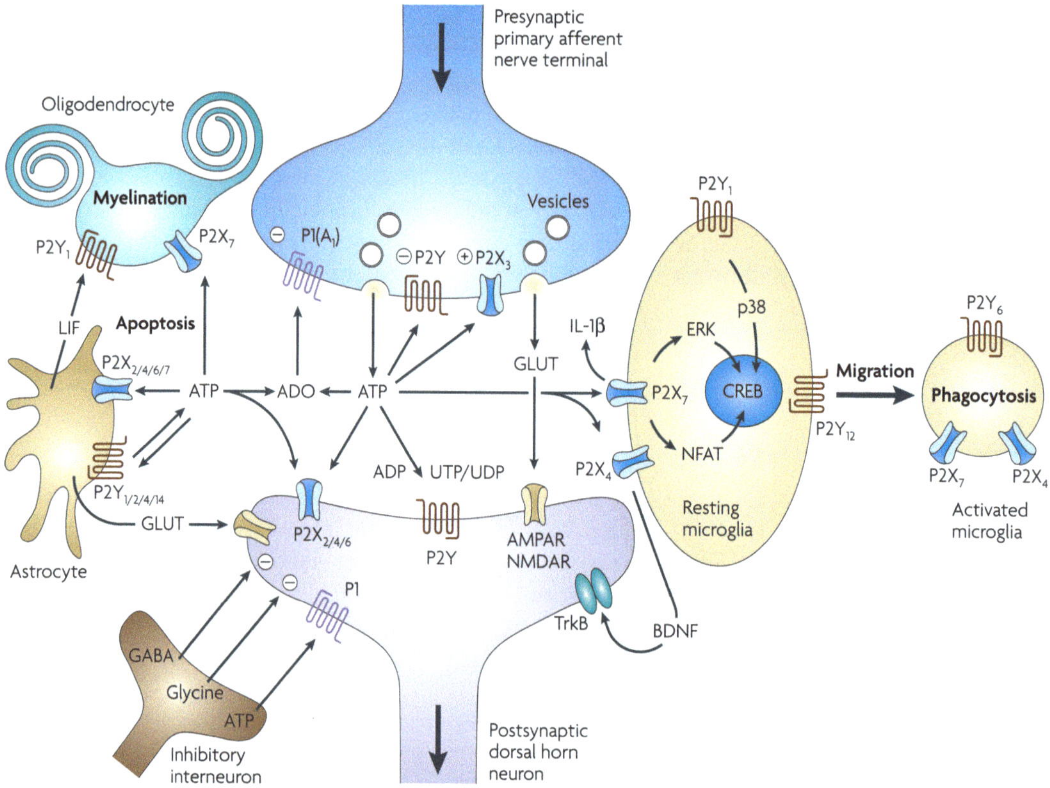

Fig. 1.2 Purinergic signalling in the spinal cord. Presynaptic primary afferent nerve terminals in the dorsal horn of the spinal cord are depicted releasing both glutamate (GLUT) and ATP as cotransmitters by exocytosis. The released ATP acts postsynaptically on $P2X_{2/4/6}$ and on various P2Y receptor subtypes activated by ADP, UTP and UDP, as well as ATP. Glutamate acts postsynaptically on α-amino-3-hydroxy-5-methyl-4-isoxazole propionic acid receptors (AMPARs) and/or *N*-methyl-d-aspartate receptors (NMDARs). ATP is broken down by ectonucleotidase to adenosine (ADO), which acts as a presynaptic inhibitory modulator through $P1(A_1)$ receptors, but ATP itself can act presynaptically either to inhibit the release of transmitter through P2Y receptors or to enhance the release of glutamate through $P2X_3$ receptors. ATP is also released from astrocytes (and probably also from microglia) together with glutamate to participate in glial–neuron interactions. Both P2X and P2Y receptor subtypes are expressed by astrocytes. Leukaemia inhibiting factor (LIF) released by astrocytes in response to ATP promotes myelination in oligodendrocytes and re-myelination through $P2Y_1$ receptors. $P2X_7$ receptors on oligodendrocytes mediate apoptosis. Resting microglia express $P2X_4$ and $P2X_7$ receptors involved in neuropathic pain. ATP, through $P2X_7$ receptors, promotes IL-1β release. Occupation of $P2X_4$ receptors leads to release of brain-derived neurotrophic factor (BDNF) to act on TrkB receptors expressed by neurons in the pain pathway. Occupation of $P2X_7$ receptors leads, through ERK and/or nuclear factor of activated T cells (NFAT), to activation of the transcription factor cAMP response element-binding protein (CREB), whereas $P2Y_1$ receptors also activate CREB, but through p38 signalling. $P2Y_{12}$ receptors on resting microglia mediate cell migration after injury, whereas $P2Y_6$ receptors that are expressed on the activated amoeboid microglia mediate phagocytosis of debris at the site of damage. Inhibitory interneurons that corelease γ-aminobutyric acid (GABA), glycine and ATP modulate the nociceptive pathway (Reproduced from Burnstock (2008). With permission from the Nature Publishing Group)

converge with glial research as it became more widely appreciated that ATP was co-released from synaptic vesicles and thus accessible to perisynaptic glia, while ATP released from glial cells could also act on neurons. This common currency for cell-cell communication opened the possibility of an intercellular signalling system that could unite glia and neurons functionally. A schematic illustrating the neuron-glial interactions is shown in Fig. 1.2.

1.3 Purinergic Signalling in Normal Behaviour

Involvement of purinergic signalling in neurotransmission and neuromodulation in the CNS is now well established, but there are relatively few studies of the involvement of purinergic signalling in behavioural pathways, apart from brainstem control of autonomic functions, although behavioural changes have been reported in pathological situations (see Burnstock 2007a). ATP and adenosine are involved in mechanisms of synaptic plasticity and memory formation (Wieraszko and Ehrlich 1994). The hypnotic/sedative (somnogenic) actions of adenosine are well known as are the central stimulant actions of methylxanthine antagonists (see Basheer et al. 2004; Dunwiddie and Masino 2001; Feldberg and Sherwood 1954). Adenosine, acting through A_1 receptors, is an endogenous, homeostatic sleep factor, mediating the sleepiness that follows prolonged wakefulness. The basal forebrain, as well as neurons in the cholinergic latero-dorsal tegmental nuclei, are essential areas for mediating the sleep inducing effects of adenosine by inhibition of wake-promoting neurons (Arrigoni et al. 2003). It has been suggested that adenosine may promote sleep by blocking inhibitory inputs on ventrolateral preoptic area sleep-active neurons (Mooradian et al. 1994). A_{2A} receptors in the subarachnoid space below the rostral forebrain, activating cells in the nucleus accumbens that increase activity of ventrolateral preoptic area neurons, may also play a role in the somnogenic effect of adenosine (Scammell et al. 2001).

The central inhibitory effects of adenosine on spontaneous locomotor activity of rodents and antagonism by caffeine have been known for some time (e.g., Barraco et al. 1983; Snyder et al. 1981). Later A_{2A} receptors on the nucleus accumbens were shown to mediate locomotor depression (Barraco et al. 1993). Modulation of striatal A_1 and A_2 receptor-mediated activity induces rotational behaviour in response to dopaminergic stimulation in intact rats (Popoli et al. 1994). Interactions between adenosine and L-type Ca^{2+} channels in the locomotor activity of rat were demonstrated (Eroglu et al. 1996). A predominant role for A_1 receptors in the motor-activity effects of acutely administered caffeine in rats has been reported (Antoniou et al. 2005). A combination of A_1 and A_{2A} receptor blocking agents induces caffeine-like spontaneous locomotor activity in mice (Kuzmin et al. 2006). It has been reported that ATP continuously modulates the cerebellar circuit by increasing the inhibitory input to Purkinje neurons, probably via $P2X_5$ and $P2Y_2$ and/or $P2Y_4$ receptor subtypes, thus decreasing the main cerebellar output activity, which contributes to locomotor coordination (Brockhaus et al. 2004). $P2X_2$ receptor immunoreactivity in the cerebellum was demonstrated and claimed to be consistent with a role for extracellular ATP acting as a fast transmitter in motor learning and coordination of movement (Kanjhan et al. 1996).

Adenosine given centrally can result in a decrease in food intake (Levine and Morley 1982). In the striatum, extracellular ATP and adenosine are involved in the regulation of the feeding-associated mesolimbic neuronal activity in an antagonistic manner (Kittner et al. 2004). It has been reported that feeding behaviour relies on tonic activation of A_{2A} receptors in the nucleus accumbens in rats (Nagel et al. 2003). NTPDase3 and 5′-ectonucleotidase regulate the levels of adenosine involved in feeding behaviour in rat brain (Belcher et al. 2006). Enhanced food intake after stimulation of hypothalamic $P2Y_1$ receptors in rats has been described (Kittner et al. 2006). Both adenosine and ATP have been implicated in mood and motivation behaviour (Fredholm 1995; Judelson et al. 2005; Krugel et al. 2004; Williams 1987).

1.4 Purinergic Pathophysiology in the CNS, Including Gliomas

There is a rapidly growing literature about the involvement of purinergic signalling in most disorders of the CNS, such as neurodegeneration diseases, including Alzheimer's, Parkinson's and Huntington's diseases and multiple sclerosis, cerebral ischaemia, migraine, neuropsychiatric and mood disorders (see Burnstock 2008 and Fig. 1.1), as well as cancer.

Glioma, the focus of this book, is a general term for malignant tumours of glial cells and includes astrocytomas, oligodendrogliomas, medulloblastomas, Schwannomas, ependymonas and glioblastomas. Ectonucleotidases were recognised early on the membranes of gliomas and detailed studies have been carried out since then (see Braganhol et al. 2009). Rat glioma C6 cells have been widely used for studies of gliomas. Adenosine uptake and ATP release from C6 cells has been demonstrated (Sinclair et al. 2000). Several P2Y receptor subtypes have been identified in C6 cells including $P2Y_1$, $P2Y_2$, $P2Y_{12}$ and perhaps $P2Y_{14}$ receptors, and implicated in the proliferation and differentiation of these cells, while $P2X_4$ and $P2X_7$ receptors also appear to be present, perhaps involved in their cell death (see Claes and Slegers 2004; Suplat-Wypych et al. 2010; Tamajusuku et al. 2010). In contrast, adenosine, after breakdown of ATP, acting via A_1 and A_{2A} and perhaps A_3 receptors, inhibits growth of C6 cells. ATP stimulates chemokine production in C6 glioma cells, where it was suggested it enhances tumour cell mobility and promotes recruitment of microglia into the developing tumours, thereby supporting tumour growth (Jantaratnotai et al. 2009). Temozolomide (TMZ) is a DNA damaging agent, which is widely used for treating primary and recurrent high-grade gliomas. It has been shown that TMZ induces an autophagy-associated ATP surge in U251 cells that protects them and may contribute to drug resistance (Katayama et al. 2007). Carnosine inhibits growth of cells isolates from human malignant glioma and recently carnosine has been shown to inhibit ATP production in both cells from freshly resected gliomas and from the T98G human glioma cell line (Renner et al. 2010).

Glioblastomas are mainly undifferentiated anaplastic cells. This is the most aggressive type of brain tumour derived from glial cells. It includes, astrocytomas, which are differentiated cells and astrogliomas, which are undifferentiated cells. ATP, acting via P2Y receptors, increased $[Ca^{2+}]_i$ in primary cultures of human glioblastoma cells and adenosine attenuated growth of mouse glioblastoma G1361 cells acting via A_1 receptors on microglia (Synowitz et al. 2006). High concentrations of ATP, probably acting via $P2Y_1$ receptors, stimulate proliferation of human astrocytoma cells. $P2Y_2$, $P2Y_6$, $P2Y_{12}$ and $P2Y_{14}$ receptors, as well as $P2X_7$ receptors, have been shown to be expressed in various astrocytoma cell lines, $P2Y_{12}$ receptors being crucial for regulating cell proliferation and differentiation, while $P2X_7$ receptor activation is associated with increase in proinflammatory factors and tumour cell migration (Ho et al. 2004; Wei et al. 2008).

References

Abbracchio MP (1997) ATP in brain function. In: Jacobson KA, Jarvis MF (eds) Purinergic approaches in experimental therapeutics. Wiley-Liss, New York

Abbracchio MP, Burnstock G (1994) Purinoceptors: are there families of P2X and P2Y purinoceptors? Pharmacol Ther 64:445–475

Abbracchio MP, Burnstock G (1998) Purinergic signalling: pathophysiological roles. Jpn J Pharmacol 78:113–145

Antoniou K, Papadopoulou-Daifoti Z, Hyphantis T, Papathanasiou G, Bekris E, Marselos M, Panlilio L, Muller CE, Goldberg SR, Ferre S (2005) A detailed behavioral analysis of the acute motor effects of caffeine in the rat: involvement of adenosine A1 and A2A receptors. Psychopharmacology (Berl) 183:154–162

Arrigoni E, Chamberlin NL, Saper CB, McCarley RW (2003) The effects of adenosine on the membrane properties of basal forebrain cholinergic neurons. Sleep 26:45–50

Ballerini P, Di Iorio P, Caciagli F, Rathbone MP, Jiang S, Nargi E, Buccella S, Giuliani P, D'Alimonte I, Fischione G, Masciulli A, Romano S, Ciccarelli R (2006) P2Y2 receptor up-regulation induced by guanosine or UTP in rat brain cultured astrocytes. Int J Immunopathol Pharmacol 19:293–308

Barberis C, McIlwain H (1976) 5′-Adenine mononucleotides in synaptosomal preparations from guinea pig neocortex: their change on incubation, superfusion and stimulation. J Neurochem 26:1015–1021

Bardoni R, Goldstein PA, Lee CJ, Gu JG, MacDermott AB (1997) ATP P2X receptors mediate fast synaptic transmission in the dorsal horn of the rat spinal cord. J Neurosci 17:5297–5304

Barraco RA, Coffin VL, Altman HJ, Phillis JW (1983) Central effects of adenosine analogs on locomotor activity in mice and antagonism of caffeine. Brain Res 272:392–395

Barraco RA, Martens KA, Parizon M, Normile HJ (1993) Adenosine A2a receptors in the nucleus accumbens mediate locomotor depression. Brain Res Bull 31:397–404

Basheer R, Strecker RE, Thakkar MM, McCarley RW (2004) Adenosine and sleep-wake regulation. Prog Neurobiol 73:379–396
Belcher SM, Zsarnovszky A, Crawford PA, Hemani H, Spurling L, Kirley TL (2006) Immunolocalization of ecto-nucleoside triphosphate diphosphohydrolase 3 in rat brain: implications for modulation of multiple homeostatic systems including feeding and sleep-wake behaviors. Neuroscience 137:1331–1346
Bo X, Burnstock G (1994) Distribution of [3H]alpha, beta-methylene ATP binding sites in rat brain and spinal cord. Neuroreport 5:1601–1604
Bowser DN, Khakh BS (2007) Vesicular ATP is the predominant cause of intercellular calcium waves in astrocytes. J Gen Physiol 129:485–491
Braganhol E, Morrone FB, Bernardi A, Huppes D, Meurer L, Edelweiss MI, Lenz G, Wink MR, Robson SC, Battastini AM (2009) Selective NTPDase2 expression modulates in vivo rat glioma growth. Cancer Sci 100:1434–1442
Brake AJ, Wagenbach MJ, Julius D (1994) New structural motif for ligand-gated ion channels defined by an ionotropic ATP receptor. Nature 371:519–523
Brambilla R, Cottini L, Fumagalli M, Ceruti S, Abbracchio MP (2003) Blockade of A2A adenosine receptors prevents basic fibroblast growth factor-induced reactive astrogliosis in rat striatal primary astrocytes. Glia 43:190–194
Brockhaus J, Dressel D, Herold S, Deitmer JW (2004) Purinergic modulation of synaptic input to Purkinje neurons in rat cerebellar brain slices. Eur J Neurosci 19:2221–2230
Buller KM, Khanna S, Sibbald JR, Day TA (1996) Central noradrenergic neurons signal via ATP to elicit vasopressin responses to haemorrhage. Neuroscience 73:637–642
Burnstock G (1972) Purinergic nerves. Pharmacol Rev 24:509–581
Burnstock G (1976) Purinergic receptors. J Theor Biol 62:491–503
Burnstock G (1978) A basis for distinguishing two types of purinergic receptor. In: Straub RW, Bolis L (eds) Cell membrane receptors for drugs and hormones: a multidisciplinary approach. Raven Press, New York
Burnstock G (1996) Purinergic neurotransmission. Semin Neurosci 8:171–174
Burnstock G (2003) Purinergic receptors in the nervous system. In: Schwiebert EM (ed) Purinergic receptors and signalling. Academic, San Diego
Burnstock G (2007a) Physiology and pathophysiology of purinergic neurotransmission. Physiol Rev 87:659–797
Burnstock G (2007b) Purine and pyrimidine receptors. Cell Mol Life Sci 64:1471–1483
Burnstock G (2008) Purinergic signalling and disorders of the central nervous system. Nat Rev Drug Discov 7:575–590
Burnstock G (2009) Purinergic cotransmission. Exp Physiol 94:20–24
Burnstock G, Kennedy C (1985) Is there a basis for distinguishing two types of P2-purinoceptor? Gen Pharmacol 16:433–440
Burnstock G, Knight GE (2004) Cellular distribution and functions of P2 receptor subtypes in different systems. Int Rev Cytol 240:31–304
Carrasquero LM, Delicado EG, Jimenez AI, Perez-Sen R, Miras-Portugal MT (2005) Cerebellar astrocytes co-express several ADP receptors. Presence of functional P2Y(13)-like receptors. Purinergic Signal 1:153–159
Claes P, Slegers H (2004) P2Y receptor activation affects the proliferation and differentiation of glial and neuronal cells: a focus on rat C6 glioma cells. Curr Neuropharmacol 2:207–220
Coco S, Calegari F, Pravettoni E, Pozzi D, Taverna E, Rosa P, Matteoli M, Verderio C (2003) Storage and release of ATP from astrocytes in culture. J Biol Chem 278:1354–1362
Cotrina ML, Lin JH, Lopez-Garcia JC, Naus CC, Nedergaard M (2000) ATP-mediated glia signaling. J Neurosci 20:2835–2844
Cunha RA, Ribeiro JA (2000) ATP as a presynaptic modulator. Life Sci 68:119–137
Diaz-Hernandez M, Pintor J, Castro E, Miras-Portugal MT (2002) Co-localisation of functional nicotinic and ionotropic nucleotide receptors in isolated cholinergic synaptic terminals. Neuropharmacology 42:20–33
Duan S, Neary JT (2006) P2X(7) receptors: properties and relevance to CNS function. Glia 54:738–746
Dubyak GR (1991) Signal transduction by P2-purinergic receptors for extracellular ATP. Am J Respir Cell Mol Biol 4:295–300
Dubyak GR (2006) ATP release mechanisms. In: Burnstock G, Arnett TR (eds) Edited monograph: nucleotides and regulation of bone cell function. Taylor & Francis, Boca Raton
Dunwiddie TV (1985) The physiological role of adenosine in the central nervous system. Int Rev Neurobiol 27:63–139
Dunwiddie TV, Masino SA (2001) The role and regulation of adenosine in the central nervous system. Annu Rev Neurosci 24:31–55
Edwards FA, Gibb AJ, Colquhoun D (1992) ATP receptor-mediated synaptic currents in the central nervous system. Nature 359:144–147
Eroglu L, Tuna R, Caglayan B (1996) Effects of nifedipine and Bay K 8644 on the R-PIA and caffeine-induced changes in the locomotor activity of rats. Pharmacol Res 33:141–144
Feldberg W, Sherwood SL (1954) Injections of drugs into the lateral ventricle of the cat. J Physiol 123:148–167

Fields RD, Burnstock G (2006) Purinergic signalling in neuron-glia interactions. Nat Rev Neurosci 7:423–436
Fredholm BB (1995) Purinoceptors in the nervous system. Pharmacol Toxicol 76:228–239
Fujii S, Sasaki H, Mikoshiba K, Kuroda Y, Yamazaki Y, Mostafa Taufiq A, Kato H (2004) A chemical LTP induced by co-activation of metabotropic and N-methyl-D-aspartate glutamate receptors in hippocampal CA1 neurons. Brain Res 999:20–28
Fumagalli M, Brambilla R, D'Ambrosi N, Volonte C, Matteoli M, Verderio C, Abbracchio MP (2003) Nucleotide-mediated calcium signaling in rat cortical astrocytes: role of P2X and P2Y receptors. Glia 43:218–203
Gibb AJ, Halliday FC (1996) Fast purinergic transmission in the central nervous system. Semin Neurosci 8:225–232
Gordon JL (1986) Extracellular ATP: effects, sources and fate. Biochem J 233:309–319
Ho MCK, Simon J, Barnard EA, Wong YH (2004) Atypical regulation of calcium signals in astrocytoma 1321N1 cells expressing the purinergic P2Y12 receptor. J Neurochem 88:84
Illes P, Alexandre Ribeiro J (2004) Molecular physiology of P2 receptors in the central nervous system. Eur J Pharmacol 483:5–17
Illes P, Zimmermann H (1999) Nucleotides and their receptors in the nervous system. Prog Brain Res 120:1–432
Illes P, Wirkner K, Nörenberg W, Masino SA, Dunwiddie TV (2001) Interaction between the transmitters ATP and glutamate in the central nervous system. Drug Dev Res 52:76–82
Inoue K, Koizumi S, Ueno S (1996) Implication of ATP receptors in brain functions. Prog Neurobiol 50:483–492
Jantaratnotai N, Choi HB, McLarnon JG (2009) ATP stimulates chemokine production via a store-operated calcium entry pathway in C6 glioma cells. BMC Cancer 9:442
Jo YH, Role LW (2002) Coordinate release of ATP and GABA at in vitro synapses of lateral hypothalamic neurons. J Neurosci 22:4794–4804
Judelson DA, Armstrong LE, Sokmen B, Roti MW, Casa DJ, Kellogg MD (2005) Effect of chronic caffeine intake on choice reaction time, mood, and visual vigilance. Physiol Behav 85:629–634
Kanjhan R, Housley GD, Thorne PR, Christie DL, Palmer DJ, Luo L, Ryan AF (1996) Localization of ATP-gated ion channels in cerebellum using P2x2R subunit-specific antisera. Neuroreport 7:2665–2669
Kanjhan R, Housley GD, Burton LD, Christie DL, Kippenberger A, Thorne PR, Luo L, Ryan AF (1999) Distribution of the P2X2 receptor subunit of the ATP-gated ion channels in the rat central nervous system. J Comp Neurol 407:11–32
Kapoor JR, Sladek CD (2000) Purinergic and adrenergic agonists synergize in stimulating vasopressin and oxytocin release. J Neurosci 20:8868–8875
Katayama M, Kawaguchi T, Berger MS, Pieper RO (2007) DNA damaging agent-induced autophagy produces a cytoprotective adenosine triphosphate surge in malignant glioma cells. Cell Death Differ 14:548–558
Kato F, Kawamura M, Shigetomi E, Tanaka J, Inoue K (2004) ATP- and adenosine-mediated signaling in the central nervous system: synaptic purinoceptors: the stage for ATP to play its "dual-role". J Pharmacol Sci 94:107–111
Khakh BS (2001) Molecular physiology of P2X receptors and ATP signalling at synapses. Nat Rev Neurosci 2:165–174
Khakh BS, Fisher JA, Nashmi R, Bowser DN, Lester HA (2005) An angstrom scale interaction between plasma membrane ATP-gated P2X2 and alpha4beta2 nicotinic channels measured with fluorescence resonance energy transfer and total internal reflection fluorescence microscopy. J Neurosci 25:6911–6920
Kittner H, Krugel U, Hoffmann E, Illes P (2004) Modulation of feeding behaviour by blocking purinergic receptors in the rat nucleus accumbens: a combined microdialysis, electroencephalographic and behavioural study. Eur J Neurosci 19:396–404
Kittner H, Franke H, Harsch JI, El-Ashmawy IM, Seidel B, Krugel U, Illes P (2006) Enhanced food intake after stimulation of hypothalamic P2Y1 receptors in rats: modulation of feeding behaviour by extracellular nucleotides. Eur J Neurosci 24:2049–2056
Kogure K, Alonso OF (1978) A pictorial representation of endogenous brain ATP by a bioluminescent method. Brain Res 154:273–284
Krugel U, Kittner H, Franke H, Illes P (2003) Purinergic modulation of neuronal activity in the mesolimbic dopaminergic system in vivo. Synapse 47:134–142
Krugel U, Spies O, Regenthal R, Illes P, Kittner H (2004) P2 receptors are involved in the mediation of motivation-related behavior. Purinergic Signal 1:21–29
Kukulski F, Sevigny J, Komoszynski M (2004) Comparative hydrolysis of extracellular adenine nucleotides and adenosine in synaptic membranes from porcine brain cortex, hippocampus, cerebellum and medulla oblongata. Brain Res 1030:49–56
Kuzmin A, Johansson B, Gimenez L, Ogren SO, Fredholm BB (2006) Combination of adenosine A1 and A2A receptor blocking agents induces caffeine-like locomotor stimulation in mice. Eur Neuropsychopharmacol 16:129–136
Lakshmi S, Joshi PG (2006) Activation of Src/kinase/phospholipase C/mitogen-activated protein kinase and induction of neurite expression by ATP, independent of nerve growth factor. Neuroscience 141:179–189
Levine AS, Morley JE (1982) Purinergic regulation of food intake. Science 217:77–79

Llewellyn-Smith IJ, Burnstock G (1998) Ultrastructural localization of P2X3 receptors in rat sensory neurons. Neuroreport 9:2545–2550
Loesch A, Burnstock G (1998) Electron-immunocytochemical localization of P2X1 receptors in the rat cerebellum. Cell Tissue Res 294:253–260
Londos C, Cooper DM, Wolff J (1980) Subclasses of external adenosine receptors. Proc Natl Acad Sci U S A 77:2551–2554
Lustig KD, Shiau AK, Brake AJ, Julius D (1993) Expression cloning of an ATP receptor from mouse neuroblastoma cells. Proc Natl Acad Sci U S A 90:5113–5117
Masino SA, Dunwiddie TV (2001) Role of purines and pyrimidines in the central nervous system. In: Abbracchio MP, Williams M (eds) Purinergic and pirimidinergic signalling. Springer, Berlin
Matsuoka I, Ohkubo S (2004) ATP- and adenosine-mediated signaling in the central nervous system: adenosine receptor activation by ATP through rapid and localized generation of adenosine by ecto-nucleotidases. J Pharmacol Sci 94:95–99
Mishra SK, Braun N, Shukla V, Fullgrabe M, Schomerus C, Korf HW, Gachet C, Ikehara Y, Sevigny J, Robson SC, Zimmermann H (2006) Extracellular nucleotide signaling in adult neural stem cells: synergism with growth factor-mediated cellular proliferation. Development 133:675–684
Montana V, Malarkey EB, Verderio C, Matteoli M, Parpura V (2006) Vesicular transmitter release from astrocytes. Glia 54:700–715
Mooradian AD, Grabau G, Bastani B (1994) Adenosine triphosphatases of rat cerebral microvessels. Effect of age and diabetes mellitus. Life Sci 55:1261–1265
Moore D, Chambers J, Waldvogel H, Faull R, Emson P (2000) Regional and cellular distribution of the P2Y(1) purinergic receptor in the human brain: striking neuronal localisation. J Comp Neurol 421:374–384
Moran-Jimenez MJ, Matute C (2000) Immunohistochemical localization of the P2Y(1) purinergic receptor in neurons and glial cells of the central nervous system. Brain Res Mol Brain Res 78:50–58
Mori M, Heuss C, Gahwiler BH, Gerber U (2001) Fast synaptic transmission mediated by P2X receptors in CA3 pyramidal cells of rat hippocampal slice cultures. J Physiol 535:115–123
Nagel J, Schladebach H, Koch M, Schwienbacher I, Muller CE, Hauber W (2003) Effects of an adenosine A2A receptor blockade in the nucleus accumbens on locomotion, feeding, and prepulse inhibition in rats. Synapse 49:279–286
Neary JT, Whittemore SR, Zhu Q, Norenberg MD (1994) Synergistic activation of DNA synthesis in astrocytes by fibroblast growth factors and extracellular ATP. J Neurochem 63:490–494
Neary JT, Rathbone MP, Cattabeni F, Abbracchio MP, Burnstock G (1996) Trophic actions of extracellular nucleotides and nucleosides on glial and neuronal cells. Trends Neurosci 19:13–18
Nieber K, Poelchen W, Illes P (1997) Role of ATP in fast excitatory synaptic potentials in locus coeruleus neurones of the rat. Br J Pharmacol 122:423–430
Nishizaki T (2004) ATP- and adenosine-mediated signaling in the central nervous system: adenosine stimulates glutamate release from astrocytes via A2a adenosine receptors. J Pharmacol Sci 94:100–102
North RA, Verkhratsky A (2006) Purinergic transmission in the central nervous system. Pflugers Arch 452:479–485
O'Connor SE, Dainty IA, Leff P (1991) Further subclassification of ATP receptors based on agonist studies. Trends Pharmacol Sci 12:137–141
Paemeleire K, Leybaert L (2000) ATP-dependent astrocyte-endothelial calcium signaling following mechanical damage to a single astrocyte in astrocyte-endothelial co-cultures. J Neurotrauma 17:345–358
Pankratov Y, Lalo U, Castro E, Miras-Portugal MT, Krishtal O (1999) ATP receptor-mediated component of the excitatory synaptic transmission in the hippocampus. Prog Brain Res 120:237–249
Pankratov Y, Lalo U, Krishtal O, Verkhratsky A (2002) Ionotropic P2X purinoreceptors mediate synaptic transmission in rat pyramidal neurones of layer II/III of somato-sensory cortex. J Physiol 542:529–536
Pankratov Y, Lalo U, Verkhratsky A, North RA (2006) Vesicular release of ATP at central synapses. Pflugers Arch 452:589–597
Pankratov Y, Lalo U, Verkhratsky A, North RA (2007) Quantal release of ATP in mouse cortex. J Gen Physiol 129:257–265
Perez MT, Bruun A (1987) Colocalization of (3H)-adenosine accumulation and GABA immunoreactivity in the chicken and rabbit retinas. Histochemistry 87:413–417
Phillis JW, Wu PH (1981) The role of adenosine and its nucleotides in central synaptic transmission. Prog Neurobiol 16:187–239
Poelchen W, Sieler D, Wirkner K, Illes P (2001) Co-transmitter function of ATP in central catecholaminergic neurons of the rat. Neuroscience 102:593–602
Popoli P, Pezzola A, de Carolis AS (1994) Modulation of striatal adenosine A1 and A2 receptors induces rotational behaviour in response to dopaminergic stimulation in intact rats. Eur J Pharmacol 257:21–25
Potter P, White TD (1980) Release of adenosine 5′-triphosphate from synaptosomes from different regions of rat brain. Neuroscience 5:1351–1356

Rathbone MP, Middlemiss PJ, Gysbers JW, Andrew C, Herman MA, Reed JK, Ciccarelli R, Di Iorio P, Caciagli F (1999) Trophic effects of purines in neurons and glial cells. Prog Neurobiol 59:663–690
Renner C, Asperger A, Seyffarth A, Meixensberger J, Gebhardt R, Gaunitz F (2010) Carnosine inhibits ATP production in cells from malignant glioma. Neurol Res 32:101–105
Richardson PJ, Brown SJ (1987) ATP release from affinity-purified rat cholinergic nerve terminals. J Neurochem 48:622–630
Robertson SJ, Ennion SJ, Evans RJ, Edwards FA (2001) Synaptic P2X receptors. Curr Opin Neurobiol 11:378–386
Rubio ME, Soto F (2001) Distinct localization of P2X receptors at excitatory postsynaptic specializations. J Neurosci 21:641–653
Scammell TE, Gerashchenko DY, Mochizuki T, McCarthy MT, Estabrooke IV, Sears CA, Saper CB, Urade Y, Hayaishi O (2001) An adenosine A2a agonist increases sleep and induces Fos in ventrolateral preoptic neurons. Neuroscience 107:653–663
Scemes E, Suadicani SO, Dahl G, Spray DC (2007) Connexin and pannexin mediated cell-cell communication. Neuron Glia Biol 3:199–208
Sinclair CJ, LaRiviere CG, Young JD, Cass CE, Baldwin SA, Parkinson FE (2000) Purine uptake and release in rat C6 glioma cells: nucleoside transport and purine metabolism under ATP-depleting conditions. J Neurochem 75:1528–1538
Snyder SH (1985) Adenosine as a neuromodulator. Annu Rev Neurosci 8:103–124
Snyder SH, Katims JJ, Annau Z, Bruns RF, Daly JW (1981) Adenosine receptors and behavioral actions of methylxanthines. Proc Natl Acad Sci U S A 78:3260–3264
Sperlagh B, Sershen H, Lajtha A, Vizi ES (1998) Co-release of endogenous ATP and [3H]noradrenaline from rat hypothalamic slices: origin and modulation by alpha2-adrenoceptors. Neuroscience 82:511–520
Stout CE, Costantin JL, Naus CC, Charles AC (2002) Intercellular calcium signaling in astrocytes via ATP release through connexin hemichannels. J Biol Chem 277:10482–10488
Suplat-Wypych D, Dygas A, Baranska J (2010) 2′, 3′-O-(4-benzoylbenzoyl)-ATP-mediated calcium signaling in rat glioma C6 cells: role of the P2Y(2) nucleotide receptor. Purinergic Signal 6:317–325
Synowitz M, Glass R, Farber K, Markovic D, Kronenberg G, Herrmann K, Schnermann J, Nolte C, van Rooijen N, Kiwit J, Kettenmann H (2006) A1 adenosine receptors in microglia control glioblastoma-host interaction. Cancer Res 66:8550–8557
Tamajusuku AS, Villodre ES, Paulus R, Coutinho-Silva R, Battasstini AM, Wink MR, Lenz G (2010) Characterization of ATP-induced cell death in the GL261 mouse glioma. J Cell Biochem 109:983–991
Valera S, Hussy N, Evans RJ, Adami N, North RA, Surprenant A, Buell G (1994) A new class of ligand-gated ion channel defined by P2x receptor for extracellular ATP. Nature 371:516–519
van Calker D, Muller M, Hamprecht B (1979) Adenosine regulates via two different types of receptors, the accumulation of cyclic AMP in cultured brain cells. J Neurochem 33:999–1005
Viscomi MT, Florenzano F, Conversi D, Bernardi G, Molinari M (2004) Axotomy dependent purinergic and nitrergic co-expression. Neuroscience 123:393–404
Wall MJ, Dale N (2007) Auto-inhibition of rat parallel fibre-Purkinje cell synapses by activity-dependent adenosine release. J Physiol 581:553–565
Webb TE, Simon J, Krishek BJ, Bateson AN, Smart TG, King BF, Burnstock G, Barnard EA (1993) Cloning and functional expression of a brain G-protein-coupled ATP receptor. FEBS Lett 324:219–225
Wei W, Ryu JK, Choi HB, McLarnon JG (2008) Expression and function of the P2X(7) receptor in rat C6 glioma cells. Cancer Lett 260:79–87
White TD (1977) Direct detection of depolarisation-induced release of ATP from a synaptosomal preparation. Nature 267:67–68
Wieraszko A, Ehrlich YH (1994) On the role of extracellular ATP in the induction of long-term potentiation in the hippocampus. J Neurochem 63:1731–1738
Williams M (1984) Mammalian central adenosine receptors. In: Lajtha A (ed) Handbook of neurochemistry. Plenum Publishing Corporation, New York
Williams M (1987) Purinergic receptors and central nervous system function. In: Meltzer HY (ed) Psychopharmacology: the third generation of progress. Raven Press, New York
Wink MR, Braganhol E, Tamajusuku AS, Lenz G, Zerbini LF, Libermann TA, Sevigny J, Battastini AM, Robson SC (2006) Nucleoside triphosphate diphosphohydrolase-2 (NTPDase2/CD39L1) is the dominant ectonucleotidase expressed by rat astrocytes. Neuroscience 138:421–432
Wittendorp MC, Boddeke HW, Biber K (2004) Adenosine A3 receptor-induced CCL2 synthesis in cultured mouse astrocytes. Glia 46:410–418
Zhang Z, Chen G, Zhou W, Song A, Xu T, Luo Q, Wang W, Gu XS, Duan S (2007) Regulated ATP release from astrocytes through lysosome exocytosis. Nat Cell Biol 9:945–953
Zimmermann H (2006) Nucleotide signaling in nervous system development. Pflugers Arch 452:573–588

Chapter 2
Adenosine Signaling in Glioma Cells

Stefania Ceruti and Maria P. Abbracchio

Abstract Purines and pyrimidines are fundamental signaling molecules in controlling the survival and proliferation of astrocytes, as well as in mediating cell-to-cell communication between glial cells and neurons in the healthy brain. The malignant transformation of astrocytes towards progressively more aggressive brain tumors (from astrocytoma to anaplastic glioblastoma) leads to modifications in both the survival and cell death pathways which overall confer a growth advantage to malignant cells and resistance to many cytotoxic stimuli. It has been demonstrated, however, that, in astrocytomas, several purinergic (in particular adenosinergic) pathways controlling cell survival and death are still effective and, in some cases, even enhanced, providing invaluable targets for purine-based chemotherapy, that still represents an appropriate pharmacological approach to brain tumors. In this chapter, the current knowledge on both receptor-mediated and receptor-independent adenosine pathways in astrocytomas will be reviewed, with a particular emphasis on the most promising targets which could be translated from *in vitro* studies to *in vivo* pharmacology.

Keywords P1 receptors • A_3 ligands • Astrocytoma • Cladribine • 2-Chloro-adenosine • Caspase-2 • Caspasc-9 • p53 mutations • CD39 • CD73 • Matrix metalloproteinases • Equilibrative nucleoside transporters

Abbreviations

ADA	Adenosine deaminase
Ado	Adenosine
AK	Adenosine kinase
APCP	α,β-Methylene ADP
2-CA	2-Chloro-adenosine
2-CdA	2-Chloro-2′-deoxyadenosine
CI-IB-MECA	2-Chloro-N^6-(3-iodobenzyl)-N-methyl-5′-carbamoyladenosine

S. Ceruti (✉) • M.P. Abbracchio
Laboratory of Molecular and Cellular Pharmacology of Purinergic Transmission,
Department of Pharmacological and Biomolecular Sciences, Università degli Studi di Milano,
9 via Balzaretti, 20133 Milan, Italy
e-mail: stefania.ceruti@unimi.it; mariapia.abbracchio@unimi.it

J. Barańska (ed.), *Glioma Signaling*, Advances in Experimental Medicine and Biology 986,
DOI 10.1007/978-94-007-4719-7_2, © Springer Science+Business Media Dordrecht 2013

CNS	Central Nervous System
8-CPT	8-Cyclo-pentyl-theophylline
DAG	Diacylglycerol
dCyd	2′-Deoxycytidine
ENT	Equilibrative nucleoside transporter
^{18}F-CPFPX	8-Cyclopentyl-3-(3-^{18}F-fluoropropyl)-1-propyl-xanthine
GSK-3β	Glycogen synthase kinase 3β
HIF-1α	Hypoxia-inducible factor 1 α subunit
IFNγ	Interferon-gamma
Ino	Inosine
IP_3	Inositol-1,4,5-trisphosphate
ITub	5-Iodotubercidin
MAP kinases	Mitogen-activated protein kinases
MMP-9	Matrix metalloproteinase-9
MRS1220	N-(9-Chloro-2-furan-2-yl-[1,2,4]triazolo[1,5-c]quinazolin-5-yl)-2-phenylacetamide
MRS1706	N-(4-Acetylphenyl)-2-[4-(2,3,6,7-tetrahydro-2,6-dioxo-1,3-dipropyl-1H-purin-8-yl) phenoxy] acetamide
MTT	3-(4,5-Dimethylthiazol-2-yl)-2,5-diphenyltetrazolium bromide
NECA	N-Ethyl-carboxamide adenosine
NTPDase	Nucleoside triphosphate diphosphohydrolase
PET	Positron emission tomography
PI	Propidium iodide
PI3K	Phosphatidylinositol 3-kinase
PKA	Protein kinase A
PKB/Akt	Protein kinase B
PKC	Protein kinase C
PLC	Phospholipase C
PLD	Phospholipase D
SAH	S-Adenosyl-homocysteine
SAM	S-Adenosyl-methionine
TNFα	Tumour necrosis factor alpha
VEGF	Vascular endothelial growth factor
zIETD-fmk	z-Ile-Glu(OMe)-Thr-Asp(OMe)-fluoromethylketone
zLEHD-fmk	z-Leu-Glu(OMe)-His-Asp(OMe)-fluoromethylketone
zVAD-fmk	z-Val-Ala-Asp(OMe)-fluoromethylketone
zVDVAD-fmk	z-Val-Asp(OMe)-Val-Ala-Asp(OMe)-fluoromethylketone.

2.1 Introduction

Due to their universal role in cell growth, differentiation and death, purinergic mechanisms are intimately involved in the regulation of tumour growth, with both positive and negative influence on cancer growth, progression and metastatisation depending on the cell type, the involvement of specific extracellular receptors and/or intracellular apoptotic pathways (for a recent commentary, see Tapper et al. (2012). Historically, the first evidences highlighting adenosine (Ado) as a regulator of cancer cell growth dates back to the 1940s, when purine derivatives started to be used as cytotoxic agents in hematologic tumours. It was then hypothesized this action to be due to impaired intracellular synthesis of nucleic acids and it was only relatively recently that the involvement of specific P1 receptors for extracellular Ado was demonstrated. Here, we summarize the available evidence implicating Ado and its receptors in the regulation of brain astrocytomas, with special emphasis on the translation of these findings to the cure of this form of cancer. It is worth mentioning that, apart from the A_3 receptor subtype, which has been the object of extensive studies as a new target for chemotherapy of brain tumours (see Sect. 2.4.2) and from the studies on the intracellular effects of Ado in astrocytoma cells (see Sect. 2.5), in many papers astrocytoma cell lines have been utilized merely as *in vitro* models of astrocytes to test the presence and the effects of Ado-mediated signaling. This has to be taken into consideration when translating data from these studies to primary normal astrocytes, which can bear significant differences with respect to malignantly transformed cells.

2.2 Adenosine Metabolism and P1 Adenosine Receptors

Ado is considered rather a neuromodulator than a true neurotransmitter, since it is neither stored as such in synaptic vesicles nor released following membrane depolarization (however, see also Melani et al. 2012). It has been estimated that basal extracellular Ado concentration is in the range of 30–200 nM (Fredholm et al. 2011), and can rapidly increase through two different mechanisms: enzymatic hydrolysis of extracellular ATP by the ecto-5′-nucleotidases CD39 and CD73 or the export via membrane bi-directional equilibrative transporters (ENTs) following its intracellular production (see also below, Fig. 2.1). Since ATP is released not only from neurons through synaptic vesicles, but also from various secretory cells (e.g., platelets, macrophages, endocrine cells; Lazarowski et al. 2011) as well as from damaged or dying cells, a delayed increase in Ado concentrations is always detected following ATP release, and, in general, Ado-mediated effects modulate the excitatory ATP effects in a compensatory inhibitory way.

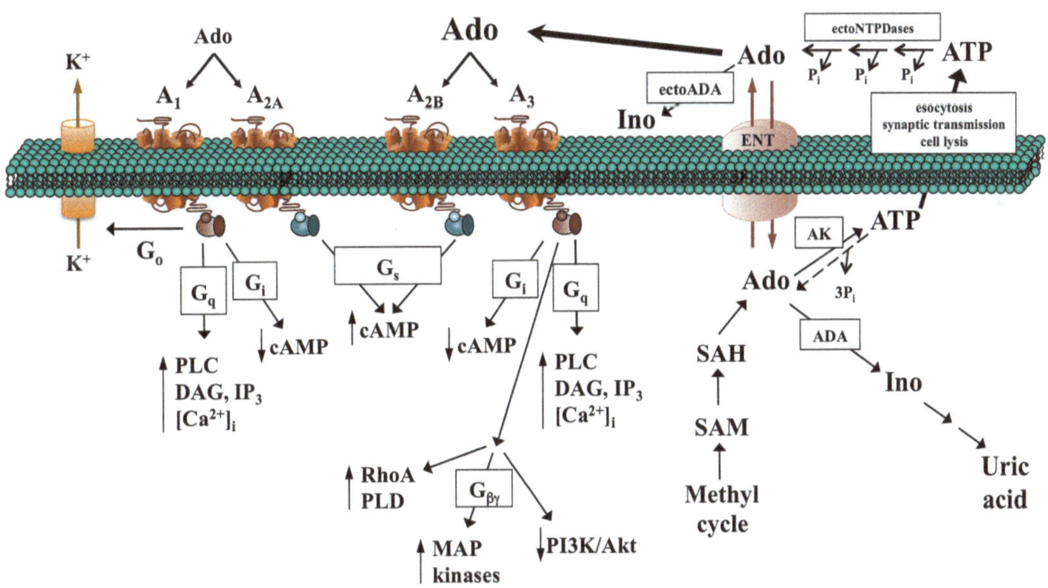

Fig. 2.1 Ado metabolism, Pl Ado receptors and corresponding signaling pathways. Ado can be generated intracellularly after SAH or ATP hydrolysis, and then transported extracellularly through ENT. Extracellular Ado concentrations can also increase thanks to ATP hydrolysis by ectoNTPDases. Low physiological Ado concentrations can activate the A_1AR and A_{2A}AR subtypes, whereas high micromolar concentrations (that can be reached following ischemic or traumatic events) are needed to recruit the A_{2B}AR and A_3AR subtypes. The classical signaling pathways activated by the four P1 receptor subtypes are shown, together with the A_3AR-mediated pathways that have been more recently identified in glioma cells. Ado can also be deaminated by ADA either extracellularly or intracellularly, and finally metabolized to uric acid. See text for details (Sects. 2.2 and 2.4.2)

In the extracellular space, Ado can activate its membrane receptors, collectively referred to as the P1 receptors. Four subtypes of G protein-coupled Ado receptors have been identified, namely the A_1 adenosine receptor (AR), A_{2A}AR, A_{2B}AR and A_3AR subtypes (Fredholm et al. 2011). Ado possesses a high affinity towards the A_1AR and A_{2A}AR subtypes, which can therefore be activated at low physiological Ado concentrations; conversely, only high Ado concentrations can effectively recruit the A_{2B}AR and A_3AR subtypes, which can thus play a key role upon pathological conditions (see also below). The A_1AR and A_3AR subtypes are either coupled to G_i proteins, which negatively regulate adenylyl cyclase activity (Fredholm et al. 2011), or G_q proteins (Linden et al. 1998; Abbracchio et al. 1995a), leading to the hydrolysis of phosphatidylinositol to generate inositol-1,4,5-trisphosphate (IP_3) and diacylglycerol (DAG). IP_3 in turn activates its receptors on the endoplasmic reticulum, and stimulates the release of calcium from the intracellular stores, whereas DAG can activate protein kinase C (PKC) and additional second messenger pathways (Fig. 2.1). Especially in the brain and heart tissues, the A_1 subtype can also directly couple to K^+ channels through the recruitment of a G_o; the opening of these channels leads to the intracellular elevation of K^+ concentrations and to membrane hyperpolarization. This latter action is at the basis of the bradycardic and anti-epileptic actions of Ado (Fredholm 2010). The A_3AR subtype can also couple to various intracellular pathways, which will be discussed in more details in Sect. 2.4.2 below. Conversely, the A_{2A}AR and A_{2B}AR subtypes are coupled to G_s and, consequently, to the activation of adenylyl cyclase and the generation of cAMP (Fredholm et al. 2011). The intracellular positive coupling to adenylyl cyclase is at the basis of the functional antagonism between the A_{2A}AR and the D2 dopamine receptors in the striatum, since the latter inhibits adenylyl cyclase functions through G_i. Alterations in this functional cross-talk between the Ado and dopamine systems plays a crucial role in the development of various neurodegenerative disorders, such as Parkinson's disease, as elegantly reviewed elsewhere (Morelli et al. 2010).

Ado is finally taken up by specific membrane transporters, which can be subdivided into equilibrative (bidirectional) or concentrative (i.e., working against concentration gradient, King et al. 2006), and phosphorylated in the cytosol by adenosine kinase (AK). Since this enzyme can operate only in the presence of high oxygen concentrations, Ado concentrations will remain high upon hypoxic conditions (such as following an ischemic event, at any site of inflammation or within a tumour mass, Vaupel et al. 2001), when no ATP can be generated; therefore, Ado-mediated actions play an important role also upon pathological conditions. Intracellular Ado can also derive from the intracellular hydrolysis of ATP or of S-adenosyl-homocysteine (SAH). The latter represents the final step of the so-called "methyl cycle", which is crucially involved in the synthesis of nucleic acids and in the methylation of proteins (Fig. 2.1). Ado can be also deaminated to inosine (Ino) by Ado deaminase (ADA), and further terminally metabolized to uric acid. Ado deamination and metabolism to uric acid can also occur extracellularly thanks to the presence of ecto-ADA enzymes on the cell surface (Fig. 2.1).

2.3 A Role for Ecto-5′-Nucleotidases CD39 and CD73 in Gliomas?

As mentioned above, Ado can be produced extracellularly from adenine nucleotides by two types of membrane-bound enzymes: CD39, the product of the ectonucleoside triphosphate diphosphohydrolase 1 gene (ENTPD1), which is responsible for the conversion of ATP to ADP and AMP, and CD73, the product of NT5E gene, which converts AMP to adenosine. Several papers have been recently published indicating that scavenging of extracellular ATP and production of adenosine have direct effects on tumour growth, spread of metastases, as well as on the adhesion, migration and homing of both cancer cells and activated immune cells mediating anti-tumour effects (for details, see Tapper et al. (2012).

Recently, Wang et al. have shown that CD73 expression inhibits anti-tumour immune responses and that, conversely, CD73-null mice bearing transplanted non-CNS tumours display a survival benefit when compared to wild type mice (Wang et al. 2011). Twelve days after tumour inoculation, higher numbers of CD8+ immune T cells are found in CD73-null mice. Authors examined in details the role of CD73 in modulating the interaction of tumour infiltrating regulatory CD4+ T cells (tumour Tregs) that can promote metastatic tumour growth, with effector T cells. Importantly, CD73-null effector T cells did not show any changes in mediating anti-tumour responses. CD4+ tumour Tregs directly inhibited CD8+ T cell functions, such as interferon-gamma (IFN-γ) production, and consequently CD8+ T cell-mediated anti-tumour effects in a CD73-dependent manner. This study therefore shows that both local and systemic production of Ado by CD73 (in tandem with CD39) results in decreased effector functions of CD8+ T cells and in a reduction of their homing to tumours. This occur by several means, including both decreased expression of adhesion molecules, direct in vivo modulation of anti-tumour responses and/or possibly induction of apoptosis (in this respect, see also Sect. 2.4). These novel data suggest that both cancer cells and host-derived cells cooperatively mediate tumour immune evasion, and that this occurs in a CD73-dependent manner.

Highly relevant to the present review, overexpression of CD73 has recently also emerged as a component of glioma cell adhesion and tumour cell-extracellular matrix interactions (Cappellari et al. 2012, see also below). In both C6 and U138MG glioma cells, treatment with 1 μM α,β-methylene ADP (APCP), a competitive CD73 inhibitor, caused a 30 % reduction in glioma cell proliferation. In addition, 100 μM Ado increased cell proliferation by 36 %, and treatment with Ado plus inhibitors of its uptake produced an additional and significant increase in cell proliferation. The inhibitory effect on cell proliferation caused by APCP was reverted by co-treatment with Ado uptake inhibitors. AMP (1 and 3 mM) decreased U138MG glioma cell proliferation by 29 and 42 %, respectively (Bavaresco et al. 2008). Taken together, these results suggest the participation of CD73 in cell proliferation, and that this process is dependent upon (i) the enzyme's production of Ado, acting here as a proliferative factor, and, (ii) removal of AMP, a toxic molecule for gliomas. These findings would therefore suggest CD73 as a new target for anti-glioma therapies. However, they are in contrast with other reports indicating that extracellular adenine nucleotides inhibit C6 glioma cell growth via Ado, which is produced by ecto-nucleotidases including CD73 at the extracellular space and then incorporated into cells by the equilibrative nucleoside transporters ENT2 (Ohkubo et al. 2007). Intracellular AMP accumulation by AK after Ado uptake would then induce C6 cell growth inhibition through pyrimidine starvation. It has to be underlined that these contrasting data have been obtained on cell lines in vitro. Therefore, at variance from other non-CNS tumours (Wang et al. 2011), elucidation of the exact role of CD39 and CD73 in gliomas still awaits direct in vivo evaluation in tumour-bearing animals where factors other than cancer cell proliferation (i.e., tumour cells' adhesion, migration and invasiveness as well as immune cell-mediated responses) can be determined (see also Conclusions).

2.4 Receptor-Mediated Effects of Adenosine on Glioma Cell Growth and Survival

The interest in the role of the purinergic system in general, and of Ado in particular, in modulating the growth and survival of various types of tumour cells has grown since the 1990s, based on the demonstration of high extracellular purine concentrations within the tumour mass. For example, concentration of Ado in the extracellular fluid of glioma tissue has been directly measured in 21 patients undergoing surgical removal of the tumour, and has been reported to be in the low micromolar range (Melani et al. 2003), meaning high enough to stimulate all the four P1 receptor subtypes. All the four Ado receptor subtypes have been identified in C6 glioma cells, functionally coupled to the modulation of adenylate cyclase activity, and quantitative real time PCR allowed to rank the expression of Ado receptors as follows: $A_1AR = A_3AR > A_{2A}AR > A_{2B}AR$ (Castillo et al. 2007). Moreover, functional ENTs (Sinclair et al. 2000) and ecto-5′-nucleotidase/CD73 (Cappellari et al. 2012) have been also found expressed by glioma cells (see above), thus suggesting that the whole machinery controlling purine metabolism is present and functional in glioma cells, and therefore cell survival and properties could be deeply influenced by intra- and extracellular Ado concentrations. Indeed, in both C6 glioma and U138MG glioblastoma cells the activity of ecto-5′-nucleotidase/CD73 increased in parallel with cell proliferation to a maximum when cells reached confluence (Bavaresco et al. 2008). This means that ATP hydrolysis to Ado will progressively augment with the growing of the tumour mass, and that Ado-mediated effects on cell growth and proliferation might become progressively more important.

Many studies have been performed on in vitro or in vivo models of cancer of different origins (such as melanoma, leukemia, prostate carcinoma etc.), or on P1 receptor-transfected cells, and they have reported different, and often opposite, effects on cell survival exerted by the various Ado receptor subtypes (Merighi et al. 2003). These contradictory effects are probably due to the various experimental models, leading to a consequent coupling of P1 receptor subtypes to different intracellular pathways controlling cell survival and proliferation depending upon the cell type, but also to the different experimental settings (e.g., hypoxic vs normoxic), and to the concentration of the agonists utilized (see Sect. 2.4.2).

The number of studies aimed at directly analyzing the effects of Ado itself or of Ado agonists on the survival of glioma cells are limited (with the exception of the A_3AR subtype; see Sect. 2.4.2), and will be reviewed in the following section.

2.4.1 A_1AR-, A_{2A}AR-, and A_{2B}AR-Mediated Effects on Glioma Cells

Based on studies performed on solid tumour cells or on transfected cells, the A_1AR subtype is generally believed to exert pro-survival and anti-apopotic actions, primarily based on its ability to activate the MAP kinases ERK1/2 and PKB, whereas the A_{2A}AR subtype can both negatively or positively affect cell survival by recruiting different intracellular signaling pathways, and the same holds true for the A_{2B}AR (Merighi et al. 2003). It is interesting to point out that the A_{2A}AR subtype has been crucially involved in the positive regulation of angiogenesis, by promoting the survival and proliferation of endothelial cells in synergy with VEGF (reviewed in Merighi et al. 2003), and its activation could therefore indirectly boost tumour growth by improving the supply of oxygen and nutrients.

At variance from the general anti-apoptotic role for the A_1AR (see above), induction of caspase-dependent cell death by Ado in RCR-1 astrocytoma cells was supposed to involve both intracellular pathways and the activation of this receptor subtype (Sai et al. 2006). Conclusions, however, appear questionable, mostly due to the extremely high Ado concentrations utilized (up to 1 mM), which more likely recall an intracellular site of action, and to the very low reversal of Ado-mediated effects exerted by the A_1AR antagonist 8-CPT.

A significant upregulation of the A_1AR has been identified in the peritumour area in animals injected with F98 or C6 glioma cells by 3D imaging reconstruction of the tumour mass (Dehnhardt et al. 2007). This upregulation was dependent upon the tumour volume. No significant changes, but rather a tendency to decrease, were instead detected within the tumour itself, possibly due to induction of cell death by necrosis. Although the functional significance of this increase in A_1AR is not known, monitoring the A_1AR expression and receptor presence around tumours could represent a valuable diagnostic and prognostic marker for glioblastoma progression. In fact, this receptor system has been studied by monitoring the binding of the radioactive molecule ^{18}F-CPFPX (a selective A_1 antagonist) by positron emission tomography (PET) *in vivo*, and described as being up-regulated during the cerebral response to glioma invasion in primary human *glioblastoma multiforme* (Bauer et al. 2005).

Specific down-regulation of the A_1AR and a parallel up-regulation of the A_{2A}R subtypes have been observed in C6 glioma cells following induction of hypoxia (Castillo et al. 2008). This effect was mimicked by Ado itself, therefore suggesting that high extracellular Ado concentrations that are generated as a consequence of hypoxic conditions (as observed within a tumour mass; see also below) can auto-regulate Ado-dependent signaling. These data further confirm that the results on receptor recruitment and activation can significantly vary depending upon the experimental setting utilized, and can even become the opposite when shifting from *in vitro* to *in vivo* conditions.

The only paper dealing with a putative role for the $A_{2A}AR$ in controlling glioma survival shows a protective role for this receptor, at least under the experimental conditions utilized. In fact, authors have demonstrated that Ado, at concentrations ranging from 125 to 500 μM, significantly protected C6 glioma cells from NO-dependent cell death induced by the incubation with a mix of cytokines (Isakovic et al. 2008). In this study, the possible involvement of the $A_{2A}R$ (although the Ado concentrations utilized here are extremely high) and of the $A_{2B}AR$ has been only postulated, based on the inability of an inhibitor of ENTs to prevent Ado-mediated effects and on previous data demonstrating that overexpression of the $A_{2A}R$ and inhibition of AK (which in turn increases extracellular Ado concentrations) in C6 glioma cells significantly inhibited NF-κB activation and cytokine-induced iNOS upregulation (Lee et al. 2005; Sands et al. 2004).

Functional $A_{2B}AR$ were identified in human astrocytoma ADF cells, (a cell line directly derived from a cancer patient; Malorni et al. 1994), which underwent desensitization following prolonged agonist exposure due to phophorylation at thereonine residues (Trincavelli et al. 2004). Interestingly, the pro-inflammatory cytokine TNFα potentiated $A_{2B}AR$ coupling and inhibited its desensitization, so that only in its presence could the A_2 receptor agonist NECA promote $A_{2A}AR$-mediated elongation of numerous thin cellular processes (Fig. 2.2; Trincavelli et al. 2004). These morphological changes generally reflect the commitment of astrocytes towards a cell differentiation program, which could be also accompanied by an increased resistance to cell death (see below, Sect. 2.4.2). It should be noted, however, that in cancer cells any stimulus promoting differentiation would conflict with their intrinsic tendency towards uncontrolled proliferation. These contradictory messages could therefore promote the decision of the cell to undergo apoptotic cell death. Thus, activation of the $A_{2B}AR$ in the presence of pro-inflammatory cytokines could prove useful in glioma therapy.

Moreover, in human U373 *glioblastoma multiforme* cells activation of the $A_{2B}AR$ recruited a series of intracellular pathways, including p38 MAP kinase and PKC delta and epsilon, which in turn promoted IL6 synthesis and release (Fiebich et al. 2005). Since this protein has been associated to the development of a malignant and aggressive phenotype in various types of cancer, such as breast carcinoma, caution should be kept in evaluating the final outcome of $A_{2B}AR$ activation in brain tumours. More experimental data are therefore needed to clearly determine the role of this receptor subtype in modulating the survival and proliferation of glioma cells.

2.4.2 The A_3 Receptor Subtype as a New Pharmacological Target for Innovative Chemotherapic Approaches to Gliomas

The A_3AR subtype is the only member of the P1 receptor family that was cloned before its pharmacological characterization. As knowledge on its structure were accumulating, it appeared that this receptor would have been difficult to characterize, mostly based on the low level of interspecies homology, reflecting in highly different pharmacological profiles, especially in terms of antagonist binding and affinities (Gessi et al. 2008). Moreover, also receptor distribution showed significant differences among species, with high receptor expression detected in the lung, spleen, testis, brain, heart and liver (Gessi et al. 2008). Interestingly, a very high levels of A_3AR protein were found in a variety of cancer cell lines, including astrocytoma and glioblastoma, suggesting a role for this receptor as a tumour marker (Gessi et al. 2008) or as a target for new chemotherapeutic approaches (see below).

The analysis of receptor dynamics in ADF astrocytoma cells has led to discover a fast desensitization response to agonist stimulation, followed by agonist-mediated internalization and recycling to the cell surface within 120 min. A more prolonged exposure to a selective A_3 agonist (Cl-IB-MECA) led to a

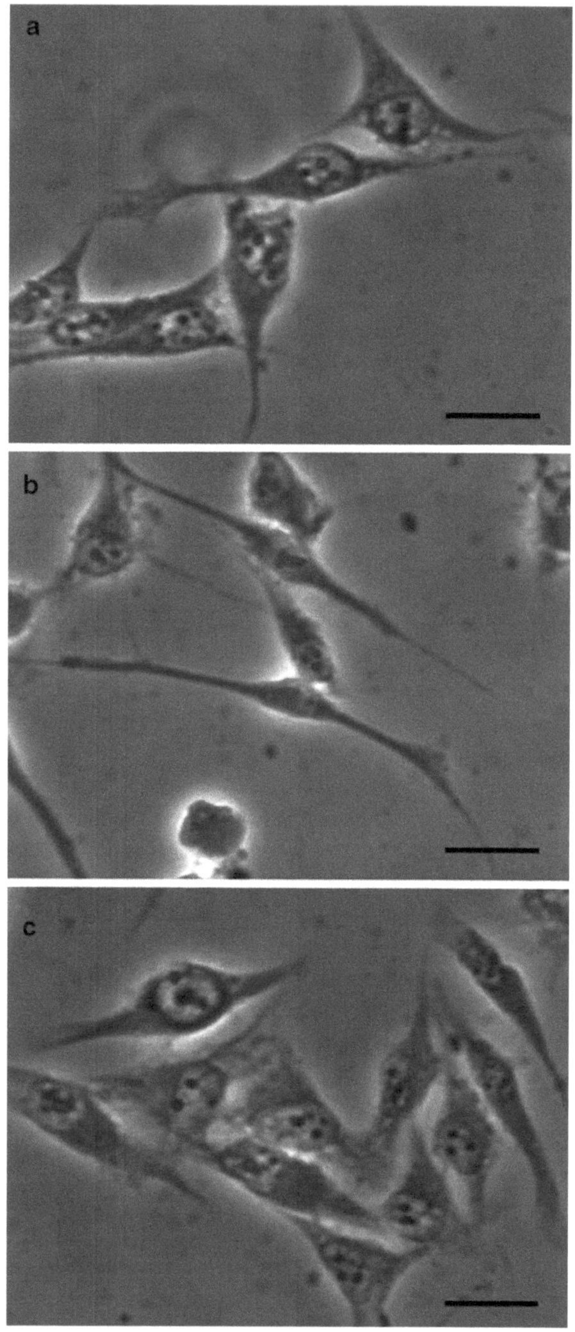

Fig. 2.2 In human astrocytoma ADF cells exposed to TNFα, NECA promotes the emission of long and thin cellular processes through the activation of the A_{2B}AR subtype. (**a**) Typical morphology of ADF cells exposed to TNFα alone for 24 h (control). (**b**) A 30-min incubation of TNFα-treated ADF cells with the non-selective agonist NECA (1 μM) induced clear morphological changes, with the emission of long processes (**c**) The full reversion of NECA-mediated effects was obtained by pre-incubating cells with the A_{2B}AR-selective antagonist MRS1706 (10 nM). Magnification: 32×; scale bars: 30 μm (Reproduced from Trincavelli et al. (2004). With permission from John Wiley and Sons)

significant down-regulation of receptor expression, with slow recovery (Trincavelli et al. 2002). This rapid desensitization kinetics is possibly due to the presence of a higher number of serine and threonine residues at the receptor C-terminal region with respect to A_1AR and A_{2A}AR subtypes (Palmer and Stiles 2000).

Apart from the "classical" second messenger systems (see Sect. 2.2 above), the A_3AR can also couple to the activation of the small G protein RhoA, which in turn stimulates phospholipase D (PLD) (Gessi et al. 2008), and to the MAP kinase signaling pathways through the release of the βγ subunit

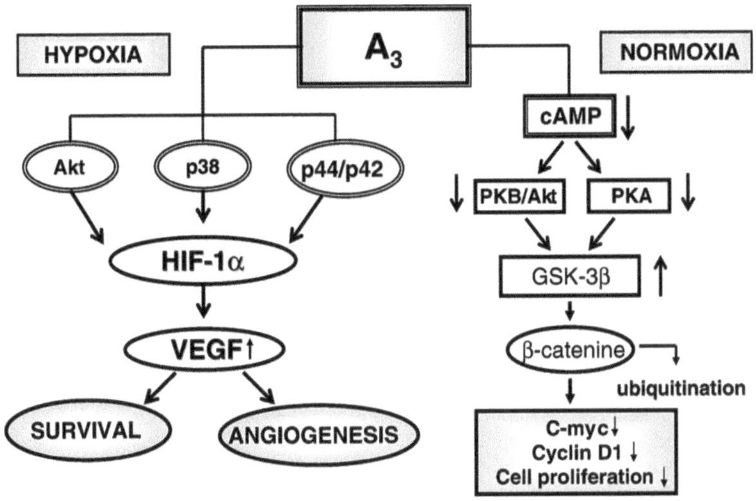

Fig. 2.3 Opposite effects exerted by A_3AR activation in tumour cells upon hypoxic and normoxic conditions. In cancer cells under normoxic conditions, activation of the A_3AR leads to a reduced expression of PKA and PKB, which in turn promotes the activation of GSK-3β, leading to inhibition of cell proliferation. Conversely, upon hypoxic conditions (such as within a solid tumour mass), A_3AR-mediated activation of MAPKinases promotes the synthesis of HIF-1α with a consequent increased cell survival and angiogenesis (Reproduced from Gessi et al. (2008). With permission from Elsevier)

from pertussis toxin-sensitive G_i proteins (Schulte and Fredholm 2000), or to the PI3K/Akt pathway (see Fig. 2.1). Concerning the latter, inhibition of PKA and PKB/Akt was demonstrated in melanoma cells upon activation of the A_3AR; this in turn maintained glycogen synthase kinase 3β (GSK-3β) in its active form, with a consequent deregulation of the Wnt pathway, increase in β-catenin degradation, and finally inhibition of tumour cell proliferation (Fishman et al. 2002). These observations, together with the high expression of the A_3AR in tumour cells and with its pro-apoptotic effects on various lymphoma, melanoma and prostate cell lines in vitro and in mouse xenografts in vivo (Fishman et al. 2002), have led to the hypothesis that A_3AR agonists might be useful as chemotherapic agents. Clinical studies are currently ongoing on the use of CF102 (Cl-IB-MECA) in patients with advanced hepatocellular carcinoma (Fredholm et al. 2011). Taken together, these results suggest an inhibitory action of the A_3AR on cancer cell growth and proliferation.

The opposite, however, seems to hold true for glioblastoma cells. In fact, the selective A_3AR antagonist, MRS1220, blocked Ado-mediated glioma proliferation (Morrone et al. 2003). Upon hypoxic conditions, activation of the A_3AR upregulated the transcription factor HIF-1α, leading to increase in VEGF production, with consequent angiogenesis (Fig. 2.3, Merighi et al. 2006). Hypoxia-induced chemoresistance of human glioblastoma cells was also demonstrated to depend upon the activation of the A_3AR linked to the PKB/Akt pathway, which in turn mediated the phosphorylation and consequent inactivation of the pro-apopototic member of the Bcl2 protein family Bad, leading to cell survival (Merighi et al. 2007). Indeed, upregulation of matrix metalloproteinase-9 (MMP-9) expression and increased tumour migration was also observed upon activation of the A_3AR in human U87MG glioblastoma cells (Gessi et al. 2010). MMPs degrade the extracellular matrix and consequently promote migration of glioma cells towards brain areas protected by an intact blood-brain barrier. Unfortunately, no differences between the migratory biology of transformed and normal brain cells have been identified, which renders this potential target difficult to be exploited (Westphal and Lamszus 2011). Based on the above-mentioned evidence, inhibition of A_3AR functions with selective antagonists could therefore represent an appealing therapeutic approach to solid tumours in general (which are characterized by an hypoxic core) and to gliomas in particular, not only to inhibit MMPs

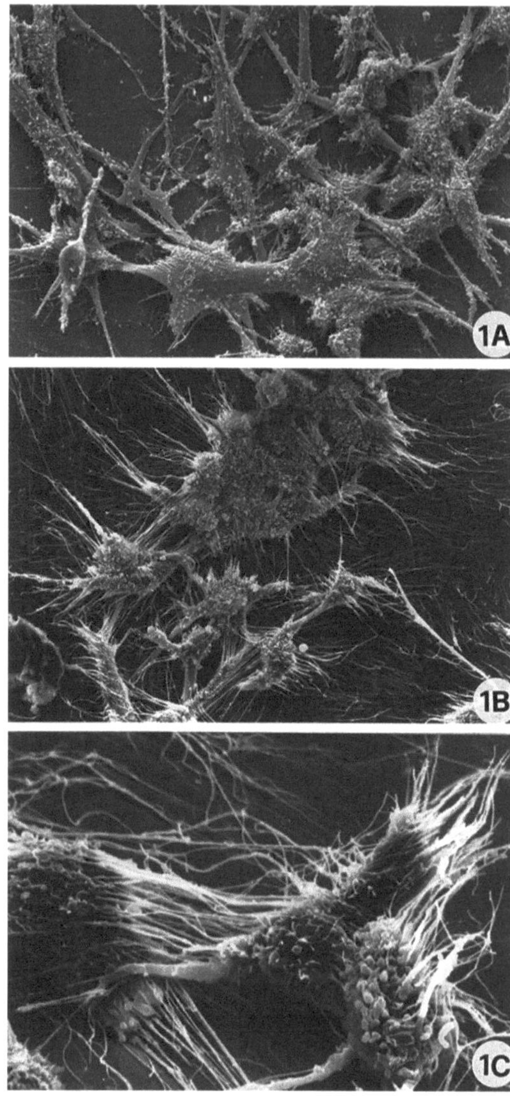

Fig. 2.4 Morphological changes induced by nanomolar concentrations of the A_3AR agonist Cl-IB-MECA in human astrocytoma ADF cells. Scanning electron micrographs of ADF cells grown for 72 h under control conditions (**a**) or in the presence of 100 nM Cl-IB-MECA (**b**, **c**). A typical *bipolar* shape with a relatively low number of cell protrusions was observed in Control cells, whereas exposure to Cl-IB-MECA induced marked morphological changes with an increased of the number and length of cellular processes. Original magnification: **a**, **b** 1000×; **c** 3300× (Reproduced from Abbracchio et al. (1997). With permission from Elsevier)

upregulation, but also to affect cancer cell growth and properties at various molecular levels, thus increasing the possibility of a more effective chemotherapy (Fig. 2.3). In this respect, possible lead compounds could emerge from the development of a series of pyrazol[3,4-d]pyrimidines derivatives with drug-like physicochemical properties, which demonstrated a high affinity and selectivity for the A_3AR and were able to potently inhibit Cl-IB-MECA- and IB-MECA-stimulated glioblastoma cell growth in vitro (Taliani et al. 2010). Taken together, data obtained in various cell cancer models suggest that A_3AR receptor stimulation exerts opposite effects on cancer cell growth and survival when taking place under normoxic or hypoxic situations, with anti-tumour or tumorigenic actions, respectively (Fig. 2.3, Gessi et al. 2008).

In line with the pro-survival role of the A_3AR in glioma cells, our group has demonstrated that in human astrocytoma ADF cells, nanomolar Cl-IB-MECA concentrations produced clear morphological changes (i.e., emission of numerous long filaments, Fig. 2.4), which were accompanied by the appearance of stress fibers, a typical hallmark of cytoskeletal rearrangement, and by the redistribution of the anti-apoptotic protein Bcl-X_L towards cell processes (Abbracchio et al. 1997), through the specific involvement of the small G protein Rho (Abbracchio et al. 2001). Overall, these changes increased the ability of the cell to adhere to the culture substrate, which in turn reflected in a reduced sensitivity to cell death, thus confirming that the A_3AR could contribute to improve cancer cell survival.

An opposite effect was, however, exerted by high Cl-IB-MECA concentrations, since a significant cytotoxicity, reduction of cell number and induction of apoptosis were detected both in C6 glioma (Appel et al. 2001) and in human astrocytoma ADF cells (Abbracchio et al. 1998) upon treatment with high micromolar concentrations of this selective A_3 agonist. These data were in line with previous results showing a dual opposite effect on ischemia-induced tissue damage exerted by either acute or chronic A_3AR agonist administration (reviewed in Jacobson 1998), suggesting an opposite outcome on cell survival depending upon the "intensity" of receptor activation. Although the selective recruitment of the A_3AR subtype by such high agonist concentrations could be questioned, these apparent discrepancies in A_3AR-mediated effects on cell survival are in line with its enigmatic (and yet-to-be fully clarified) role as "Dr. Jekill and Mr. Hyde" (Gessi et al. 2008).

A further confirmation of the possible anti-proliferative role of the A_3AR in glioma comes from a study showing a significant upregulation of the mRNAs encoding for the A_3AR and for ENTPDase/CD73 upon exposure of C6 and U138-MG glioma cells to indomethacin (Bernardi et al. 2007). Increases in CD73 mRNA was accompanied by a significantly higher ratio of ATP hydrolysis; although actual Ado concentration was not measured in this paper, it can be envisaged that it reaches values high enough to activate the A_3AR subtype. Indomethacin also significantly inhibited glioma cell proliferation, and this effect was reverted by the A_3AR agonist MRS1220, but not by antagonists acting at other P1 receptors (Bernardi et al. 2007), thus suggesting a specific involvement of this receptor subtype in indomethacin-mediated effects.

Finally, it should be also noted that A_3ARs can modulate the immune system by activating NK cells and, possibly, NK-mediated disruption of tumour cells. Conversely, within the environment of a solid tumour Ado has been demonstrated to exert an immunosuppressive role through the A_3AR, which could interfere with the recognition of tumour cells by T killer cells (Gessi et al. 2008, see also above Sect. 2.3). Therefore this double immunomodulatory effect can positively or negatively influence tumour growth, progression, and metastatisation depending upon the cell environment.

2.5 Receptor-Independent Effects of Adenosine Analogues in Glioma Cell Growth and Survival

Apart from the more recently identified receptor-mediated effects, the intracellular cytotoxic actions of purine and pyrimidine derivatives have been known since the 1940s and are at the basis of their use as chemotherapic and anti-metabolites. For example, 2-CdA is utilized as the chemotherapic drug of choice in hairy cell leukemia and in other haematological malignancies, with the name of Cladribine (Golomb 2011). The molecular basis of the effects exerted by nucleoside analogues were initially mainly ascribed to their ability to alter intracellular purine and pyrimidine pools, leading to impaired synthesis of nucleic acids, or to directly interfere with the enzymes involved in DNA synthesis, thanks to their structural similarities with endogenous nucleosides (Dighiero 1996). In the 1990s, however, it was demonstrated that nucleoside analogues could also activate specific pathways of apoptotic cell death. For example, in haemathological disorders, Cladribine was demonstrated to recruit both the extrinsic caspase-8-dependent pathway of cell death (Nomura et al. 2000), and the intrinsic mitochondrial pathway of death, with the involvement of caspase-9 (Genini et al. 2000). Their actions on CNS cancer cells have not, however, been investigated properly.

The first clues of a possible cytotoxic role for Ado analogues on astrocytes came from the demonstration that the exposure of primary rat astrocytes to 2-chloro-adenosine (2-CA), a non-selective

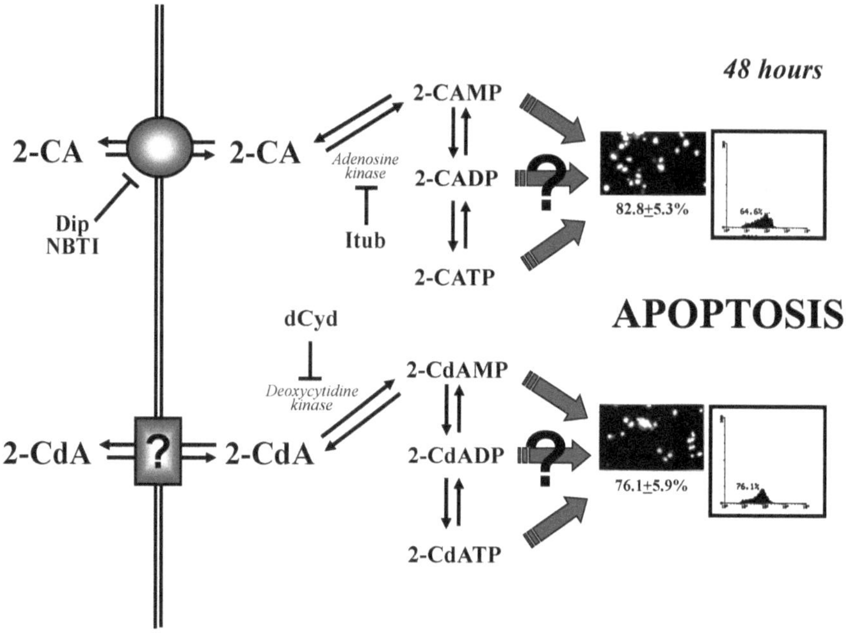

Fig. 2.5 Different intracellular pathways of apoptotic cell death activated by 2-CA and 2-CdA in human astrocytoma ADF cells. 2-CA and 2-CdA are prodrugs, since they have to be intracellularly phosphorylated/activated by two distinct kinases to exert their toxic effects. This has been demonstrated by inhibiting 2-CA- and 2-CdA-mediated effects with two specific inhibitors: Itub (acting on adenosine kinase) and dCyd (acting on deoxycytidine kinase), in the case of 2-CA and 2-CdA, respectively. Thus, the active toxic species are represented by the corresponding 2-chloronucleotide or deoxy-nucleotide derivatives, which led to induction of cell death by apoptosis through complex intracellular pathways (see text for details and Fig. 2.6). Apoptosis has been demonstrated by: (1) staining of cells floating in the culture supernatants with the chromatin fluorescent dye Hoechst 33258 (see pictures), and (2) flow-cytometric analysis of PI-stained cultures (see histograms on the *right*). The former technique demonstrated chromatin condensation and nuclear fragmentation in a highly significant percentage of detached cells (see micrographs and numbers *below*), whereas the appearance of hypodiploid DNA after a 48-h exposure to either adenosine analogue was shown by flow-cytometry (see numbers on histograms). Numbers represent the percentage of apoptotic cell death. For more details, see text (Reproduced from Ceruti et al. (2003a). With permission from John Wiley and Sons)

agonist of P1 receptors, induced a dramatic reduction of cell number (Abbracchio et al. 1994), due to the induction of cell death by apoptosis (Abbracchio et al. 1995b; Ceruti et al. 1997). To our surprise, Ado receptor antagonists did not inhibit 2-CA-mediated effects, thus suggesting an intracellular site of action for the Ado analogue. These results prompted us to analyze the possible cytotoxic effects of 2-CA on human astrocytoma cells (ADF cells, see above), in comparison with 2-CdA, in order to verify the possible use of Ado analogues as antitumour agents also in solid brain tumours. First of all, we demonstrated a high cytotoxic activity exerted by the two Ado analogues on ADF cells, due to the induction of apoptosis (Fig. 2.5, Ceruti et al. 2000, 2003a). 2-CdA resulted to be more potent than 2-CA in inducing ADF cell death, with significant effects already detected at 5 μM concentration. Both molecules exerted their effects after their intracellular phosphorylation to the corresponding nucleotide derivatives, which occurred through the recruitment of two independent biochemical pathways, as demonstrated by the complete inhibition of 2-CA-mediated actions by Itub (an inhibitor of AK) and of 2-CdA-induced effects by dCyd, which competes for the active site of deoxycytidine kinase (Fig. 2.5, Ceruti et al. 2000, 2003a). Only in the case of Cladribine, was a cell cycle block demonstrated to precede induction of apoptosis, further confirming that, although closely related chemically, these two compounds indeed recruited diverging pathways of death (Ceruti et al. 2000).

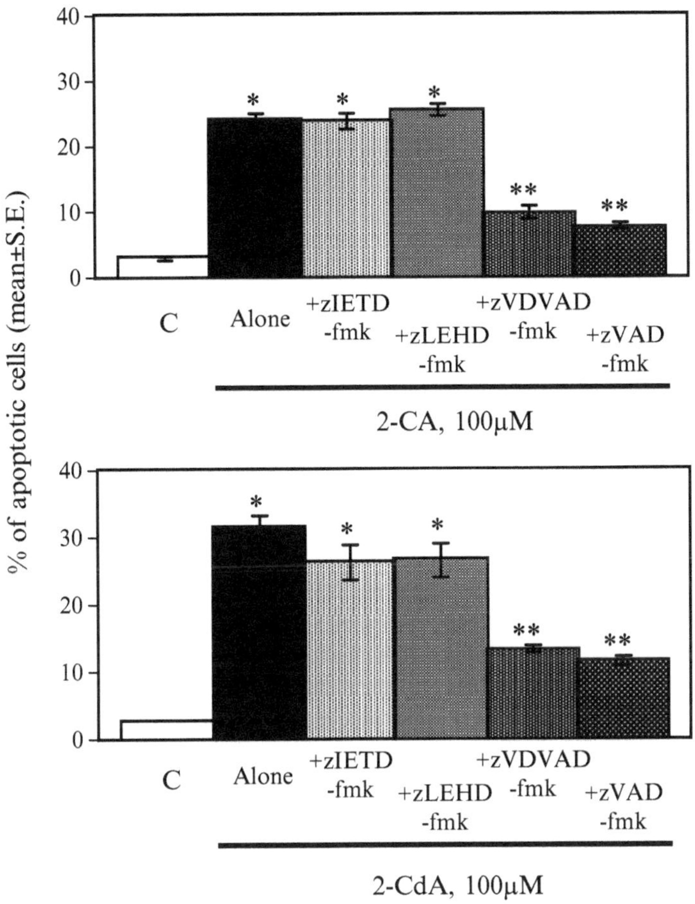

Fig. 2.6 Specific involvement of caspase-2 as the upstream initiator caspase in 2-CA- and 2-CdA-induced cell death of ADF cells. Human astrocytoma ADF cells were exposed for 24 h to either 100 μM 2-CA or 2-CdA, alone or in combination with selective inhibitors of caspase-8 (zIETD-fmk), caspase-9 (zLEHD-fmk), caspase-2 (zVDVAD-fmk) or with the pan-caspase inhibitor zVAD-fmk (all utilized at 3 μM), as indicated. At the end of the incubation, the percentage of apoptotic nuclei was determined by flow cytometric analysis of PI-stained nuclei. Only the selective caspase-2 inhibitors and the pan-caspase inhibitor were able to significantly protect cells from Ado analogue-induced death. See text for details. Data represent the mean ± S.E.M. of three independent experiments. $^{*}p<0.05$ with respect to control, and $^{**}p<0.05$ with respect to control and to corresponding Ado analog alone, one-way ANOVA (Scheffe' F test) (Reproduced from Ceruti et al. (2003b). With permission from ASPET)

Nevertheless, the two Ado analogues converged on a common and rather new pathway of apoptosis; in fact, both Ado analogues activated an atypical caspase pathway in ADF cells, with the recruitment of caspase-2 as initiator caspase, followed by the contemporary consequent activation of caspase-8 and caspase-9, and of caspase-3 as the main effector caspase (Fig. 2.6, Ceruti et al. 2003b). This was the first demonstration of the possible activation of a caspase-2-dependent pathway of cell death in astrocytoma cells, which was later demonstrated also following gamma-irradiation and overexpression of the pro-apoptotic molecule Smac/Diablo (Giagkousiklidis et al. 2005). A key contribution of the MAP kinase signaling pathways upon exposure of ADF cells to 2-CA and 2-CdA was also demonstrated (Ceruti et al. manuscript in preparation).

It is well known that tumorigenesis is linked to the acquisition of several genetic defects, which render malignant cells resistant to the normal growth inhibitory and apoptotic signals, with a parallel deregulation of proliferation leading to out-of-control expansion of damaged cells (Kelly and Strasser 2011). In this respect, we have detected various genetic alterations borne by ADF cells. First of all, ADF cells are insensitive to apoptotic triggers (e.g., betulinic acid, potassium cyanide, and 2-deoxy-ribose) that recruit the "classical" intrinsic pathway of cell death, e.g. the mitochondria/cytochrome C/ caspase-9 pathway (Ceruti et al. 2005). This is possibly due to the expression of a mutated form of caspase-9, bearing a single C-to-T nucleotide substitution in position 83 of the coding sequence, leading to a single amino acid substitution (A-to-V) in position 28 of the caspase-9 protein. This mutation occurs in the caspase pro-domain region, which allows the proper assembly of the apoptosome complex (Bratton and Salvesen 2010); it can be anticipated that a highly ramified amino acid as valine substituting for alanine will negatively influence the generation of the proper protein-to-protein interactions. Moreover, ADF cells also express a truncated splice variant of caspase-9, the so-called caspase-9β, which lacks a relevant part of the large subunit of the protein, including the catalytic site (Ceruti et al. 2005). Caspase-9β is therefore inactive and it is generally believed to behave as a dominant negative of the full-length caspase-9. Based on these observations, we therefore postulated that ADF cells couldn't efficiently activate the intrinsic pathway of cell death, which could explain their resistance to several known cytotoxic agents.

We also identified a single mutation in the p53 protein expressed by ADF cells (Ceruti et al. 2006): a single G-to-A nucleotide substitution in position 797 of the coding sequence, with a consequent single amino acid substitution (G-to-E) at position 266 of the p53 protein, belonging to the DNA-binding domain. This mutation leads to an inactive p53 isoform (Ceruti et al. 2006), meaning that ADF cells cannot activate the cell death pathways that depend upon the transcriptional activity of p53.

Despite the presence of the above-mentioned multiple genetic alterations of different cell death pathways, both 2-CA and 2-CdA have proved able to induce massive ADF cell death by recruiting caspase-2 and, in the case of Cladribine, also by interfering with the cell cycle (see above, Ceruti et al. 2000, 2003b). This unveils the cytotoxic potential of these derivatives also in the presence of resistance to various pharmacological approaches, and suggests the use of Ado derivatives as second line drugs in astrocytomas refractory to currently utilized chemotherapic regimens.

It is also important to note that both 2-CA and 2-CdA were ineffective in inducing cell death on primary neuronal cultures when utilized at concentrations, which were already almost maximally effective on ADF cells (Fig. 2.7, Ceruti et al. 2000). Although the mechanism at the basis of this selective sparing of neuronal cells (despite the marked apoptotic effect observed on cancerous glial cells) has not been clarified, it could represent an important additional therapeutic advantage in the case of in vivo administration of these drugs. Moreover, a specific carrier for Ado at the level of the blood-brain barrier has been demonstrated (Pardridge et al. 1994), and since 2-CA and Ado utilize the same ENT to permeate cells, it may be hypothesized that 2-CA may easily cross the blood-brain barrier. Conversely, for 2-CdA, concentrations in cerebrospinal fluid of leukemic patients have been demonstrated to represent 25 % of plasma concentrations (Liliemark 1997), suggesting a high tropism toward nervous tissues. Moreover, an early disruption of the blood-brain barrier is known to occur in several CNS pathologies that would favour accumulation of these antitumour agents in brain tissues.

Despite the above-mentioned encouraging preclinical data, a few unsuccessful clinical studies have been performed on the use of 2-CdA in malignant gliomas. A Phase II study performed on seven patients with recurrent glioma has shown no significant effects by Cladribine (Rajkumar et al. 1999). A Clinical Trial entitled "Chemotherapy Followed by Radiation Therapy in Treating Patients with Malignant Glioma: Cladribine in combination with radiotherapy in patients with high grade glioma" is available on the USA ClinicalTrials.gov website (#NCT00019071). The study has been completed, but neither publications nor results have been posted on the website, thus making it impossible to evaluate its outcome. No data are currently available on the possible effects exerted by Cladribine, which is now considered one of the drugs of choice in multiple sclerosis for its immunomodulatory activities (Killestein et al. 2011), as a second-line drug on refractory gliomas (see above).

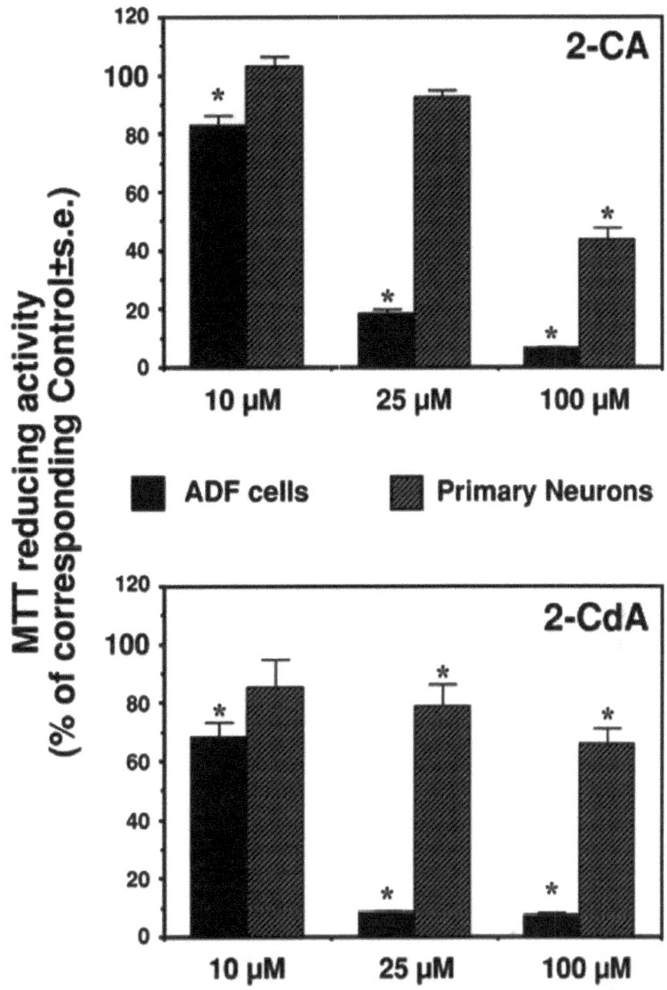

Fig. 2.7 Selective sparing of primary neurons by concentrations of 2-CA and 2-CdA inducing significant cytotoxicity in human astrocytoma ADF cells. ADF cells and primary cortical neurons were exposed to increasing 2-CA (*upper panel*) or 2-CdA concentrations (*lower panel*) for 48 h. At the end of the incubation period the percentage of cell death was evaluated by the ability of cultures to metabolize the MTT dye, a typical feature of healthy cells. Data are expressed as percentage of the MTT metabolizing activity in the corresponding control cultures and represent the mean ± S.E.M. of five independent experiments run in triplicate. *$p < 0.05$ with respect to corresponding control, one way ANOVA (Scheffe's F-test) (Reproduced from Ceruti et al. (2000). With permission from John Wiley and Sons)

More recently, as already mentioned above, Ado itself has been demonstrated to have inhibitory effects on the growth of C6 glioma cells by activating an intracellular pathway involving its uptake through ENT2 followed by phosphorylation by AK (Ohkubo et al. 2007). Authors hypothesize the induction of pyrimidine starvation by excessive intracellular AMP concentrations, since extracellular uridine completely reversed Ado-mediated effects. These results suggest that, in glioma cells, not only can elevated extracellular Ado concentrations activate its membrane receptors (see Sect. 2.4 above), but that the nucleoside can also directly modulate cell proliferation through intracellular signaling pathways.

2.6 Conclusions and Future Perspectives

The ubiquitous distribution and heterogeneity of the various components of Ado-mediated signaling (i.e., metabolizing enzymes, various subtypes of membrane receptors and of membrane transporters, direct actions on intracellular enzymes involved in cell survival, etc.), and the multiplicity of effects described upon its recruitment, with sometimes opposite outcomes, have made it difficult to exploit this system as a new pharmacological target for brain tumours. Nevertheless, the data collected so far and reviewed in this article strongly suggest that the Ado system plays a fundamental role in controlling the proliferation, migration, and survival of glioma cells, where a marked hypoxic environment is generated. Not only Ado receptors, but also nucleotide metabolizing enzymes could represent important and druggable targets in the fight against brain tumours. Moreover, the ability of Ado analogues to recruit atypical intracellular pathways of death in tumours where the "classical" apoptotic triggers are useless and to selectively spare surrounding healthy cells would represent additional therapeutic advantages, which should be further explored. Contrasting results can often be reconciled by taking into consideration the various experimental settings, the different cell lines, and the agonists' concentrations utilized. While many studies have been addressed to assess the potential of Ado analogs as anti-glioma agents *in vitro*, we feel that an accurate evaluation of these compounds in appropriate *in vivo* models is still missing. We therefore envisage that well-designed *in vivo* studies could shed light on the current inconsistencies between different results, and more clearly highlight the possible advantages of targeting the Ado system for the therapy of glioma.

References

Abbracchio MP, Saffrey MJ, Höpker V, Burnstock G (1994) Modulation of astroglial cell proliferation by analogues of adenosine and ATP in primary cultures of rat striatum. Neuroscience 59:67–76

Abbracchio MP, Brambilla R, Ceruti S, Kim HO, von Lubitz DKJE, Jacobson KA, Cattabeni F (1995a) G protein-dependent activation of phospholipase C by adenosine A_3 receptors in rat brain. Mol Pharmacol 48:1038–1045

Abbracchio MP, Ceruti S, Barbieri D, Franceschi C, Malorni W, Biondo L, Burnstock G, Cattabeni F (1995b) A novel action for adenosine: apoptosis of astroglial cells in rat brain primary cultures. Biochem Biophys Res Commun 213:908–915

Abbracchio MP, Rainaldi G, Giammarioli AM, Ceruti S, Brambilla R, Cattabeni F, Barbieri D, Franceschi C, Jacobson KA, Malorni W (1997) The A3 adenosine receptor mediates cell spreading, reorganization of actin cytoskeleton, and distribution of Bcl-XL: studies in human astroglioma cells. Biochem Biophys Res Commun 241:297–304

Abbracchio MP, Ceruti S, Brambilla R, Barbieri D, Camurri A, Franceschi C, Giammarioli AM, Jacobson KA, Cattabeni F, Malorni W (1998) Adenosine A_3 receptors and viability of astrocytes. Drug Dev Res 45:379–386

Abbracchio MP, Camurri A, Ceruti S, Cattabeni F, Falzano L, Giammarioli AM, Jacobson KA, Trincavelli L, Martini C, Malorni W, Fiorentini C (2001) The A_3 adenosine receptor induces cytoskeleton rearrangement in human astrocytoma cells via a specific action on Rho proteins. Ann NY Acad Sci 939:63–73

Appel E, Kazimirsky G, Ashkenazi E, Kim SG, Jacobson KA, Brodie C (2001) Roles of BCL-2 and caspase 3 in the adenosine A_3 receptor-induced apoptosis. J Mol Neurosci 17:285–292

Bauer A, Langen KJ, Bidmon H, Holschbach MH, Weber S, Olson RA, Coenen HH, Zilles K (2005) ^{18}F-Cyclopentyl-3(3-^{18}F-Fluoropropyl)-1-Propyl-xanthine PET identifies changes in cerebral A1Adenosine receptor density caused by glioma invasion. J Nucl Med 46:450–454

Bavaresco L, Bernardi A, Braganhol E, Cappellari AR, Rockenbach L, Farias PF, Wink MR, Delgado-Cañedo A, Battastini AM (2008) The role of ecto-5′-nucleotidase/CD73 in glioma cell line proliferation. Mol Cell Biochem 319:61–68

Bernardi A, Bavaresco L, Wink MR, Jacques-Silva MC, Delgado-Cañedo A, Lenz G, Battastini AM (2007) Indomethacin stimulates activity and expression of ecto-5′-nucleotidase/CD73 in glioma cell lines. Eur J Pharmacol 569:8–15

Bratton SB, Salvesen GS (2010) Regulation of the Apaf-1-caspase-9 apoptosome. J Cell Sci 123:3209–3214

Cappellari AR, Vasques GJ, Bavaresco L, Braganhol E, Battastini AM (2012) Involvement of ecto-5′-nucleotidase/CD73 in U138MG glioma cell adhesion. Mol Cell Biochem 359:315–322

Castillo AC, Albasanz JL, Fernández M, Martìn M (2007) Endogenous expression of adenosine A_1, A_2 and A_3 receptors in rat C6 glioma cells. Neurochem Res 32:1056–1070

Castillo CA, León D, Ruiz MA, Albasanz JL, Martín M (2008) Modulation of adenosine A_1 and A_{2A} receptors in C6 glioma cells during hypoxia: involvement of endogenous adenosine. J Neurochem 105:2315–2329

Ceruti S, Barbieri D, Veronese E, Cattabeni F, Cossarizza A, Giammarioli AM, Malorni W, Franceschi C, Abbracchio MP (1997) Different pathways of apoptosis revealed by 2-chloro-adenosine and deoxy-D-ribose in mammalian astroglial cells. J Neurosci Res 47:372–383

Ceruti S, Franceschi C, Barbieri D, Malorni W, Camurri A, Giammarioli AM, Ambrosini A, Racagni G, Cattabeni F, Abbracchio MP (2000) Apoptosis induced by 2-chloro-adenosine and 2-chloro-2′-deoxy-adenosine in a human astrocytoma cell line: differential mechanisms and possible clinical relevance. J Neurosci Res 60:388–400

Ceruti S, Mazzola A, Beltrami E, Passera D, Piantoni E, Cattabeni F, Abbracchio MP (2003a) Intracellular phosphorylation of chloro-adenosine analogs is a pre-requisite for activation of caspase-3 and induction of apoptosis in human astrocytoma cells. Drug Dev Res 58:396–404
Ceruti S, Beltrami E, Matarrese P, Mazzola A, Cattabeni F, Malorni W, Abbracchio MP (2003b) A key role for caspase-2 and caspase-3 in the apoptosis induced by 2-chloro-2′-deoxy-adenosine (cladribine) and 2-chloro-adenosine in human astrocytoma cells. Mol Pharmacol 63:1437–1447
Ceruti S, Mazzola A, Abbracchio MP (2005) Resistance of human astrocytoma cells to apoptosis induced by mitochondria-damaging agents: possible implications for anticancer therapy. J Pharmacol Exp Ther 314:825–837
Ceruti S, Mazzola A, Abbracchio MP (2006) Proteasome inhibitors potentiate etoposide-induced cell death in human astrocytoma cells bearing a mutated p53 isoform. J Pharmacol Exp Ther 319:1424–1434
Dehnhardt M, Palm C, Vieten A, Bauer A, Pietrzyk U (2007) Quantifying the A_1AR distribution in peritumoural zones around experimental F98 and C6 rat brain tumours. J Neurooncol 85:49–63
Dighiero G (1996) Adverse and beneficial immunological effects of purine nucleoside analogues. Hematol Cell Ther 38:S75–S81
Fiebich BL, Akundi RS, Biber K, Hamke M, Schmidt C, Butcher RD, van Calker D, Willmroth F (2005) IL-6 expression induced by adenosine A2b receptor stimulation in U373 MG cells depends on p38 mitogen activated kinase and protein kinase C. Neurochem Int 46:501–512
Fishman P, Bar-Yehuda S, Madi L, Cohn I (2002) A3 adenosine receptor as a target for cancer therapy. Anti-Cancer Drugs 13:437–443
Fredholm BB (2010) Adenosine receptors as drug targets. Exp Cell Res 316:1284–1288
Fredholm BB, Ijzerman AP, Jacobson KA, Linden J, Müller CE (2011) International Union of Basic and Clinical Pharmacology. LXXXI. Nomenclature and classification of adenosine receptors – an update. Pharmacol Rev 63:1–34
Genini D, Adachi S, Chao Q, Rose DW, Carrera CJ, Cottam HB, Carson DA, Leoni LM (2000) Deoxyadenosine analogs induce programmed cell death in chronic lymphocytic leukemia cells by damaging the DNA and by directly affecting the mitochondria. Blood 96:3537–3543
Gessi S, Merighi S, Varani K, Leung E, Mac Lennan S, Borea PA (2008) The A3 adenosine receptor: an enigmatic player in cell biology. Pharmacol Ther 117:123–140
Gessi S, Sacchetto V, Fogli E, Merighi S, Varani K, Baraldi PG, Tabrizi MA, Leung E, Maclennan S, Borea PA (2010) Modulation of metalloproteinase-9 in U87MG glioblastoma cells by A3 adenosine receptors. Biochem Pharmacol 79:1483–1495
Giagkousiklidis S, Vogler M, Westhoff MA, Kasperczyk H, Debatin KM, Fulda S (2005) Sensitization for gamma-irradiation-induced apoptosis by second mitochondria-derived activator of caspase. Cancer Res 65:10502–10513
Golomb HM (2011) Fifty years of hairy cell leukemia treatments. Leuk Lymphoma 52:3–5
Isakovic A, Harhaji L, Dacevic M, Trajkovic V (2008) Adenosine rescues glioma cells from cytokine-induced death by interfering with the signaling network involved in nitric oxide production. Eur J Pharmacol 591:106–113
Jacobson KA (1998) Adenosine A3 receptors: novel ligands and paradoxical effects. Trends Pharmacol Sci 19:184–191
Kelly GL, Strasser A (2011) The essential role of evasion from cell death in cancer. Adv Cancer Res 111:39–96
Killestein J, Rudick RA, Polman CH (2011) Oral treatment for multiple sclerosis. Lancet Neurol 10:1026–1034
King AE, Ackley MA, Cass CE, Young JD, Baldwin SA (2006) Nucleoside transporters: from scavengers to novel therapeutic targets. Trends Pharmacol Sci 27:416–425
Lazarowski ER, Sesma JI, Seminario-Vidal L, Kreda SM (2011) Molecular mechanisms of purine and pyrimidine nucleotide release. Adv Pharmacol 61:221–261
Lee JK, Won JS, Singh AK, Singh I (2005) Adenosine kinase inhibitor attenuates the expression of inducible nitric oxide synthase in glial cells. Neuropharmacology 48:151–160
Liliemark J (1997) The clinical pharmacokinetics of cladribine. Clin Pharmacokinet 32:120–131
Linden J, Auchampach JA, Jin X, Figler RA (1998) The structure and function of A1 and A2B adenosine receptors. Life Sci 62:1519–1524
Malorni W, Rainaldi G, Rivabene R, Santini MT (1994) Different susceptibilities to cell death induced by t-butylhydroperoxide could depend upon cell histotype-associated growth features. Cell Biol Toxicol 10:207–218
Melani A, De Micheli E, Pinna G, Alfieri A, Corte LD, Pedata F (2003) Adenosine extracellular levels in human brain gliomas: an intraoperative microdialysis study. Neurosci Lett 346:93–96
Melani A, Corti F, Stephan H, Müller CE, Donati C, Bruni P, Vannucchi MG, Pedata F (2012) Ecto-ATPase inhibition: ATP and adenosine release under physiological and ischemic in vivo conditions in the rat striatum. Exp Neurol 233:193–204
Merighi S, Mirandola P, Varani K, Gessi S, Leung E, Baraldi PG, Tabrizi MA, Borea PA (2003) A glance at adenosine receptors: novel target for antitumor therapy. Pharmacol Ther 100:31–48
Merighi S, Benini A, Mirandola P, Gessi S, Varani K, Leung E, Maclennan S, Borea PA (2006) Adenosine modulates vascular endothelial growth factor expression via hypoxia-inducible factor-1 in human glioblastoma cells. Biochem Pharmacol 72:19–31

Merighi S, Benini A, Mirandola P, Gessi S, Varani K, Leung E, Maclennan S, Baraldi PG, Borea PA (2007) Hypoxia inhibits paclitaxel-induced apoptosis through adenosine-mediated phosphorylation of bad in glioblastoma cells. Mol Pharmacol 72:162–172
Morelli M, Carta AR, Kachroo A, Schwarzschild MA (2010) Pathophysiological roles for purines: adenosine, caffeine and urate. Prog Brain Res 183:183–208
Morrone FB, Jacques-Silva MC, Horn AP, Bernardi A, Schwartsmann G, Rodnight R, Lenz G (2003) Extracellular nucleotides and nucleosides induce proliferation and increase nucleoside transport in human glioma cell lines. J Neurooncol 64:211–218
Nomura Y, Inanami O, Takahashi K, Matsuda A, Kuwabara M (2000) 2-Chloro-2′-deoxyadenosine induces apoptosis through the Fas/Fas ligand pathway in human leukemia cell line MOLT-4. Leukemia 14:299–306
Ohkubo S, Nagata K, Nakahata N (2007) Adenosine uptake-dependent C6 cell growth inhibition. Eur J Pharmacol 577:35–43
Palmer TM, Stiles GL (2000) Identification of threonine residues controlling the agonist-dependent phosphorylation and desensitization of the rat A(3) adenosine receptor. Mol Pharmacol 57:539–545
Pardridge WM, Yoshikawa T, Kang YS, Miller LP (1994) Blood-brain barrier transport and brain metabolism of adenosine and adenosine analogs. J Pharmacol Exp Ther 268:14–18
Rajkumar SV, Burch PA, Nair S, Dinapoli RP, Scheithauer B, O'Fallon JR, Etzell PS, Leitch JM, Morton RF, Marks RS (1999) Phase II North Central Cancer Treatment Group study of 2-chlorodeoxyadenosine in patients with recurrent glioma. Am J Clin Oncol 22:168–171
Sai K, Yang D, Yamamoto H, Fujikawa H, Yamamoto S, Nagata T, Saito M, Yamamura T, Nishizaki T (2006) A(1) adenosine receptor signal and AMPK involving caspase-9/-3 activation are responsible for adenosine-induced RCR-1 astrocytoma cell death. Neurotoxicology 27:458–467
Sands WA, Martin AF, Strong EW, Palmer TM (2004) Specific inhibition of nuclear factor-kappaB-dependent inflammatory responses by cell type-specific mechanisms upon A2A adenosine receptor gene transfer. Mol Pharmacol 66:1147–1159
Schulte G, Fredholm BB (2000) Human adenosine A(1), A(2A), A(2B), and A(3) receptors expressed in Chinese hamster ovary cells all mediate the phosphorylation of extracellular-regulated kinase 1/2. Mol Pharmacol 58:477–482
Sinclair CJ, LaRivière CG, Young JD, Cass CE, Baldwin SA, Parkinson FE (2000) Purine uptake and release in rat C6 glioma cells: nucleoside transport and purine metabolism under ATP-depleting conditions. J Neurochem 75:1528–1538
Taliani S, La Motta C, Mugnaini L, Simorini F, Salerno S, Marini AM, Da Settimo F, Cosconati S, Cosimelli B, Greco G, Limongelli V, Marinelli L, Novellino E, Ciampi O, Daniele S, Trincavelli ML, Martini C (2010) Novel N2-substituted pyrazolo[3,4-d]pyrimidine adenosine A3 receptor antagonists: inhibition of A_3-mediated human glioblastoma cell proliferation. J Med Chem 53:3954–3963
Tapper E, Schmelzle M, Eltzschig HK, Robson SC (2012) Erratum to: role of CD73 and extracellular adenosine in disease. Purinergic Signal 8:339
Trincavelli ML, Tuscano D, Marroni M, Falleni A, Gremigni V, Ceruti S, Abbracchio MP, Jacobson KA, Cattabeni F, Martini C (2002) A3 adenosine receptors in human astrocytoma cells: agonist-mediated desensitization, internalization, and down-regulation. Mol Pharmacol 62:1373–1384
Trincavelli ML, Marroni M, Tuscano D, Ceruti S, Mazzola A, Mitro N, Abbracchio MP, Martini C (2004) Regulation of A_{2B} adenosine receptor functioning by tumour necrosis factor a in human astroglial cells. J Neurochem 91:1180–1190
Vaupel P, Thews O, Hoeckel M (2001) Treatment resistance of solid tumors: role of hypoxia and anemia. Med Oncol 18:243–259
Wang L, Fan J, Thompson LF, Zhang Y, Shin T, Curiel TJ, Zhang B (2011) CD73 has distinct roles in non hematopoietic and hematopoietic cells to promote tumor growth in mice. J Clin Invest 121:2371–2382
Westphal M, Lamszus K (2011) The neurobiology of gliomas: from cell biology to the development of therapeutic approaches. Nat Rev Neurosci 12:495–508

Chapter 3
Cross-Talk in Nucleotide Signaling in Glioma C6 Cells

Dorota Wypych and Jolanta Barańska

Abstract The chapter is focused on the mechanism of action of metabotropic P2Y nucleotide receptors: $P2Y_1$, $P2Y_2$, $P2Y_{12}$, $P2Y_{14}$ and the ionotropic $P2X_7$ receptor in glioma C6 cells. $P2Y_1$ and $P2Y_{12}$ both respond to ADP, but while $P2Y_1$ links to PLC and elevates cytosolic Ca^{2+} concentration, $P2Y_{12}$ negatively couples to adenylate cyclase, maintaining cAMP at low level. In glioma C6, these two P2Y receptors modulate activities of ERK1/2 and PI3K/Akt signaling and the effects depend on physiological conditions of the cells. During prolonged serum deprivation, cell growth is arrested, the expression of the $P2Y_1$ receptor strongly decreases and $P2Y_{12}$ becomes a major player responsible for ADP-evoked signal transduction. The $P2Y_{12}$ receptor activates ERK1/2 kinase phosphorylation (a known cell proliferation regulator) and stimulates Akt activity, contributing to glioma invasiveness. In contrast, $P2Y_1$ has an inhibitory effect on Akt pathway signaling. Furthermore, the $P2X_7$ receptor, often responsible for apoptotic fate, is not involved in Ca^{2+}elevation in C6 cells. The shift in nucleotide receptor expression from $P2Y_1$ to $P2Y_{12}$ during serum withdrawal, the cross talk between both receptors and the lack of $P2X_7$ activity shows the precise self-regulating mechanism, enhancing survival and preserving the neoplastic features of C6 cells.

Keywords $P2Y_1$, $P2Y_2$, $P2Y_{12}$, $P2Y_{14}$, $P2X_7$ nucleotide receptors • Serum withdrawal • $P2Y_1$/$P2Y_{12}$ cross-talk • cAMP • ERK1/2, PI3K/Akt activity • Glioma C6 cells

---

D. Wypych (✉)
Nencki Institute of Experimental Biology, Polish Academy of Sciences, 3 Pasteur St, PL 02-093 Warsaw, Poland

Laboratory of Molecular and Cellular Pharmacology of Purinergic Transmission, Department of Pharmacological and Biomolecular Sciences, Università degli Studi di Milano, 9 via Balzaretti, 20133 Milan, Italy
e-mail: d.wypych@nencki.gov.pl

J. Barańska
Nencki Institute of Experimental Biology, Polish Academy of Sciences, 3 Pasteur St, PL 02-093 Warsaw, Poland
e-mail: j.baranska@nencki.gov.pl

J. Barańska (ed.), *Glioma Signaling*, Advances in Experimental Medicine and Biology 986, DOI 10.1007/978-94-007-4719-7_3,

Abbreviations

Akt	Protein kinase B/Akt kinase
Ap_4A	p^1,p^4-Di(adenosine-5′)tetraphosphate
BzATP	2′,3′-*O*-(4-Benzoylbenzoyl)-ATP
$[Ca^{2+}]_i$	Intracellular calcium concentration
DAG	Diacylglycerol
ER	Endoplasmic reticulum
ERK1/2	Extracellular signal-regulated kinases 1/2
Gap1	GTP-ase activating protein 1
GBM	*Glioblastoma multiforme*
GEF	Guanine nucleotide exchange factor
GFAP	Glial fibrillary acidic protein
GFP	Green fluorescent protein
GPCR	G protein-coupled receptor
IP_3	Inositol-1,4,5-trisphosphate
LPS	Lipopolysacharide
MEK	MAP kinase-ERK kinase
2MeSADP	2-Methylthio ADP
OxATP	Periodate-oxidized ATP
PDK1	Phosphoinositide-dependent protein kinase 1
PI3K	Phosphatidylinositol 3-kinase
PIP_2	Phosphatidylinositol-4,5-bisphosphate
PKA	Protein kinase A
PKC	Protein kinase C
PLC	Phospholipase C
PPADS	Pyridoxal-phosphate-6-azophenyl-2′,4′disulfonic acid
PTEN	Phosphatase and tensin homolog
PTX	Pertussis toxin
RTK	Receptor tyrosine kinase

3.1 Introduction

Gliomas are the most common brain tissue primary tumors that originate from glial cells. One of the most malignant form of gliomas, resistant to therapeutic intervention, is human *glioblastoma multiforme* (GBM) characterized by a marked increase in cell proliferation and invasiveness (Barth 1998; Collins 1998; Grobben et al. 2002; Salcman 1995). There are many models used in studies of glioma cell biology. One of them, widely used in GBM research, is rat glioma C6 cells. This tumoral cell line was originally produced in Wistar-Furth rats exposed to N,N′-nitroso-methylurea (Benda et al. 1968). Glioma C6 cells have oligodendrocytic as well as astrocytic progenitor properties and are also often used as a biochemical model for studies related to astrocytes (Brismar 1995). Moreover, the cells injected into rodent brain become morphologically similar to GBM (Auer et al. 1981). They are used in both *in vivo* and *in vitro* studies of this kind of tumor (see Chap. 5) since brain cancer treatment are still needed.

Recent evidence suggests that glioma development is determined by ATP signaling (White and Burnstock 2006). However, not only ATP but also UTP as well as their metabolites: ADP, UDP and adenosine have an influence on cell metabolism by stimulation of specific receptors present on the plasma membrane (see Chap. 1). This stimulation regulates proliferation, invasiveness and cell death. Thus, nucleotide receptors might be useful therapeutic targets (Burnstock 2002).

As described in Chap. 1, the large family of nucleotide receptors is divided into two major classes: P1, responding to adenosine, and P2, responding to ATP, UTP, ADP, UDP and UDP-sugar derivatives. To the P2 class belong: the intrinsic ligand-gated ion channels – P2X receptors and the metabotropic G-protein coupled P2Y receptors (Abbracchio and Burnstock 1994). To date, eight mammalian P2Y receptors have been cloned and pharmacologically characterized (see Fig. 1.1). Among them, $P2Y_2$ is activated by ATP and UTP, whereas $P2Y_1$ and $P2Y_{12}$ both respond to ADP. $P2Y_1$ and $P2Y_2$ are coupled to phospholipase C (PLC) and are responsible for Ca^{2+} mobilization from intracellular stores, while the $P2Y_{12}$ receptor is negatively coupled to adenylate cyclase (Ralevic and Burnstock 1998). However, one receptor can activate more than one signaling pathway. Such an example is a novel $P2Y_{14}$ receptor that specifically responds to UDP-glucose and related sugar-nucleotides (Abbracchio et al. 2003). Its stimulation leads both to calcium mobilization and to inhibition of adenylate cyclase. $P2Y_1$, $P2Y_2$, $P2Y_4$, $P2Y_6$, $P2Y_{12}$, $P2Y_{13}$ and $P2Y_{14}$ are expressed in glioma C6 cells (Barańska et al. 2004; Braganhol et al. 2009; Czajkowski and Barańska 2002; Krzemiński et al. 2008; Van Kolen and Slegers 2006b). Contrary to P2Y receptors, the expression and activity of P2X receptors in glioma C6 cells have not been intensively examined. However, it was shown that the $P2X_7$ receptor, which responds to ATP is not active in this cell line (Supłat-Wypych et al. 2010).

This chapter is focused on the cross-talk between the metabotropic $P2Y_1$ and $P2Y_{12}$ nucleotide receptors. Their stimulation modulates activities of extracellular signal-regulated kinases (ERK1/2) and phosphatidylinositol 3-kinase/Akt kinase (PI3K/Akt) (Barańska et al. 2004). $P2Y_{12}$ not only reduces the level of cAMP in the cell but is also responsible for activating ERK1/2 kinase phosphorylation (a known cell proliferation regulator) and stimulating Akt activity (Czajkowski et al. 2004; Krzemiński et al. 2007). Thus, $P2Y_{12}$ acts in favor of cell growth and proliferation. In contrast, $P2Y_1$ has an inhibitory effect on the Akt signaling pathway. The above effects depend on the physiological conditions. During prolonged serum deprivation, $P2Y_1$ receptor expression is strongly decreased while that of $P2Y_{12}$ is distinctly increased in C6 cells. The intense pathological proliferation and invasiveness of glioma cells seems to depend on the constitutive low level of cAMP (Stork and Schmitt 2002) and up-regulation of the PI3K/Akt pathway (Kubiatowski et al. 2001). Thus, the shift in expression from $P2Y_1$ to $P2Y_{12}$ during serum starvation suggests the existence of a self-sufficient mechanism for survival of C6 cells under unfavorable conditions. Furthermore, it is well known that prolonged activation of the $P2X_7$ receptor is a common signal for apoptosis and cell death due to massive Ca^{2+} entry in many cells (Ferrari et al. 1997). The lack of such $P2X_7$ receptor activity in glioma C6 cells supports the hypothesis that gliomas are strongly oriented to pro-survival mechanism and intensive proliferation.

3.2 Properties of P2 Receptors

ATP release from neuronal cells and synaptic terminals occurs *via* exocytosis. However, recent findings indicate that both adenine and uridine nucleotides can be released from non-neuronal cells *via* regulated processes such as: exocytosis, secretion of granules, efflux, changes in osmolarity or mechanical perturbation. Moreover, in primary astrocytes, astrocytoma and glioma cells ATP release is accompanied by enhanced accumulation of extracellular UTP and this phenomenon is a result of the UDP-glucose release (for review see Lazarowski 2006).

ATP released to the extracellular space can exert its effects *via* activation of the plasma membrane P2 nucleotide receptors, both P2X and P2Y. Afterwards, ATP can also be degraded to ADP, AMP and adenosine by ecto-enzymes (ectonucleotidases) present on the cell surface (Zimmermann 2000; see also Chap. 5). To date, four subtypes of P1 adenosine receptors have been cloned: A_1, A_{2A}, A_{2B} and A_3. They all belong to the G protein-coupled metabotropic receptor superfamily and their effects are presented and discussed in Chap. 2 of this book. Here, we present the description of P2 receptors, whose activation plays a crucial role in glioma biology.

3.2.1 P2X Receptors

To date, seven mammalian P2X receptor subtypes ($P2X_1$ – $P2X_7$) have been cloned (for review see Burnstock 2007; Ralevic and Burnstock 1998). Ionotropic P2X receptors are ATP-gated, nonselective cation channels formed by three subunits (for review see North and Surprenant 2000; North 2002). Each subunit consists of two transmembrane domains with short intracellular NH_2 and COOH termini. These two segments are associated by a large extracellular N-glycosylated hydrophilic loop. One type or more than one type of subunits can form homotrimeric or heterotrimeric receptors, respectively. Activation of P2X receptors by ATP, but not UTP, produces rapid Ca^{2+} and Na^+ influx coupled with K^+ efflux (Fig. 3.1) and results in depolarization of cells. This, in turn, leads to opening of additional plasma membrane voltage-dependent calcium channels and Ca^{2+} flows in from the extracellular space. Thus, upon P2X receptor activation, opening of cation channels results in elevation of intracellular Ca^{2+} concentration ($[Ca^{2+}]_i$) and in reduction of the membrane potential. Such effects are usually detected by properties of the unitary currents of ion channels in outside-out patches and by measuring increase of $[Ca^{2+}]_i$. The time course of calcium response is one of the main criteria distinguishing ionotropic (P2X) form metabotropic (P2Y) receptors. P2X are characterized by rapid, within milliseconds, $[Ca^{2+}]_i$ elevation while P2Y receptors response is much slower and occurs in minutes (North and Surprenant 2000; Ralevic and Burnstock 1998).

The P2X receptor family shows many pharmacological differences, however ATP is the main agonist of all P2X subunits and neither UTP nor UDP lead to their activation. Another ligand of several P2X receptor subunits is an analog of ATP, α,β-methylene ATP, although it is able to activate receptors in a different range (for more details see Burnstock 2007). Among members of the P2X receptor family the homomeric $P2X_7$ is atypical. It needs ATP concentration an order of magnitude higher (milimolar) than other P2X family members and is 10–30 times more sensitive to the ATP analog, 2′,3′-*O*-(4-benzoylbenzoyl)-ATP (BzATP). Furthermore, activation of $P2X_7$ by milimolar concentrations of ATP for prolonged time generates the conversion of this cation channel to a large nonselective transmembrane pore. In consequence, massive Ca^{2+} influx is observed and small molecules (up to 900 Da in size) enter the cell. Due to such properties, $P2X_7$ receptors may be responsible for an apoptotic fate of the cell (Anderson and Nedergaard 2006; North 2002; White and Burnstock 2006).

P2X receptors are widely expressed in the body. They are present not only in excitable cells, such as neurons and skeletal muscle cells, but also in non-excitable ones, for example: smooth muscle, epithelial, endothelial or glial cells. Due to rapid response, they are largely viewed as mediators of fast, short-term neuronal signaling and cell-to-cell communication. P2X receptors also play an

Fig. 3.1 Schematic diagram of nucleotide signaling pathways activated in glioma C6 cells. *Upper part*: Effects of extracellular nucleotides on metabotropic (P2Y) and ionotropic (P2X) receptors. ATP stimulates P2X receptors and together with UTP acts on the $P2Y_2$ receptor. Ectoenzymes hydrolyze ATP to ADP and to adenosine, which acts on P1 receptors. ADP stimulates both $P2Y_1$ and $P2Y_{12}$ receptors. $P2Y_1$ and $P2Y_2$ activate PLC *via* protein G_q. (**a**) PLC converts PIP_2 to IP_3 and DAG, which activates PKC. (**b**) IP_3 binds to its receptors in the endoplasmic reticulum (ER) and causes release of Ca^{2+} to cytosol. Calcium inhibits adenylate cyclase (AC) activity. (**c**) The $P2Y_{12}$ receptor coupled to G_i protein inhibits AC activity, decreases intracellular level of cAMP and inhibits PKA activity. *Lower part*: Visualization of signaling cascades in glioma C6 cells described above. (**a**) PIP_2 breakdown activated by ADP *via* $P2Y_1$ receptor activation. PIP_2 is bound to GFP and fluorescence microscope pictures show the same cell *before* and *after* 5 min of ADP stimulation. *Arrows* mark points of PIP_2 breakdown. (**b**) Effect of ADP, ATP and UTP on $[Ca^{2+}]_i$ measured in Fura-2 loaded cells. Experiments performed in the presence of extracellular Ca^{2+}. Each trace represents the response of an individual cell. Notice a rapid rise in $[Ca^{2+}]_i$ (the first phase of the signal) upon addition of agonists (*arrows*), followed by a sustained elevation of $[Ca^{2+}]_i$ (the second phase of the signal). (**c**) ADP-induced inhibition of the isoproterenol-elicited cAMP response. Notice a very low cAMP level in control cells in the absence of isoproterenol. The inhibitory effect of ADP was reversed by PTX, a specific inhibitor of G_i protein, responsible for the inhibition of AC. For details, see text (Adapted from Czajkowski et al. (2002) (**a**) with permission from Elsevier and from Sabała et al. 2001 (**b** and **c**) with permission from John Wiley and Sons)

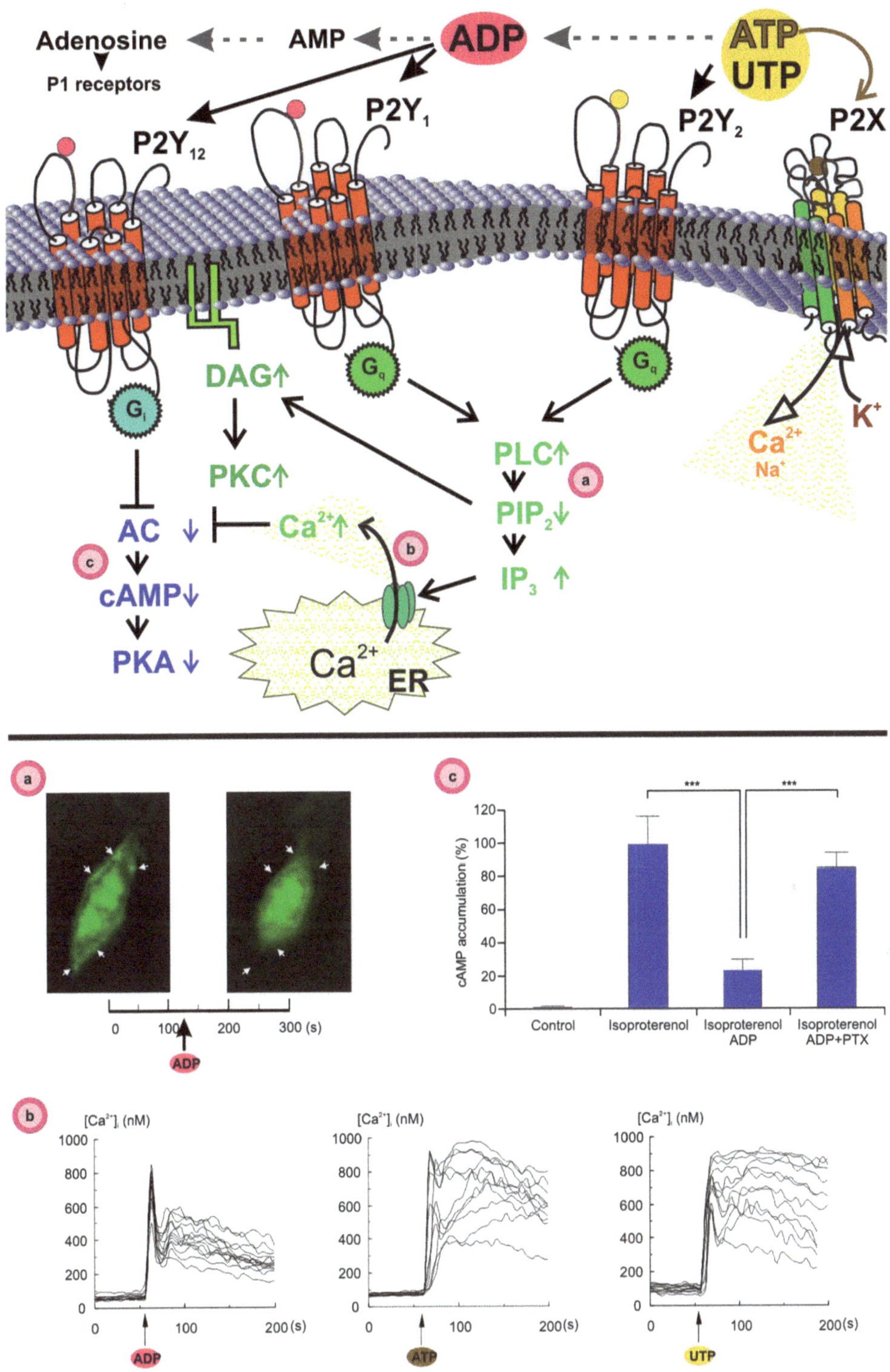
Adenosine
P1 receptors
AMP
ADP
ATP
UTP
P2Y12
P2Y1
P2Y2
P2X
Gi
Gq
Gq
DAG↑
PKC↑
PLC↑
PIP2↓
IP3 ↑
AC ↓
cAMP↓
PKA ↓
Ca2+↑
Ca2+
ER
K+
Ca2+
Na+
a
b
c
0
100
200
300 (s)
ADP
cAMP accumulation (%)
120
100
80
60
40
20
0
***
***
Control
Isoproterenol
Isoproterenol ADP
Isoproterenol ADP+PTX
[Ca2+]i (nM)
1000
800
600
400
200
0
0
100
200 (s)
ADP
ATP
UTP

important role in muscle development and fiber contraction. Ryten et al. demonstrated that in satellite cells activation of $P2X_5$ receptors inhibited proliferation and stimulated expression of muscle cell differentiation markers (Ryten et al. 2002). Similarly, results from our laboratory showed that in the C2C12 cell line, $P2X_5$ is involved in regulation of the shift between myoblast proliferation and differentiation (Banachewicz et al. 2005).

Fumagalli et al. presented evidence that primary rat cortical astrocytes express all cloned P2X receptors except for the $P2X_6$ subtype (Fumagalli et al. 2003). In contrast, glioma C6 cells growing *in vitro* express mRNA for the $P2X_2$, $P2X_4$ and $P2X_5$ subtypes. However, after implantation of the cells to rat brain and then their further growth as primary culture (*ex vivo* C6 model) only $P2X_4$ mRNA can be detected (Braganhol et al. 2009). Expression of $P2X_4$ was also reported in C6 glioma cells implanted into rat brain, where the presence of tumor cells was associated with microglia activation (Guo et al. 2004). $P2X_7$ receptors expressed in neuronal populations, due to their ability to massively increase the intracellular Ca^{2+} level, play a role in excitotoxicity. In various CNS injury models, e.g. *in vivo* spinal cord trauma, activation of $P2X_7$ receptors by a high level of released ATP, leads to death of neurons (Anderson and Nedergaard 2006). On the other hand, in glial cells with an immunocompetent role in CNS (microglia), $P2X_7$ receptors participate in controlling proliferation by supporting cell cycle progression (Bianco et al. 2006). The expression and functional activity of the $P2X_7$ receptors in glioma C6 cells will be discussed in further sections of this Chapter.

3.2.2 P2Y Receptors

Up to now, the mammalian P2Y receptor family comprises eight subtypes: $P2Y_1$, $P2Y_2$, $P2Y_4$, $P2Y_6$, $P2Y_{11}$, $P2Y_{12}$, $P2Y_{13}$ and $P2Y_{14}$ (Abbracchio et al. 2003; Boarder and Webb 2001; Communi et al. 2001; Harden et al. 1998; Hollopeter et al. 2001; King et al. 2000; Zhang et al. 2002). The missing numbers represent receptors cloned from non-mammalian vertebrates and those not fully characterized, called "orphan" receptors.

Similarly to other metabotropic receptor proteins, P2Y contain seven transmembrane domains linked to each other by extra- and intra-cellular hydrophilic loops, the extracellular NH_2 terminus and intracellular COOH terminus (Burnstock 1997, 2007; Ralevic and Burnstock 1998). The schematic structure of the receptors is presented in Fig. 3.1. P2Y receptors act *via* binding to a single α,β,γ – heterotrimeric G protein. It is suggested that positively charged amino acid residues present in transmembrane regions 3, 6 and 7 form a ligand-docking pocket and are involved in electrostatic interaction with phosphate groups of exogenous ATP. This binding is affected by functional diversity of the intracellular loops and the COOH terminus (Ralevic and Burnstock 1998).

From a phylogenetic and structural point of view, the P2Y receptor family can be divided into two subgroups with relatively high structural divergence. The first one contains $P2Y_1$, $P2Y_2$, $P2Y_4$, $P2Y_6$ and $P2Y_{11}$ receptors, while to the other one belong $P2Y_{12}$, $P2Y_{13}$ and $P2Y_{14}$ receptors (Abbracchio et al. 2003). Each subgroup is characterized by specific nucleotide-binding motifs in transmembrane domains 6 and 7, considered as important for ligand binding and crucial for agonist activity (Abbracchio et al. 2003). All P2Y receptors classified to the first subgroup are coupled to $G_{q/11}$ protein. Their stimulation, *via* G_q protein, activates phospholipase C (PLC), which leads to hydrolysis of phosphatidylinositol-4,5-bisphosphate (PIP_2) and to formation of diacylglicerol (DAG) and inositol-1,4,5-trisphosphate (IP_3) (Boarder and Hourani 1998) (Fig. 3.1). DAG activates protein kinase C (PKC), while IP_3 binds to its specific receptors on the endoplasmic reticulum (ER) membrane, evoking Ca^{2+} release from the ER stores (Berridge 1995; Putney and Bird 1993). Taken together, stimulation of $P2Y_1$, $P2Y_2$, $P2Y_4$ and $P2Y_6$ receptors leads to mobilization of intracellular Ca^{2+}. The last member of this subgroup – $P2Y_{11}$ can be coupled to either $G_{q/11}$ or G_s proteins. In consequence, it is positively linked to two signaling pathways by activating either PLC or adenylate cyclase, respectively (Communi et al. 1997; Ralevic and Burnstock 1998). $P2Y_{12}$, $P2Y_{13}$ and $P2Y_{14}$ receptors, phylogenetically and structurally classified to the second subgroup of P2Y receptors, are all coupled to G_i protein and their stimulation negatively affects adenylate cyclase and decreases cAMP level (Abbracchio et al. 2003; Moore et al. 2003) (Fig. 3.1). Thus, a slower response of P2Y receptors to agonists than that of P2X is a result of the involvement of the second messenger systems.

From a pharmacological point of view, the P2Y receptor family is characterized by different sensitivities to adenine and/or uridine nucleotides (see Table 3.1). It has been found that ADP is the most potent natural agonist of $P2Y_1$ receptors, while UTP and UDP are not effective (Leon et al. 1997). However, the stable analog of ADP, 2-methylthio ADP (2MeSADP), is even a more potent $P2Y_1$ agonist. $P2Y_1$ is blocked by suramin, pyridoxal-phosphate-6-azophenyl-2′,4′ disulfonic acid (PPADS), MRS2179 and MRS2269. The two last compounds were identified as selective antagonists of this receptor (Burnstock 2007) (Table 3.1). $P2Y_2$ responds to both ATP and UTP with approximately equal potency and it is insensitive to or only weakly activated by ADP or 2MeSADP (Nicholas et al. 1996). It is worth to add that two rodent receptors: $P2Y_2$ and $P2Y_4$ are equipotently activated by ATP and UTP. They can be distinguished by their proper antagonists, since suramin blocks $P2Y_2$ while Reactive blue 2 blocks $P2Y_4$ receptors (Burnstock 2007). Contrary to rat, the human $P2Y_4$ receptor is highly more selective to UTP than ATP (Nicholas et al. 1996). Furthermore, the $P2Y_6$ receptor is UDP-sensitive and is activated weakly, if at all, by UTP, ATP or ADP (Nicholas et al. 1996). The $P2Y_{11}$ receptor was cloned from human placenta and has been shown to be activated by ATP and ADP but not by UTP or UDP (Communi et al. 1997). $P2Y_{12}$ and $P2Y_{13}$ receptors are phylogenetically and structurally related (Communi et al. 2001). They are selective for ADP and 2MeSADP and blocked by ARC- compounds. The $P2Y_{13}$ receptor can be distinguished from $P2Y_{12}$ by the use of PPADS, which acts as an antagonist of $P2Y_{13}$ (but not of $P2Y_{12}$), or by the use of p^1,p^4-di(adenosine-5′)tetraphosphate (Ap_4A), which acts as an agonist of $P2Y_{12}$ and as an antagonist of the $P2Y_{13}$ receptor (Claes et al. 2001). The recently cloned $P2Y_{14}$ receptor is activated by UDP-glucose, UDP-galactose, UDP-glucuronic acid and UDP-*N*-acetyl-glucosamine, while no significant responses are detected when ATP, ADP or UTP are used (Abbracchio et al. 2003; Harden et al. 2010).

As mentioned before (see Sect. 3.2.2), all receptors belonging to the $P2Y_{12}$ like subfamily are exclusively coupled to $G_{i/o}$ protein and decrease cAMP level in cells (Harden et al. 2010). However, it has also been suggested that the $P2Y_{14}$ receptor coupled to $G_{i/o}$ protein can activate PLCβ *via* β/γ subunits of this protein (Bianco et al. 2005). A similar activation of PLCβ *via* $G_{i/o}$ protein coupling might occur in the case of other P2Y receptors and among them $P2Y_2$ (Burnstock 2007).

Interestingly, for the last several years a couple of reports appeared describing a change in the downstream pathways activated by nucleotide receptors after binding integrins. It was shown that the arginine-glycine-aspartic acid (RGD) domain is responsible for selectively attaching $\alpha_V\beta_3/\beta_5$ integrins to $P2Y_2$ receptors in human astrocytoma cells (Erb et al. 2001). Thus, UTP is able to stimulate G_0 and G_{12} proteins which lead to the activation of small GTPases: Rac and Rho – proteins crucial for actin cytoskeletal rearrangements required for chemotaxis (Bagchi et al. 2005; Liao et al. 2007). Moreover, in primary rat astrocytes, UTP-stimulated cell migration was inhibited by anti α_V integrin antibodies (Wang et al. 2005). All those data suggest that in regions of strong substratum adhesion, $P2Y_2$ receptors may transduct signals through subunits other than G_q, but also that $P2Y_2$-associated pathways may be important factors regulating migration-dependent events like astrogliosis in brain disorders. The above issues will be precisely described in Chap. 6 of this book.

P2Y receptors are widely distributed in nearly all human and rodent tissues. Several neuronal and glial cell lines have been derived from human or rodent brain (see Chap. 1 and, for review, see Sak and Illes 2005).

It has been suggested that in CNS, P2Y receptors may take a part in many basic functions, such as cellular metabolism, proliferation and differentiation, both under physiological and pathological conditions (Abbracchio and Verderio 2006; Di Virgilio et al. 2001). In rat cortical astrocytes,

Table 3.1 The main properties of P2Y receptors

Subtype	Chromosomal location of the gene (*H. sapiens*)	Number of amino acids	Agonists activity	Antagonists	The main activated pathways
$P2Y_1$	3q25.2	373	2-MeSADP > 2-MeSATP = ADP > ATP	MRS 2279, MRS 2179, PPADS, suramin	$G_{q/11}$ ⇨ Ca^{2+}↑
$P2Y_2$	11q13.5-q14.1	377	UTP = ATP	suramin	$G_{q/11}$ ⇨ Ca^{2+}↑
$P2Y_4$	Xq13	365	UTP ≥ ATP	Reactive blue 2, PPADS	$G_{q/11}$ ⇨ Ca^{2+}↑
$P2Y_6$	11q13.5	328	UDP > UTP >> ATP	Reactive blue 2, PPADS, suramin	$G_{q/11}$ ⇨ Ca^{2+}↑
$P2Y_{11}$	19p13.2	374	AR-C67085MX > BzATP ≥ ATPγS > ATP	suramin, reactive blue 2,	$G_{q/11}$ ⇨ Ca^{2+}↑ G_s ⇨ cAMP↑
$P2Y_{12}$	3q24-q25	342	ADP = 2-MeSADP	AR-C67085MX, cangrelor	G_i ⇨ cAMP↓
$P2Y_{13}$	3q24	354*	ADP = 2-MeSADP >> ATP & 2-MeSATP	AR-C67085MX, cangrelor	G_i ⇨ cAMP↓
$P2Y_{14}$	3q24-q25.1	338	UDP-glucose = UDP-galactose	–	$G_{i/0}$ ⇨ cAMP↓; Ca^{2+}↑

Designations

2-MeSADP, 2-Methylthioadenosine- 5′- O- diphosphate; 2-MeSATP, 2- Methylthioadenosine- 5′- O-triphosphate; ATPγS, adenosine 5′-0(3-thiotriphosphate); AR-C67085MX, 2-Propylthio-D-β,γ-dichloromethylene-ATP; cangrelor, (known also as: AR-C69931MX) – [(2R,3S,4R,5R)-3,4-dihydroxy-5-[6-(2-methylsulfanylethylamino)-2-(3,3,3-trifluoropropylsulfanyl)purin-9-yl]oxolan-2-yl]methyl dihydrogen phosphate; BzATP, 2′,3′-*O*-(4-benzoylbenzoyl)-adenosine 5′-triphosphate; MRS 2179, 3′-Adenylic acid,2′-deoxy-N-methyl-, 5′-(dihydrogen phosphate); MRS 2279, (1R,2S,4S,5S)-1-[(phosphato)methyl]-4-(2- chloro-6-aminopurin-9-yl) bicyclo [3.1.0]-hexane-2- phosphate; PPADS, Pyridoxal phosphate-6-azophenyl-2′,4′-disulfonic acid

↑ increase

↓ decrease

*49 % identity in amino acid sequence with $P2Y_{12}$ receptor

Fumagalli et al. described the expression of $P2Y_1$, $P2Y_2$, $P2Y_4$, $P2Y_6$, $P2Y_{12}$ and $P2Y_{14}$ receptors and suggested their participation in astrocyte-to-astrocyte communications (Fumagalli et al. 2003). This phenomenon occurs due to the release of ATP, which may act as an autocrine and paracrine messenger initiating purinergic signaling processes. A similar role of these receptors in communication between astrocytes and neurons was also reported (Bezzi and Volterra 2001).

The expression of P2Y receptors was examined in mouse microglial cell line N9 and in primary microglia derived from rat brain, cultivated either in resting conditions or exposed to bacterial lipopolysacharide (LPS) (Bianco et al. 2005). Long-term exposure to LPS resulted in the increase of $P2Y_6$ and $P2Y_{14}$ receptors activity. This was explained as a part of a protective mechanism that minimizes sensitivity of microglia to external cell death-inducing signals (Bianco et al. 2005). These data are in agreement with the supposition that the $P2Y_{14}$ receptor may play an important neuroimmune function (Abbracchio et al. 2003). Similarly, in 1321N1 astrocytoma cells, activation of $P2Y_6$ by UDP protects cells against TNF-α induced apoptosis (Kim et al. 2003). Moreover, $P2Y_6$ mRNA expression appeared only after implantation of glioma C6 cells into rat brain and further growth in culture (*ex vivo* model), while in C6 control cell line these receptors were not detected. In general, glioma C6 cells express multiple P2Y receptor subtypes: $P2Y_1$, $P2Y_2$, $P2Y_4$, $P2Y_6$, $P2Y_{12}$, $P2Y_{13}$ and $P2Y_{14}$ (Barańska et al. 2004; Braganhol et al. 2009; Czajkowski and Barańska 2002; Krzemiński et al. 2008; Van Kolen and Slegers 2006b).

In rat native cerebellar astrocytes, Carrasquero et al. described expression of $P2Y_1$ and $P2Y_{13}$ receptors, while $P2Y_{12}$ was not detected (Carrasquero et al. 2005). The authors suggested that in these cells $P2Y_{13}$ substitutes for $P2Y_{12}$ and cooperates with $P2Y_1$, similarly as it is observed in platelets, where $P2Y_1$ and $P2Y_{12}$ take part in ADP-induced platelet aggregation (Gachet 2006; Hollopeter et al. 2001). The $P2Y_{13}$ receptor was found to be highly expressed not only in human and mouse brain tissue but also in rat primary astrocytes, where it was proposed to participate in reactive astrogliosis (Fumagalli et al. 2004). In glioma C6 cells, the $P2Y_{12}$ receptor is highly expressed and functionally active. Moreover, its involvement in signaling processes is often opposite to that of the $P2Y_1$ receptor (Baranska et al. 2004; Krzemiński et al. 2007).

The expression and functional activity of $P2Y_1$ and $P2Y_{12}$ receptors in glioma C6 cells and the cross-talk between them will be discussed in further sections of this Chapter.

3.3 $P2Y_1$, $P2Y_2$ and $P2Y_{12}$ Receptor Expression and Functionality in Glioma C6 Cells

In the 1990s of the twentieth century, extracellular ATP was shown to generate IP_3 formation in glioma C6 cells, indicating the presence of purinergic receptors coupled to phosphoinositide turnover and hydrolysis (Hirano et al. 1991; Kitanaka et al. 1992; Lin and Chuang 1993, 1994).

While the presence of ATP-responding, PLC-activated $P2Y_2$ receptors in this cell line was generally accepted, the presence of $P2Y_1$ receptors that stimulate PLC activity was a subject of a serious debate. The studies concerning accumulation of IP_3, both in C6-2B and in C6 cells, suggested the presence of $P2Y_1$-like receptors affecting not IP_3 formation but inhibition of adenylate cyclase. Moreover, these receptors were not blocked by the $P2Y_1$ receptor antagonist, PPADS (Boarder and Hourani 1998; Schachter et al. 1996, 1997). Therefore, the existence of two different ADP-responding receptors has been proposed: one ($P2Y_1$), coupled to PLC and blocked by PPADS and another one, termed P2Y-C6 or $P2Y_{AC}$, present in glioma C6 cells and negatively coupled to adenylate cyclase (Claes et al. 2001; Grobben et al. 2001). On the other hand, Webb and others reported that the cloned rat $P2Y_1$ receptor DNA and that derived from glioma C6 cells were 100 % identical in the entire coding region (Webb et al. 1996). Thus, the authors suggested that in different cell types the same $P2Y_1$ receptor might either activate PLC or inhibit adenylate cyclase.

These confusing data were finally clarified by experiments performed in our laboratory in which the activity of $P2Y_1$ receptors was estimated by their ability to increase $[Ca^{2+}]_i$. We have previously shown that glioma C6 cells belong to the type of non-excitable cells (Barańska et al. 1995). In such cells, the voltage-dependent Ca^{2+} channels are absent and the biphasic, capacitative Ca^{2+} signaling is mediated by the inositol phosphate system (Berridge 1995; Putney and Bird 1993; see also Chap. 4). In this process, the initial rise of $[Ca^{2+}]_i$ (the first phase of the cell response) results from a direct effect of IP_3 on the ER calcium stores and can be observed even in the absence of external Ca^{2+}. The second phase of the calcium signal is generated by an opening of voltage-independent plasma membrane Ca^{2+} channels and Ca^{2+} inflow into the cell as a consequence of the depletion of the ER stores. We have shown that in glioma C6 cells, ADP as well as ATP and UTP initiate a calcium response typical for non-excitable cells and compatible with the capacitative model of Ca^{2+} influx (Sabala et al. 2001). All the above mentioned nucleotides induce $[Ca^{2+}]_i$ elevation which starts with an initial peak response (the first phase) and is followed by the second phase – a long, sustained plateau (Fig. 3.1b). Moreover, soon afterwards, cells treated with ATP failed to respond to UTP and *vice versa*, whereas similar cross-desensitization between ADP and UTP, or ADP and ATP did not occur. These results indicate that ATP and UTP act on the same common subtype of P2Y receptors ($P2Y_2$), whereas ADP is recognized by another one (Sabała et al. 2001). Treatment of cells with PPADS and MRS2179, the $P2Y_1$ receptor specific antagonists, abolish ADP-induced Ca^{2+} mobilization. The $P2Y_2$ receptor is insensitive to these compounds. Furthermore, 2MeSADP, a selective agonist of high potency for the $P2Y_1$ receptor, is more efficient than ADP in increasing $[Ca^{2+}]_i$ (Czajkowski et al. 2002; Sabała et al. 2001).

As it is shown in Fig. 3.1b, the kinetics of ADP-evoked $[Ca^{2+}]_i$ changes differs from those generated by ATP and UTP. The response is more transient and quickly returns to the basal level. Therefore, one could expect that ADP-evoked IP_3 accumulation was also transient and too short to be estimated. This could be a reason for the former belief that in glioma C6 cells, ADP-responding receptor stimulation did not lead to PLC activation. However, using a direct visualization of GFP coupled PLCδ-PH domain of PIP_2 in C6 cells, we revealed that after stimulation with ADP, the PIP_2-bound GFP fluorescence was reduced at the cell surface, indicating the breakdown of PIP_2 by activated PLC (Fig. 3.1a) (Czajkowski et al. 2002).

In contrast to PLC activation, ADP-evoked adenylate cyclase inhibition has been documented for a long time (Boyer et al. 1993). Thus, in glioma C6 cells ADP has two activities: it stimulates PLC and inhibits adenylate cyclase. The latter activity can be shown by inhibition of the isoproterenol-elicited cAMP accumulation by 80 % in the presence of ADP (Fig. 3.1c). This effect is reversed by the treatment of cells with pertussis toxin (PTX), indicating G_i protein involvement, and is insensitive to PPADS (Sabała et al. 2001).

Knockdown of the $P2Y_1$ receptor with antisense oligonucleotides, as well as treatment of cells with MRS2179 or PPADS, caused loss of receptor-mediated PLC activity but were without effect on ADP- induced adenylate cyclase inhibition. All of these indicate that there exist two different receptors in glioma C6 cells, both responding to ADP (Czajkowski et al. 2002). At the same time, a P2Y receptor subtype, previously called $P2T_{AC}$, that is coupled to the inhibition of adenylate cyclase in rat blood platelets, was cloned and designated as $P2Y_{12}$ (Hollopeter et al. 2001; Savi et al. 2001). The presence of this receptor was also reported in glioma C6-2B cells (Jin et al. 2001). Its identity and the presence in glioma C6 cells, as well as the presence of $P2Y_1$ and $P2Y_2$ receptors, was verified and confirmed by polymerase chain reaction using reverse-transcribed total RNA (RT-PCR) and Western blot analysis (Barańska et al. 2004; Czajkowski et al. 2002, 2004; Krzemiński et al. 2007; Sabała et al. 2001) (Fig. 3.2).

Thus, in glioma C6 cells the $P2Y_1$, $P2Y_2$ and $P2Y_{12}$ receptors are expressed. ATP and UTP act as agonists on $P2Y_2$, whereas ADP stimulates both $P2Y_1$ and $P2Y_{12}$ receptors. The molecular identity, pharmacological analysis and the typical biphasic Ca^{2+} response showed that $P2Y_2$ and $P2Y_1$ are linked to PLC and Ca^{2+} release, whereas $P2Y_{12}$ is negatively coupled to adenylate cyclase through G_i protein.

In conclusion, it is worth to add that one of the most characteristic features of glioma C6 cells is the massive expression of the $P2Y_{12}$ receptor, long known to exist in this cell line and previously denominated as P2Y-C6 or $P2Y_{AC}$. High expression of this receptor leads to a very low constitutive basal level of cAMP (see Fig. 3.1c). The other characteristic feature is the instability of the $P2Y_1$ receptor expression. The lack of activity of the $P2Y_1$ receptor reported previously (Grobben et al. 2001; Schachter et al. 1997) might be explained not only by the transient and very short stimulation of PLC and IP_3 accumulation but also, and most likely, by conditions of cell culturing and handling. It has been recently shown that the long-term absence of serum in the C6 cell culture medium specifically affects the $P2Y_1$ receptor, strongly decreasing its expression and, simultaneously, its activity (Barańska et al. 2004; Czajkowski et al. 2004; Krzemiński et al. 2007 and see Chap. 4). The effect of serum withdrawal on $P2Y_1$ and $P2Y_{12}$ receptor expression is presented in detail in the next Section.

3.4 Serum Withdrawal

3.4.1 Effect on $P2Y_1$ and $P2Y_{12}$ Receptor Expression

It is well established that proliferation of cells is induced by numerous substances present in animal serum such as growth factors, cytokines or bioactive lipids. Among them, an important serum component is lysophosphatidic acid, bound to albumin and carried in the blood stream (Jalink et al. 1994; Kranenburg and Moolenaar 2001). Therefore, media used for cells cultivation are usually supplemented with sera originating from various sources. However, it is known that for some *in vitro* studies, such as mitogenic signaling, it is better to use withdrawal cells (Grobben et al. 2001; Tu et al. 2000). Such proceeding should also be applied for glioma C6 cells in the case of mitogen-activated protein kinase studies, including extracellular signal-regulated kinases 1/2 (ERK1/2). In cells cultivated in the presence of serum the basal activity of ERK1/2 is so high that nucleotides, ADP and UTP, are unable to induce additional stimulation of ERK1/2 activity (Barańska et al. 2004; Czajkowski et al. 2004).

Sometimes, differences in cell culture conditions may be misleading. The group of Slegers was unable to detect activation of PLC and calcium release after ADP stimulation in glioma C6 cells grown in chemically defined serum-free medium. In consequence, the authors suggested the lack of $P2Y_1$ receptor expression in those cells (Grobben et al. 2001). In contrast to the results of Slegers' group, studies performed on cells growing in the medium supplemented with serum revealed the presence and functional activity of the classic $P2Y_1$ receptors (Czajkowski et al. 2002; Sabała et al. 2001). These data show that variations in the composition of the culture medium, such as the presence or absence of serum, may have important implications. Therefore, comparison of the effects of these two cell culture conditions seemed to be essential (Czajkowski et al. 2004; Krzemiński et al. 2007).

Figure 3.2a shows the mRNA level of $P2Y_1$, $P2Y_2$ and $P2Y_{12}$ receptors in glioma C6 cells, either serum-deprived or cultured in the presence of 10 % fetal bovine serum. In cells starved for 48 h in the serum-free medium, the level of $P2Y_1$ mRNA strongly decreased, whereas $P2Y_{12}$ mRNA expression slightly increased. When C6 cells were cultured for 98 h in chemically defined serum-free medium, as described by Grobben et al. (Grobben et al. 2001), the same effect was observed (Czajkowski et al. 2004) (not shown in Fig. 3.2). In contrast, serum deprivation was without any effect on $P2Y_2$ mRNA expression (Fig. 3.2a). A similar pattern of changes was demonstrated for proteins (Krzemiński et al. 2007). Figure 3.2b shows the level of $P2Y_1$ and $P2Y_{12}$ receptor protein in cells cultivated for 24, 48, 72 and 98 h in the serum-free medium. The level of the $P2Y_1$ receptor gradually decreased and was very low after 72 h, and particularly after 96 h, while the level of $P2Y_{12}$ increased. However, even after 96 h of serum deprivation, when the medium was replaced by that containing serum, the level of the $P2Y_1$ receptor could be restored (Fig. 3.2c).

The interrelation between $P2Y_1$ and $P2Y_{12}$ expression led to the hypothesis, that increased level of $P2Y_{12}$ receptors, observed during long-time serum starvation, might be a consequence of $P2Y_1$ decrease. Up to now, there are no literature data about the regulation of transcription of $P2Y_1$ and

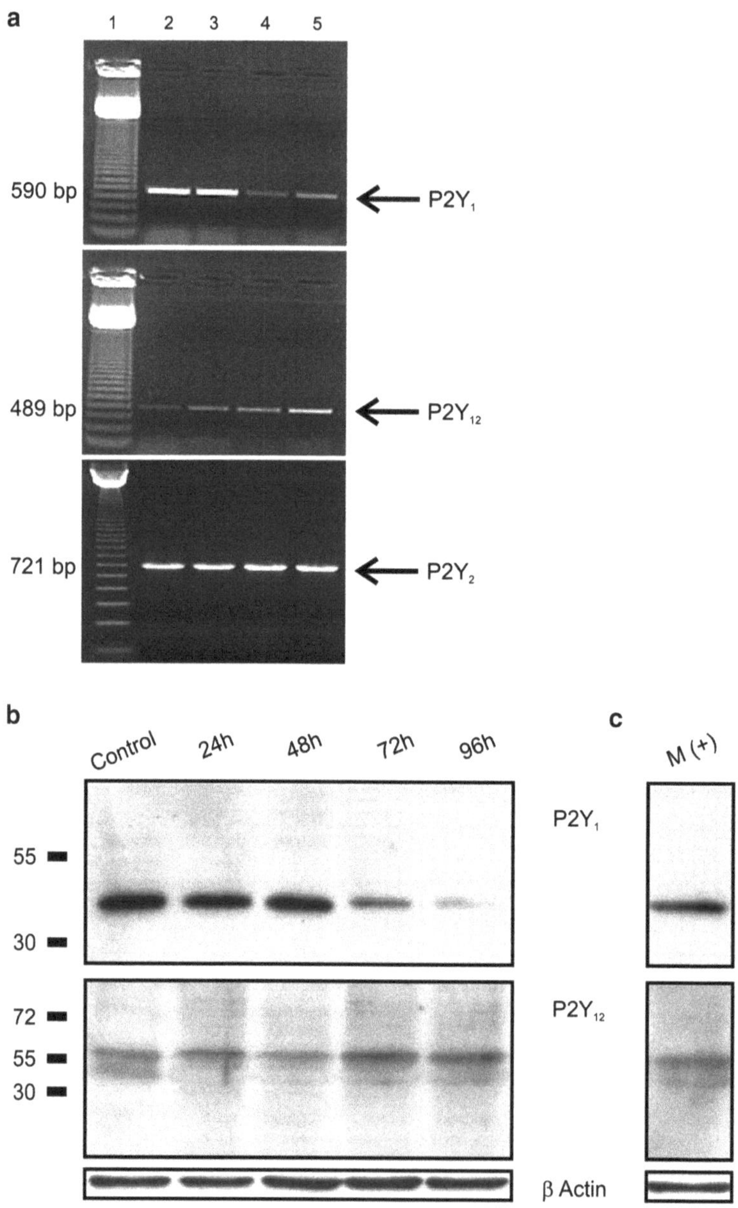

Fig. 3.2 Effect of serum deprivation on mRNA and protein levels of nucleotide receptors in glioma C6 cells. (**a**) Detection of $P2Y_1$, $P2Y_2$ and $P2Y_{12}$ mRNA by RT-PCR reaction. Amplification products of 590, 489 and 721 bp correspond to $P2Y_1$, $P2Y_{12}$ and $P2Y_2$ mRNAs, respectively. Lane 1, 123 DNA ladder; lanes 2 and 3, cells grown in the medium supplemented with 10 % serum; lanes 4 and 5, cells after 48 h serum deprivation. (**b**) Changes in the protein level of $P2Y_1$ and $P2Y_{12}$ receptors detected using Western blot in non-starved cells (control) and cells cultivated without serum for 24, 48, 72 and 96 h, respectively. (**c**). Cells cultivated in serum-free medium for 96 h followed by 24 h recovery in fresh medium with serum (M+) (Reproduced from Barańska et al. (2004) (**a**) with permission from Elsevier and from Krzemiński et al. 2007 (**b** and **c**) with permission from John Wiley and Sons)

$P2Y_{12}$ genes. Furthermore, rat genome analysis in the Ensembl database showed no essential differences between $P2Y_1$ and $P2Y_{12}$ gene promoters in transcription factor binding sites, which were additional arguments in favor of the hypothesis. The development and spreading of the siRNA technique encouraged its verification.

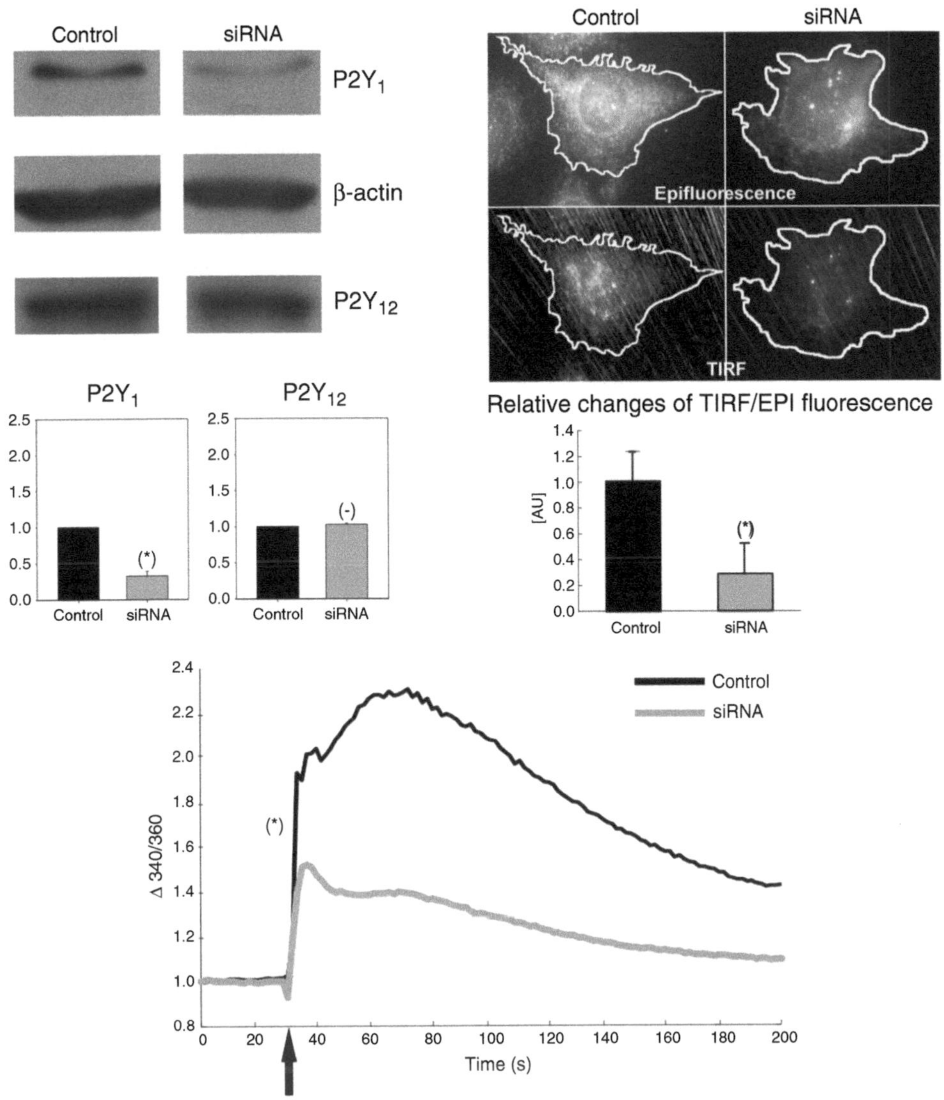

Fig. 3.3 Selective exclusion of $P2Y_1$ receptors by siRNA in glioma C6 cells. (**a**) A typical Western blot experiment. The level of each receptor protein in total protein extract was measured densitometrically and is presented with reference to a control protein (β-actin) (number of analyzed experiments, $N=3$). The protein level in the control is assumed as 1 arbitrary unit (AU). The analysis shows a significant decrease of $P2Y_1$ receptor protein 0.32 ± 0.07 AU ($p<0.05$) and no change in the amount of $P2Y_{12}$ receptor protein 1.02 ± 0.02 AU. (**b**) Immunocytochemical staining of $P2Y_1$ receptors visualized in epifluorescence (*upper pictures*) and in Total Internal Reflection Fluorescence Microscopy (TIRF) module (*lower pictures*). The relative ratio between $P2Y_1$ receptors on the plasma membrane to the total amount of receptors in control cells is 1 ± 0.22 AU (number of analyzed cells, $n=14$) and after siRNA transfection it decreases to 0.29 ± 0.23 AU ($n=18$). The change is statistically significant ($p<0.05$). (**c**) Functional study of $P2Y_1$ receptors – the mean calcium response of control (*black*) and siRNA transfected cells (*grey*) evoked by 10 μM 2MeSADP as indicated by *arrow*, in buffer with 2 mM $CaCl_2$. The change was statistically significant ($p<0.001$)

The selective exclusion of the $P2Y_1$ receptor was confirmed by Western blot and TIRFM (Total Internal Reflection Fluorescence Microscopy). Both analyses showed a decreased level of $P2Y_1$ receptors in total protein extracts and in the plasma membrane, respectively (Fig. 3.3a, b). Moreover, studies of the activity of the $P2Y_1$ receptor confirmed this result (Fig. 3.3c). However, further experiments showed that decrease of the $P2Y_1$ protein level did not cause an increase of the level of $P2Y_{12}$ receptors

(Fig. 3.3a). Similarly, the inhibition of $P2Y_{12}$ expression by siRNA did not lead to the increase of $P2Y_1$ (not shown). In other words, the increased level of $P2Y_{12}$ receptors is not the consequence of inhibition of $P2Y_1$ receptor expression, but is the result of changes in cell culture medium, i.e. serum withdrawal (unpublished results). It can be deemed that the lack of factor/factors present in serum causes changes in expression of both these receptors independently. Nevertheless, the change in expression of $P2Y_1$ and $P2Y_{12}$ during long-term serum starvation is very strong, comparable with a selective experimental blockade by siRNA. Hence, long-term serum starvation is a good model for functional studies of both these receptors.

In conclusion, the long-term serum deprived glioma C6 cells are characterized by very low expression of the $P2Y_1$ receptor. Under such conditions, the $P2Y_{12}$ receptor strongly predominates and is primarily responsible for the ADP-evoked signal transduction. Consequently, the decrease in $P2Y_1$ mRNA and protein expression is reflected by the distinctly lower ADP-evoked calcium response when compared with cells incubated in the presence of serum (Barańska et al. 2004).

In the very instructive review article Sak and Illes reported that in neuronal cancer cell lines of both human and rodent origin, the $P2Y_1$ subtype of P2Y receptors appears to be absent (Sak and Illes 2005). This phenomenon seems to be a general property of neuronal malignant cells. In contrast, in healthy neuronal cells, the $P2Y_1$ receptor predominates (Illes and Alexandre Ribeiro 2004). In an inhospitable environment created by prolonged serum starvation, glioma C6 cells demonstrate similar malignant-like behavior.

3.4.2 Effect on C6 Cell Morphology, Growth and Differentiation

Glioma C6 cells normally have an irregular, flattened shape. They are usually characterized by a typical, bipolar, fibroblast-like morphology (see Chap. 4). However, during prolonged (up to 96 h) serum deprivation 75 % of cells gradually adopted a rounded, astrocyte-like shape. Nevertheless, after 96 h of serum deprivation, when the medium was replaced with one containing 10 % serum, the cells returned to their typical fibroblast-like morphology and the amount of altered cells decreased to about 25 % (Krzemiński et al. 2007).

Long-term serum withdrawal also induced cell cycle arrest. Such effect was demonstrated in many cell lines including astrocytes (Chou and Langan 2003). Tu et al. reported that in glioma C6 cells, 48 h serum deprivation induced growth arrest synchronized in the G_0/G_i phase of cell cycle and the cells minimally incorporated [^{3}H]tymidine, a marker of proliferation (Tu et al. 2000). It was shown that 96 h of serum deprivation arrested 95 % of C6 cells in the G_0/G_i phase (Krzemiński et al. 2007). Again, addition of medium containing serum almost completely restored cells to the control pattern. These data indicate that the observed alterations are not permanent and cells are still ready to be activated. Furthermore, serum deprivation did not cause a significant loss of viability or apoptotic features of glioma C6 cells (Krzemiński et al. 2007).

Similarly to the long-term (days) serum deprivation, the same morphological changes could be observed when cells were incubated only for 10 min in a medium without serum and then, still in the absence of serum, treated with agonists that increased intracellular concentration of cAMP (Oey 1975). Furthermore, Koschel and Tas showed that the treatment of glioma C6 cells with isoproterenol, an agonist of β-adrenergic receptor positively coupled to adenylate cyclase, resulted in a rapid increase in cAMP and subsequently induced a change in the cell morphology from a flat, fibroblast-like to a rounded, astrocyte-like shape (Koschel and Tas 1993). This change was completed in about 30–40 min and occurred only in the absence of serum in cell culture media. The presence of serum prevented these changes and its addition to the morphologically changed cells, restored the fibroblast-like appearance. The component of serum responsible for this effect was identified as lysophosphatidic acid (Koschel and Tas 1993). Sphingosine 1-phosphate, another bioactive lipid present in serum, was also found to be able to reverse this change (Tas and Koschel 1998). It has been further found that lysophosphatidic acid and sphingosine 1-phosphate receptors are widely expressed in numerous cell types. They can be coupled to multiple G proteins: G_i, G_q and $G_{12/13}$ and mediate variety of responses, among them a long-term stimulation of cell proliferation (Kranenburg and Moolenaar 2001; Moolenaar 1999; van Koppen et al. 1996). In fibroblasts, lysophosphatidic acid acting on a receptor coupled with G_i protein inhibits, similarly as the $P2Y_{12}$ receptor, adenylate cyclase, and leads to a decrease of intracellular cAMP level (van Corven et al. 1989).

In glioma C6 cells growing in chemically defined serum-free medium, agents activating β-adrenergic receptor or membrane-permeable cAMP analogues stimulating adenylate cyclase activity (such as dibutyryl cAMP), induced the synthesis of glial fibrillary acidic protein (GFAP) concomitantly with growth arrest (Claes et al. 2004; Roymans et al. 2001). GFAP is widely expressed in astrocytes and is used as a marker of the induction of differentiation into an astrocyte type II (Dahl 1981; Messens and Slegers 1992). Together with changes in GFAP expression, C6 cells, more rounded than cells growing in the presence of 10 % serum (Messens and Slegers 1992), became even more astrocyte-like with a stellate shape (Claes et al. 2004). Furthermore, this cAMP-induced differentiation of C6 cells could be inhibited by activation of the $P2Y_{12}$ receptor (Claes et al. 2004).

In contrast to control conditions, long-term serum starvation, although it imposed growth arrest and astrocyte-like appearance on C6 cells, did not induce differentiation into astrocytes and oligodendrocytes. Under such conditions, expression of GFAP and NG2 proteoglycan (a marker of oligodendrocyte progenitors) was not increased either in the presence or absence of serum in the culture medium (Krzemiński et al. 2007). Taken together, these data pointed to the role of the $P2Y_{12}$ receptor in cell survival. Different expression of $P2Y_1$ and $P2Y_{12}$ receptors in serum-starved and non-starved cells seems to have an important biological significance. The inhibitory effect of $P2Y_{12}$ on adenylate cyclase activity might maintain the low intracellular cAMP level and, due to that, block cAMP-induced differentiation of C6 cells, partially compensating for the lack of serum in the culture medium.

3.4.3 Effect on the $P2Y_{14}$ Receptor

As it was already mentioned above, the $P2Y_{14}$ receptor responds to sugar nucleotides, such as UDP-glucose, UDP-galactose, UDP-glucoronic acid and UDP-N-acetyl-glucosamine. Phylogenetically it belongs to the group of $P2Y_{12}$ and $P2Y_{13}$ receptors (Abbracchio et al. 2003; Harden et al. 2010). $P2Y_{14}$ shares 45 % homology with the $P2Y_{12}$ receptor and, similarly to $P2Y_{12}$ and $P2Y_{13}$, appears to be coupled to PTX-sensitive $G_{i/o}$ protein. The studies on HEK 293 cells co-transfected with $P2Y_{14}$ and different chimeric G α subunits recognized and confirmed that the $P2Y_{14}$ receptor is coupled to $G_{i/o}$ proteins (Moore et al. 2003). Indeed, in murine T-lymphocytes and human neutrophils UDP-glucose induces inhibition of adenylate cyclase (Moore et al. 2003; Scrivens and Dickenson 2005, 2006). However, in rat cortical astrocytes, human monocytes and N9 microglial cells, UDP-glucose also stimulates elevation of intracellular Ca^{2+} level (Bianco et al. 2005; Fumagalli et al. 2003; Skelton et al. 2003).

Our studies showed the presence of the $P2Y_{14}$ receptor on glioma C6 cells. Western blot technique and the treatment with deglycosylation enzyme, N-glycosidase F, demonstrated that in these cells the $P2Y_{14}$ receptor exists in two forms: glycosylated, which highly predominates, and non-glycosylated one (Krzemiński et al. 2008). Binding of UDP-glucose induced two responses sensitive to PTX: an increase of calcium concentration and an inhibition of adenylate cyclase. This suggests participation of the β/γ subunits of $G_{i/o}$ protein in PLCβ activation, as it was previously proposed by Bianco et al. (2005). However, this result does not exclude the possibility that two UDP-glucose receptors exist: one coupled to PLC and another one negatively coupled to adenylate cyclase (Krzemiński et al. 2008). It is worth to add that recently UDP was found to be a potent agonist of both human and rodent $P2Y_{14}$ receptors (Carter et al. 2009; Fricks et al. 2008, 2009). Harden et al. hypothesized that $P2Y_{14}$ and

$P2Y_6$, co-existing in many tissues, might be coordinately activated by UDP, resulting in G_q-dependent activation of PLC by $P2Y_6$ and G_i-dependent inhibition of adenylate cyclase by $P2Y_{14}$, similarly as in the case of $P2Y_1$ and $P2Y_{12}$ receptors activated by ADP (Harden et al. 2010).

Since in glioma C6 cells long-term serum deprivation distinctly affects expression and activity of $P2Y_1$ and $P2Y_{12}$, the same procedure was applied for studies concerning the $P2Y_{14}$ receptor. It was found that during long-term serum deprivation expression of the glycosylated form strongly decreased, whereas that of the non-glycosylated form increased. Simultaneously, calcium response induced by UDP-glucose was reduced in serum-starved cells. Moreover, in 96 h serum-deprived cells, inhibition of isoproterenol-induced cAMP accumulation was two times higher than in non-starved cells. These data show that the glycosylated form of the $P2Y_{14}$ receptor is primarily involved in intracellular calcium mobilization, while the non-glycosylated one in adenylate cyclase inhibition (Krzemiński et al. 2008).

What seems to be the most interesting is the response of glioma C6 cells to the long-term serum deprivation. Serum starvation favors an increase in the expression of receptors, whose activation inhibits cAMP accumulation, such as $P2Y_{12}$ and the non-glycosylated form of $P2Y_{14}$. Since cAMP has been shown to inhibit growth and induce differentiation of glioma cells (Dugan et al. 1999; Kim et al. 2001), such cross-talk between $P2Y_1$ and $P2Y_{12}$ receptors and between glycosylated and non-glycosylated forms of the $P2Y_{14}$ receptor seems to be an important self-regulating process acting against cells differentiation and maintaining proliferation and survival in unfavorable conditions.

3.5 Cyclic AMP Effect on Cell Proliferation, Growth and Differentiation

Cell proliferation and survival is *inter alia* regulated by growth factors *via* receptor tyrosine kinases (RTK). Activation of these cell-surface receptors lead to phosphorylation of ERK1/2 and/or activation of PI3 kinase. These two signaling pathways may be regulated by cyclic AMP.

cAMP is a common second messenger that influences many cellular processes and gene transcription *via* protein phosphorylation (Sutherland 1972). Elevated levels of cAMP lead to activation of different cAMP targets, such as the cAMP-dependent protein kinase A (PKA), considered as the main target of cAMP in eukaryotic cells. However, Epac, the guanine exchange factor (GEF) directly activated by cAMP, was also shown to be a receptor for this compound (Bos 2003; de Rooij et al. 1998; Kopperud et al. 2003). Epac specifically activates a small G protein, Rap1, a member of the Ras-like family of small GTPases (see Fig. 3.4). Depending on PKA and Epac abundance, distribution and localization, these two agents may act independently, synergistically or in opposition to each other in regulating specific cellular functions (for more details see Cheng et al. 2008)

In various types of cells, cyclic AMP may cause different effects. It is involved in regulating various cellular features of differentiation, such as neurite outgrowth in PC-12 cells, or adipocyte formation from 3 T3-L1 fibroblasts (Cheng et al. 2008; Wan et al. 2011). In epithelial cells, hepatocytes, keratinocytes, pancreatic islet β cells or Schwann cells it stimulates proliferation. On the contrary, an increased level of cAMP in fibroblasts, smooth muscle, lymphoid cells or epithelial tumor cells diminishes cell growth (Dugan et al. 1999). These divergent effects depend on whether cAMP activates or inhibits ERK1/2, whose signaling is important for cellular proliferation, differentiation and survival.

In central nervous system-derived neurons, cAMP activates ERK1/2 thus stimulating their growth and survival (Dugan et al. 1999). In spinal motor neurons cultured *in vitro*, elevation of intracellular cAMP extended neuronal survival in serum-free medium for as much as 1 week (Hanson et al. 1998). On the opposite, in primary astrocyte and glioma C6 cells, cAMP inhibits ERK1/2 activity and simultaneously inhibits cells growth (Dugan et al. 1999; Kurino et al. 1996; Qiu et al. 2000; Wang et al. 2001). Proposed mechanisms of these processes are presented in Fig. 3.4.

Thus, in a variety of cell types multiple distinct mechanisms regulate cAMP inhibition or activation of ERK1/2 signaling (for review see Stork and Schmitt 2002). Figure 3.4 shows simplified and

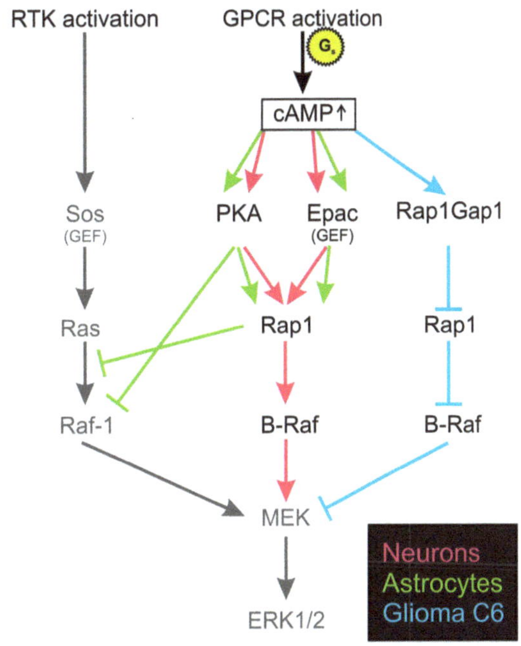

Fig. 3.4 Cyclic AMP-mediated regulation of ERK1/2 signaling in neurons (*pink*), astrocytes (*green*) and glioma C6 cells (*blue*). For details, see text

schematic models of the canonical receptor tyrosine kinase (RTK) signaling pathway (gray color) and those regulated by cAMP in neurons (pink color), astrocytes (green color) and glioma C6 cells (blue color).

RTK signaling occurs *via* a small, monomeric protein G, Ras, and a serine-threonine protein kinase, Raf-1 (Fig. 3.4, gray color). After growth factor stimulation, receptors undergo activation by dimerization and autophosphorylation on tyrosine residues. They, in turn, form a signaling complex with adaptor proteins and a GEF protein (Sos), which catalyzes guanine nucleotide exchange (GDP/GTP) on Ras. Activated Ras (Ras-GTP) recruits Raf-1 to the plasma membrane. Association with Ras results in Raf-1 activation and subsequent phosphorylation of protein kinase MEK (MAP kinase-ERK kinase), which further phosphorylates ERK1/2 both on tyrosine and threonine residues. Doubly phosphorylated ERK1/2 undergoes dimerization and translocation to the nucleus, where it serves as an activator of early response genes. In the case of receptor stimulation by the epidermal growth factor (EGF), ERK activation is an absolute requirement for cell proliferation. The role of P2Y receptors in the cross-talk between growth factor receptors and ERK1/2 activation was reviewed by Van Kolen and Slegers (2006b).

However, activation of MEK and further ERK1/2 may occur due to an alternative pathway, in which not RTK, but G-protein coupled receptors linked to G_s and responsible for intracellular cAMP accumulation take part. In neurons, elevated cAMP level activates PKA or/and Epac, which in turn activates small G protein Rap1 (Fig. 3.4, pink color). This activation occurs either due to direct cAMP binding to Epac or due to cAMP-induced phosphorylation of PKA. Epac belongs to a family of cAMP-binding nucleotide exchange factor proteins (GEFs) that catalyze GDP/GTP exchange on Rap1. Contrary to Ras, Rap1 activates the serine-threonine kinase B-Raf. Activated B-Raf phosphorylates MEK which, in turn, phosphorylates and activates ERK1/2. This pathway was described in PC12 cells (Vossler et al. 1997) and was found to occur in cells that possess the B-Raf protein. B-Raf mRNA was detected in brain, and in neurons; B-Raf seems to be the major MEK activator (Qiu et al. 2000). In contrast, astrocytes do not express this protein (Dugan et al. 1999).

In neurons, when the cAMP level is not increased, growth factors recruit Ras and activate ERK1/2. When cAMP level is elevated, Rap1 is recruited and activated by cAMP (Fig. 3.4, pink color). In astrocytes lacking B-Raf, the increased level of cAMP inhibits ERK1/2 (Fig. 3.4, green color). This inhibition may be attributed to a direct PKA-mediated phosphorylation of Raf-1, which inhibits Ras binding and Raf-1 activity. It is also suggested that Rap1 antagonizes Ras by interfering in Ras-effectors function. Activated Rap1 (when bound to GTP) competes with Ras for Raf-1 binding and thereby blocks Ras-induced Raf-1 activation (Fig. 3.4, green color) (Dugan et al. 1999; Wang et al. 2001). These observations led to the hypothesis that an increased cAMP level stimulates ERK1/2 and induces proliferation in B-Raf expressing cells, but inhibits this cascade in B-Raf negative cells (Dugan et al. 1999).

In glioma C6 cells, the increased level of cAMP also inhibits ERK1/2 activity and cell growth. Dugan et al. suggested that in these cells, the growth-inhibitory effect of cAMP is mediated similarly as in astrocytes (Dugan et al. 1999). However, Wang et al. detected the B-Raf protein in C6 glioma cells and found that, in contrast to astrocytes, cAMP inhibits ERK1/2 by inhibiting Rap1 (Fig. 3.4, blue color) (Wang et al. 2001). Consequently, neither PKA nor Epac were involved in the cAMP-dependent inactivation of ERK1/2. The authors suggested that in C6 cells the cAMP-dependent Rap1 and ERK1/2 inhibition occurred due to Rap1Gap1 activity modulation. Gap1 (GTPase-activating protein 1) is regulated by G_i protein α subunit, which also inhibits adenylate cyclase activity (Zwartkruis and Bos 1999). Thus, an increased level of cAMP leads to Rap1Gap1 stimulation, resulting in Rap1 inhibition. Since the B-Raf protein was detected, Wang et al. proposed that cAMP inhibits ERK1/2 by inhibiting the Rap1/B-Raf/MEK/ERK signaling cascade (Wang et al. 2001). Therefore, the mechanism of this process seems to be different from the one suggested for astrocytes. However, another possibility is a direct cAMP inhibition of some GEF-Rap1 proteins, resulting in Rap1 inhibition (Wang et al. 2001).

There are, nevertheless, several examples where cAMP can inhibit cell proliferation without inhibiting ERK1/2 activity (McKenzie and Pouyssegur 1996). In embryonic fibroblast cells, Ras-1 gene removal does not affect ERK activation, although further development is blocked due to increased apoptosis. Moreover, cAMP inhibition of ERK1/2 may be time-dependent, as in PC12 cells, where cAMP blocks early activation but enhances late activation of ERK1/2 (Arslan and Fredholm 2000).

In many cells, cAMP inhibits not only ERK1/2 but also the phosphatidylinositol 3-kinase (PI3K)/protein kinase B (PKB, also called Akt) signaling cascade. Upon RTK activation, PI3K catalytic subunit is recruited to the plasma membrane and increases the levels of phosphatidylinositol-3,4,5-trisphosphate and inositol-3,4-bisphosphate. They, in turn, induce translocation of two kinases to the plasma membrane, Akt and PDK1 (phosphoinositide (PI3K) – dependent kinase 1), by binding to the plekstrin (PH) homology domain. In the membrane, PDK1 phosphorylates and activates Akt, which further affects phosphorylation of various targets. Among them are BAD and caspase-9, two known members of the pro-apoptotic Bcl-family. Their inactivation by Akt phosphorylation results in the inhibition of apoptosis. A similar result, the rescue from apoptosis, also occurs when cells express constitutively active PI3K and ERK1/2 (Brazil and Hemmings 2001; Kim et al. 2001; Wang et al. 2001).

The inhibitory effect of cAMP on the PI3K/Akt pathway was reported in Swiss 3T3, HEK293, Cos, Rat2 and glioma C6 cells (Wang et al. 2001). In the latter cell line, inhibition of Rap1 by the Rap1Gap1 protein causes a decrease in Akt phosphorylation, suggesting that cAMP influences the Akt pathway similarly to the ERK1/2 pathway, *via* inhibition of Rap1 activity. According to Wang et al. Akt inhibition is a consequence of PI3K inhibition (Wang et al. 2001). Kim et al. explained it showing that cAMP attenuates the PI3K/Akt pathway by decreasing the lipid kinase activity of PI3K and, due to that, the level of phosphatidylinositol-3,4,5-trisphosphate (Kim et al. 2001). In consequence, PDK1 translocation to the membrane is blocked and therefore Akt cannot be phosphorylated and activated.

As mentioned before, an increased level of intracellular cAMP results in inhibition of Akt activity. It leads to growth arrest and differentiation of glioma C6 cells into type II astrocytes, accompanied by induction of GFAP synthesis and changes in cell morphology (Aerts et al. 2011; Claes et al. 2004; Van Kolen and Slegers 2004). In addition, Aerts et al. reported that these cAMP-mediated effects inhibit the expression of ecto-nucleotide pyrophosphatase/phosphodiesterase 1, an ecto-enzyme bound to the cell-surface, which regulates short-term and long-term cellular functions in the central nervous system (Aerts et al. 2011; Neary and Zimmermann 2009). However, these cAMP-dependent effects might be reversed by the activation of $P2Y_{12}$ receptors, negatively bound to the adenylate cyclase (Van Kolen and Slegers 2004).

Thus, high concentration of intracellular cAMP is dangerous for C6 cell metabolism. In consequence of the increased cAMP level, ERK1/2 and PI3K/Akt signaling pathways are inhibited and thus, C6 cell proliferation is stopped, growth is arrested and differentiation is induced. To prevent such effects, the maintenance of the low cAMP level in this cell line is essential. The cells use various mechanisms to block cAMP synthesis. One of them, described by Weber in rapidly growing hepatoma cells, is to steadily increase phosphodiesterase activity concurrently with a decrease in adenylate cyclase activity (Weber 2002). The other, reported by Chiono et al. is to inhibit type VI adenylate cyclase, which predominates in gliomas, by increasing $[Ca^{2+}]_i$ (Chiono et al. 1995). Glioma C6 cells are characterized by a very low constitutive level of cAMP (see Fig. 3.1c), which may be caused by the high activity of $P2Y_{12}$ receptors. Accordingly, during serum deprivation $P2Y_{12}$ expression is preferentially increased, most probably as the result of the same, self-regulating mechanism maintaining a low level of intracellular cAMP.

3.6 Effect of Extracellular Nucleotides on ERK1/2 and PI3K/Akt Activity: $P2Y_1$/ $P2Y_{12}$ Cross-Talk

The effect of the $P2Y_2$ receptor on ERK1/2 activity in glioma C6 cells was examined by Tu et al. (2000). ATP- and UTP-induced stimulation of this receptor resulted in an increase in $[^3H]$tymidine incorporation and p42/p44 ERK1/2 phosphorylation. $P2Y_2$-mediated stimulation of proliferation occurred through the Ras/Raf-1/MEK/ERK signaling cascade, which was described in details in the previous Section. The pathway was modulated by PLC activation, enhanced activity of protein kinase C and Ca^{2+} release. Thus, it could be concluded that the positive effect of ATP and UTP on C6 cell proliferation and DNA synthesis involved a cross-talk between $G\alpha_q$-coupled $P2Y_2$ receptors signaling and the receptor tyrosine kinase–dependent cascade (Tu et al. 2000). The mechanism by which the $P2Y_1$ receptor stimulates ERK1/2, most probably involves a PLC-dependent pathway (Czajkowski et al. 2004), similar to that described by Tu et al. for the $P2Y_2$ receptor (Tu et al. 2000).

However, the other data showed that in transiently transfected HEK 293 cells, free β/γ subunits of heterotrimeric G_i protein could also stimulate proliferation *via* the Ras/C-Raf/MEK/ERK pathway (Ito et al. 1995). Moreover, in contrast to Tu et al. and Wang et al. (Tu et al. 2000; Wang et al. 2001), who proposed that activation of ERK1/2 in glioma C6 cells might occur through Ras/Raf-1/MEK or Rap1/B-Raf/MEK, respectively (see Sect. 3.5), the group of Slegers suggested an alternative signaling cascade in the same cell line (Grobben et al. 2001). It involved another small G protein, a monomeric GTPase, RhoA, and the signaling pathway through RhoA/PCK/Raf/MEK/ERK was proposed (Grobben et al. 2001). It was further shown that an atypical PKCζ was involved in the RhoA-dependent mitogenic signaling and the proliferation was enhanced by $P2Y_{AC}$ ($P2Y_{12}$) receptor activation. It occurred through $G\alpha_i$ (but not β/γ subunits)-dependent, RhoA-dependent/PKCζ/MEK/ERK pathway (Van Kolen and Slegers 2006a). On the other hand, Soulet et al. presented evidence that the stimulated human $P2Y_{12}$ receptor, stably expressed in Chinese hamster ovary cells, activated two major signaling mechanisms: dependent and independent on G_i protein (Soulet et al. 2004). The PTX-sensitive mechanism, in which G_i protein was involved, led to cell proliferation *via* activation of both the ERK1/2 and PI3K/Akt kinase signaling. The other, PTX-insensitive, independent of G_i, occurred *via* the activation of RhoA and Rho-kinase, resulting in actin cytoskeleton reorganization, but not in cell proliferation. The authors suggested that in the latter mechanism the $G_{12/13}$ heterotrimeric protein might be involved (Soulet et al. 2004). Thus, the above data concerning mechanisms of $P2Y_{12}$-G_i protein-coupled-receptor signaling inducing ERK1/2 activation are rather divergent. The reason for that may be different cell lines used in experiments, different subclones of these lines formed during different passages, or various cell culture conditions. Nevertheless, the experiments concerning the activity of $P2Y_{12}$ receptors are usually performed using ADP or 2MeSADP as agonists. However, the above-mentioned compounds act not only on G_i-dependent $P2Y_{12}$, but also on G_q-dependent $P2Y_1$ receptors. Therefore, the effects of stimulation of both receptors should be investigated.

As it was already mentioned (see Sect. 3.4.1), glioma C6 cells cultivated in the presence of serum are characterized by a high constitutive activity of ERK1/2 (Barańska et al. 2004; Czajkowski et al. 2004). Similar high basal activity of PI3K/Akt is a characteristic feature of gliomas and other tumor cells (Furnari et al. 1998). The latter fact is usually explained as a result of a loss or a very low level of a lipid phosphatase (PTEN). This phosphatase has an activity antagonistic to that of PI3K since it dephosphorylates the 3′ position of phosphoinositides (Cantley and Neel 1999; Kubiatowski et al. 2001). Therefore, investigations concerning ERK1/2 and PI3K/Akt activities are usually performed in media devoid of sera. This is also why 48 and 96 h serum deprived glioma C6 cells were used in studies on ERK1/2 and PI3K/Akt activities. Under these conditions, cells were growth-arrested and incorporated [^{3}H]tymidine, a marker of proliferation, only minimally (Czajkowski et al. 2004; Krzemiński et al. 2007; Tu et al. 2000).

Czajkowski et al. showed that in 48 h serum-starved C6 cells, both p44 ERK1 and p42 ERK2 were activated in a time-dependent manner in response to ADP and 2MeSADP (Czajkowski et al. 2004). ERK phosphorylation was reduced by removal of the extracellular Ca^{2+} and by specific antagonists of $P2Y_1$ and $P2Y_{12}$ receptors. The inhibitory effect of MRS2179, a $P2Y_1$ receptor antagonist, on ADP-stimulated ERK1/2 activity was approximately 40 %. ARC69931MX, a $P2Y_{12}$ receptor selective antagonist, inhibited ADP-induced ERK1/2 phosphorylation by 85 %. Moreover, the effects of both antagonists were additive (Czajkowski et al. 2004). The same effects were observed for the cells cultivated for 96 h in serum-free medium (Krzemiński et al. 2007). Extracellular Ca^{2+} removal inhibited ADP-induced ERK1/2 activation by about 70 %, whereas the one induced by UTP (the agonist of $P2Y_2$ receptors) was completely blocked (Czajkowski et al. 2004). These data point out the participation of both $P2Y_1$ and $P2Y_{12}$ receptors in ERK1/2 activation, but also show that the effect of ADP on ERK1/2 kinase activation and cell proliferation is primarily mediated by $P2Y_{12}$ receptors.

Stimulation of PI3K activity evoked by ADP *via* $P2Y_1$ and $P2Y_{12}$ receptors was examined in C6 cells grown either in the presence or in the absence of serum in the cultivation medium (Barańska et al. 2004; Czajkowski et al. 2004). The non-starved cells were characterized by a high constitutive activity of the enzyme and their stimulation by ADP resulted in decreased PI3K activity. On the contrary, in cells cultivated for 48 h without serum, ADP evoked a gradual increase in PI3K activity. Furthermore, preincubation of 48 h serum-starved cells with MRS2179 and then, still in its presence, stimulation of the cells with ADP, markedly increased PI3K activity. It indicated that blocking of $P2Y_1$ receptors had a stimulatory effect on the kinase (Czajkowski et al. 2004). Furthermore, Krzemiński et al. demonstrated that even after 96 h cultivation in serum-free medium, glioma C6 cells were able to respond to 2MeSADP and showed enhanced Akt phosphorylation, whereas $P2Y_{12}$ antagonist, ARC69931MX, completely suppressed Akt phosphorylation (Krzemiński et al. 2007).

Figure 3.5 shows changes of constitutive ERK1/2 and Akt phosphorylation in unstimulated glioma C6 cells during long-term (for up to 96 h) serum deprivation (Krzemiński et al. 2007). As it is shown, the level of p42 and p44 ERK1/2 phosphorylation was greatly reduced. This is convergent with inhibition of cells proliferation. After 96 h of serum deprivation, the replacement of the medium for a fresh one, either serum-free (M-) or containing serum (M+), restored ERK1/2 phosphorylation. On the other hand, prolonged serum deprivation (up to 72 h) increased Akt phosphorylation and this process was stopped only after very long (96 h) serum starvation. Moreover, after this time the replacement of the medium for a fresh one, but still serum-free (M-), increased Akt phosphorylation. In contrast, fresh medium with serum (M+) had an opposite effect, as the phosphorylation of Akt was decreased.

These data show the cross-talk between PI3K/Akt activity and $P2Y_1$ and $P2Y_{12}$ receptor expression (see Sect. 3.4.1). When cells are grown in the presence of serum, $P2Y_1$ receptors signaling

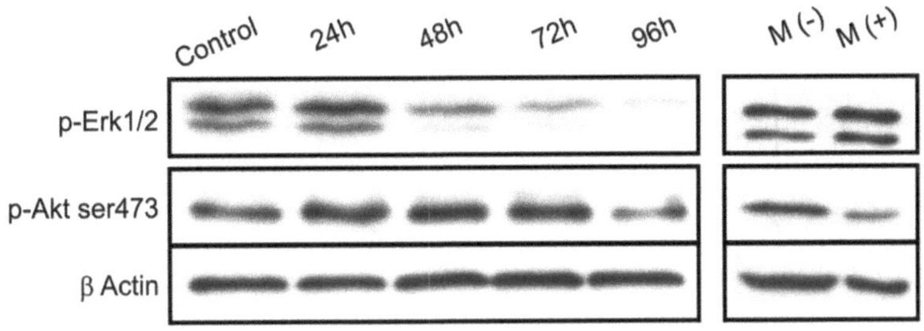

Fig. 3.5 Changes in constitutive ERK1/2 and Akt phosphorylation of unstimulated glioma C6 cells during long-term serum starvation. C6 cells were grown in the presence of 10 % fetal bovine serum (control), or were cultivated in serum-free medium for 24, 48, 72 or 96 h. (M−), cells cultivated in serum-free medium for 96 h followed by 24 h recovery in fresh medium without serum. (M+), cells cultivated in serum-free medium for 96 h followed by 24 h recovery in fresh medium with 10 % serum. The level of p42, p44 ERK1/2 and Akt phosphorylation was monitored by Western blot analysis. β-actin was used as a reference protein (Reproduced from Krzemiński et al. 2007 with permission from John Wiley and Sons)

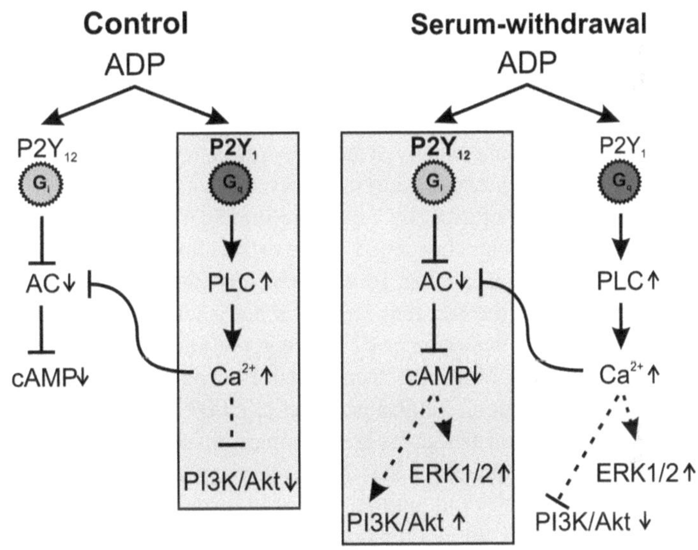

Fig. 3.6 Comparison of signal transduction pathways activated by $P2Y_1$ and $P2Y_{12}$ receptors in C6 glioma cells growing in the presence of serum (*control*) or cultivated in serum-free medium (*serum-withdrawal*). During long-term serum-starvation, the $P2Y_1$ expression strongly decreases and $P2Y_{12}$ receptor predominates. *Dashed lines* symbolize indirect stimulatory or inhibitory actions. For details, see text

predominates and inhibits P13K activity. On the other hand, when the cells are serum-deprived the expression of $P2Y_1$ receptors is strongly decreased and, in parallel, Akt phosphorylation is increased. Moreover, after 96 h of serum deprivation, when the medium is replaced for the one containing serum the protein level of $P2Y_1$ receptors increases and phosphorylation of Akt decreases again (Fig. 3.5). The inhibitory effect of the $P2Y_1$ receptor on PI3K/Akt activity is still poorly understood. On the other hand, the stimulatory effect of $P2Y_{12}$ receptors, which regulate intracellular cAMP level, on PI3K/Akt activity may occur *via* Rap1 (Aerts et al. 2011; Kim et al. 2001; Wang et al. 2001).

Thus, in C6 cells, $P2Y_{12}$ receptors are predominantly involved in ERK1/2 activation. They have a stimulatory effect on PI3K/Akt activity, whereas the effect of $P2Y_1$ receptors is inhibitory. Therefore, it seems that $P2Y_{12}$ receptors primarily contribute to glioma C6 cell growth and proliferation. The high constitutive activity of Akt, which increases further during prolonged serum starvation, seems to be also important for cell survival and enhanced invasiveness.

Signaling pathways activated by ADP-stimulation in control and serum-deprived glioma C6 cells are illustrated in Fig. 3.6.

3.7 The $P2X_7$ Receptor

As previously mentioned (see Sect. 3.2.1), the $P2X_7$ receptor is an atypical member of the P2X receptor family. It differs from other P2X receptors in that it is characterized by a relative low affinity for ATP and is activated by high, milimolar ATP concentrations. ATP binding to the $P2X_7$ opens the channel *via* a ligand-gated mechanism and evokes a fast influx of Ca^{2+} as well as Na^+ and the efflux of K^+ (North and Surprenant 2000; North 2002). Another characteristic that distinguishes $P2X_7$ from other P2X receptors is its ability to open nonspecific pores upon repeated or prolonged ATP application. These pores are responsible for the massive influx of Ca^{2+}. However, they are permeable not only to cations, but also to small molecules and high molecular weight dyes. It was reported that activation of $P2X_7$ and subsequent opening of large pores might be responsible for ATP-mediated release of amino acids in cortical astrocytes (Duan et al. 2003) as well as the release of ATP and other nucleotides in glioma C6 cells (De Vuyst et al. 2007). More recently, it was suggested that the ATP-induced ion flow through $P2X_7$ led to the activation of pannexin-1 hemichannel. In consequence, pre-interleukin-1β was prematurely processed and interleukin-1β was released from macrophages (Pelegrin and Surprenant 2006). Hemichannels are regulated by the extracellular Ca^{2+} level, they are closed at the physiological, milimolar Ca^{2+} concentration and open in response to its lowering (Srinivas et al. 2006), till apoptotic or necrotic cell death occurs (Chen and Brosnan 2006; Locovei et al. 2007).

As it was already described (see Sect. 3.2.1), activation of $P2X_7$ receptors results in reduction of membrane potential and in elevation of $[Ca^{2+}]_i$. After the opening of a receptor-channel, extracellular Ca^{2+} flows in and promotes the opening of the plasma membrane voltage-dependent calcium channels. These phenomena are usually detected by measurements of the unitary currents passing through single ion channels (patch-clamp) and measurements of $[Ca^{2+}]_i$ (North and Surprenant 2000). It is worth to add that activation of $P2X_7$ by a high, milimolar ATP concentration raises problems. First, 1 mM ATP is so acidic that it can change pH of the extracellular medium. That is why $P2X_7$ receptors are extensively activated by its analog, BzATP, which is 10–30 times more potent than ATP. The other problem is connected with the fact that BzATP is not a selective agonist of $P2X_7$, as at the same micromolar concentration it activates other P2X receptors (Anderson and Nedergaard 2006; North and Surprenant 2000; North 2002). Moreover, there is also evidence that BzATP may activate P2Y metabotropic receptors (Carrasquero et al. 2009; Fischer et al. 2009; Supłat-Wypych et al. 2010; Wildman et al. 2003). Glioma C6 cells and astrocytes are examples of such discrepancies (see below).

As mentioned in Sect. 3.2.1, $P2X_7$ receptors are widely expressed in neuronal cells. In cortical astrocytes Fumagali et al. and Bianco et al. provided evidence that $P2X_7$ receptors are functionally active (Bianco et al. 2009; Fumagalli et al. 2003). Nevertheless, using the same cultured cortical astroglia, Fischer et al. came to a different conclusion (Fischer et al. 2009). The latter authors found that BzATP caused $[Ca^{2+}]_i$ increase not only in the presence, but also in the absence of extracellular Ca^{2+}, which suggests that it activates both ionotropic and metabotropic receptors. Moreover, Carrasquero et al., measuring calcium responses in rat cerebellar astrocytes, proposed that both $P2Y_{13}$ and $P2X_7$ receptors were activated by BzATP (Carrasquero et al. 2009). However, the authors suggested that the first transient phase of calcium response was induced by the $P2Y_{13}$ receptor, whereas the second, sustained response was generated by $P2X_7$. This result seemed to be controversial since one of the main criteria to distinguish ionotropic from metabotropic receptors is the time course of the response. For ionotropic receptors, it occurs within milliseconds while the metabotropic ones respond within minutes (North and Surprenant 2000; Ralevic and Burnstock 1998).

In glioma C6 cells, neither Braganhol et al., nor De Vuyst et al., found $P2X_7$ mRNA or protein (Braganhol et al. 2009; De Vuyst et al. 2007). However, Wei et al. suggested the presence of functionally active $P2X_7$ receptors in these cells (Wei et al. 2008). This assumption was based on the fact that BzATP led to the increase of $[Ca^{2+}]_i$ and that the response was blocked by periodate-oxidized ATP (OxATP), commonly used as $P2X_7$ antagonist. On the contrary, Supłat-Wypych et al. provided evidence that $P2X_7$ receptors were not activated in C6 cells (Supłat-Wypych et al. 2010). In the absence of extracellular calcium, BzATP generated an increase of $[Ca^{2+}]_i$ *via* Ca^{2+} release from intracellular stores. In the presence of extracellular calcium, BzATP established a biphasic Ca^{2+} response, in a manner typical for P2Y metabotropic receptors coupled to G_q protein that involves PLC activation and the capacitative calcium entry. Moreover, OxATP blocked not only BzATP- but also UTP- induced intracellular Ca^{2+} elevation and brilliant blue G, a selective antagonist of the rat $P2X_7$ receptor, was without effect on the kinetics of calcium signals. Furthermore, neither BzATP (300 μM) nor ATP (3 mM) opened large transmembrane pores. All the above observations, together with the lack of response of UTP-desensitized cells to BzATP and *vice versa*, provide support for the activation of P2Y receptors, most probably the $P2Y_2$ receptor, by BzATP.

Taken together, *in vitro* studies suggest that in glioma C6 cells functionally active $P2X_7$ receptors do not exist. It is well known that stimulation of $P2X_7$ receptors *via* prolonged ATP exposure causes often apoptotic fate of the cell, due to massive Ca^{2+} entry. These facts, together with the shift between $P2Y_1$ and $P2Y_{12}$ receptors expression during long-term serum starvation might belong to the same self-sufficient mechanism, important for survival, up-regulated growth and invasiveness of glioma C6 cells.

On the other hand, in the *in vivo* study, Ryu et al. using immunohistochemical analysis observed colocalization of the $P2X_7$ receptor with tumor cells and microglia 2 weeks after intrastratial injection of C6 cells into rat brains (Ryu et al. 2011). Intravenous administration of the $P2X_7$ receptor antagonist, brilliant blue G, reduced tumor volume by 52 %. These results suggest that, *in vivo*, $P2X_7$ might have a different function than opening a channel and generating Ca^{2+} influx and that the $P2X_7$ receptor could be involved in signaling pathway(s) contributing to glioma growth. Moreover, brilliant blue G administrated in such a way can also influence $P2X_7$ receptors in other cells, like microglia, the activity of which is crucial to the tumor growth and invasion (Sliwa et al. 2007, see also Sect. 7.2.2).

3.8 Concluding Remarks

Rat glioma C6 is a tumoral cell line of glial origin. It has biological features of astrocytes and gliomas. One of such features is the expression of $P2Y_1$ receptors, commonly present in astrocytes and neuronal healthy cells but not in neuronal cancer cells. In C6 cells, $P2Y_1$ receptors are present and functionally active; however, their expression is highly instable. This is especially visible during serum withdrawal, when the expression of $P2Y_1$ receptors gradually decreases, and becomes undetectable after very long time of serum starvation. In contrast, $P2Y_{12}$ receptors, coupled to G_i protein and negatively linked to adenylate cyclase, are massively expressed during unfavorable conditions of serum deprivation. Both these receptors are stimulated by the same agonists. The decrease in $P2Y_1$ receptor expression is not related to the fact that these receptors induce signaling *via* G_q protein, PLC stimulation and Ca^{2+} mobilization since the expression of $P2Y_2$ receptors, acting through the same signaling cascade, is not changed during serum deprivation. The experiments with siRNA show that the shift between $P2Y_1$ and $P2Y_{12}$ expression observed during serum withdrawal is not due to a concerted cell response, but rather to the lack of serum which differentially affects $P2Y_1$ and $P2Y_{12}$ synthesis. Nevertheless, long-term serum withdrawal seems to be a convenient tool for functional studies of these receptors.

Thus, as far as receptors are concerned, the characteristic features of glioma C6 cells are: (1) instable expression of $P2Y_1$ and (2) massive expression of $P2Y_{12}$ receptors. As to the second messengers, such characteristic features are: (1) very low basal level of cAMP, (2) constitutively active PI3K/Akt signaling pathway, (3) very active ERK1/2 signaling. All these cellular traits cooperate to maintain sustained proliferation of C6 cells.

In platelets, $P2Y_1$ and $P2Y_{12}$ receptors cooperate in ADP-induced platelet aggregation. In glioma C6 cells, such cooperation is more complicated. It takes part in the stimulation of ERK1/2 activity, although the role of $P2Y_{12}$ in this process is predominating. On the other hand, the effects of these two receptors on PI3K/Akt signaling are opposite. Both in the presence and in the absence of serum, $P2Y_{12}$ activates PI3K/Akt signaling pathway, while $P2Y_1$ inhibits it.

In C6 cells cultivated in the presence of serum, the basal activity of ERK1/2 is so high that its additional activation by extracellular nucleotides cannot be observed. During serum starvation, ERK1/2 activity strongly declines, concomitantly with cell growth arrest. On the contrary, Akt activity distinctly increases, in parallel with increasing $P2Y_{12}$ and decreasing $P2Y_1$ receptor expression. Thus, it seems that under unfavorable conditions of serum withdrawal, the most important for C6 cells metabolism is to maintain a high activity of the PI3K/Akt signaling pathway, which contributes to proliferative and invasive properties of these cells. $P2Y_{12}$ acts in favor of such activity, while $P2Y_1$ acts against it and probably therefore $P2Y_{12}$ expression prevails, whereas that of $P2Y_1$ is strongly attenuated. A similar regularity takes place in the case of the $P2Y_{14}$ receptor. In C6 cells, two forms of this receptor occur: a glycosylated one, coupled to PLC, and a non-glycosylated one, negatively coupled to adenylate cyclase. Under serum withdrawal conditions, the latter form dominates.

$P2Y_{12}$ plays also an important role in maintaining a very low cAMP level. Since high cAMP level is dangerous for glioma C6 metabolism, the cells use various mechanisms to block cAMP synthesis. One of them is the inhibition of the activity of adenylate cyclase type VI, which predominates in gliomas, by increased intracellular Ca^{2+} level. Thus, in C6 cells cultivated with serum, the increase in $[Ca^{2+}]_i$ induced by activation of $P2Y_1$ receptors may down regulate adenylate cyclase in cooperation with $P2Y_{12}$ receptors, negatively coupled to this enzyme. When the cells are serum-starved and their growth is arrested, the expression of $P2Y_1$ receptors, which inhibit the PI3K/Akt signaling pathway, decreases and the expression of $P2Y_{12}$ receptors, predominates. The latter receptors stimulate ERK1/2 and PI3K/Akt activity and during serum-starvation play an important role in the maintenance of a very low level of cAMP and sustained high activity of the PI3K/Akt cascade.

Taken together, these data show the precise intracellular machinery that act in favor of maintaining glioma growth and proliferation, as well as invasiveness. In C6 cells, nucleotide receptors are involved in all of these processes and the cross-talk between $P2Y_1$ and $P2Y_{12}$ receptors seems to be an important self-regulating mechanism that helps to preserve their tumoral features.

References

Abbracchio MP, Burnstock G (1994) Purinoceptors: are there families of P2X and P2Y purinoceptors? Pharmacol Ther 64:445–475

Abbracchio MP, Verderio C (2006) Pathophysiological roles of P2 receptors in glial cells. Novartis Found Symp 276:91–103, discussion 103–112, 275–181

Abbracchio MP, Boeynaems JM, Barnard EA, Boyer JL, Kennedy C, Miras-Portugal MT, King BF, Gachet C, Jacobson KA, Weisman GA, Burnstock G (2003) Characterization of the UDP-glucose receptor (re-named here the P2Y14 receptor) adds diversity to the P2Y receptor family. Trends Pharmacol Sci 24:52–55

Aerts I, Grobben B, Van Ostade X, Slegers H (2011) Cyclic AMP-dependent down regulation of ecto-nucleotide pyrophosphatase/phosphodiesterase 1 (NPP1) in rat C6 glioma. Eur J Pharmacol 654:1–9

Anderson CM, Nedergaard M (2006) Emerging challenges of assigning P2X7 receptor function and immunoreactivity in neurons. Trends Neurosci 29:257–262

Arslan G, Fredholm BB (2000) Stimulatory and inhibitory effects of adenosine A(2A) receptors on nerve growth factor-induced phosphorylation of extracellular regulated kinases 1/2 in PC12 cells. Neurosci Lett 292:183–186

Auer RN, Del Maestro RF, Anderson R (1981) A simple and reproducible experimental in vivo glioma model. Can J Neurol Sci 8:325–331

Bagchi S, Liao Z, Gonzalez FA, Chorna NE, Seye CI, Weisman GA, Erb L (2005) The P2Y2 nucleotide receptor interacts with alphav integrins to activate Go and induce cell migration. J Biol Chem 280:39050–39057

Banachewicz W, Supłat D, Krzemiński P, Pomorski P, Barańska J (2005) P2 nucleotide receptors on C2C12 satellite cells. Purinergic Signal 1:249–257

Barańska J, Chaban V, Czarny M, Sabała P (1995) Changes in Ca2+ concentration in phorbol ester and thapsigargin treated glioma C6 cells. The role of protein kinase C in regulation of Ca2+ entry. Cell Calcium 17:207–215
Barańska J, Czajkowski R, Sabała P (2004) Cross-talks between nucleotide receptor-induced signaling pathways in serum-deprived and non-starved glioma C6 cells. Adv Enzyme Regul 44:219–232
Barth RF (1998) Rat brain tumor models in experimental neuro-oncology: the 9L, C6, T9, F98, RG2 (D74), RT-2 and CNS-1 gliomas. J Neurooncol 36:91–102
Benda P, Lightbody J, Sato G, Levine L, Sweet W (1968) Differentiated rat glial cell strain in tissue culture. Science 161:370–371
Berridge MJ (1995) Capacitative calcium entry. Biochem J 312(Pt 1):1–11
Bezzi P, Volterra A (2001) A neuron-glia signalling network in the active brain. Curr Opin Neurobiol 11:387–394
Bianco F, Fumagalli M, Pravettoni E, D'Ambrosi N, Volonte C, Matteoli M, Abbracchio MP, Verderio C (2005) Pathophysiological roles of extracellular nucleotides in glial cells: differential expression of purinergic receptors in resting and activated microglia. Brain Res Brain Res Rev 48:144–156
Bianco F, Ceruti S, Colombo A, Fumagalli M, Ferrari D, Pizzirani C, Matteoli M, Di Virgilio F, Abbracchio MP, Verderio C (2006) A role for P2X7 in microglial proliferation. J Neurochem 99:745–758
Bianco F, Colombo A, Saglietti L, Lecca D, Abbracchio MP, Matteoli M, Verderio C (2009) Different properties of P2X(7) receptor in hippocampal and cortical astrocytes. Purinergic Signal 5:233–240
Boarder MR, Hourani SM (1998) The regulation of vascular function by P2 receptors: multiple sites and multiple receptors. Trends Pharmacol Sci 19:99–107
Boarder M, Webb TE (2001) P2Y receptors: structure and function. In: Abbracchio MP, Williams M (eds) Purinergic and pirimidinergic signalling. Springer, Berlin
Bos JL (2003) Epac: a new cAMP target and new avenues in cAMP research. Nat Rev Mol Cell Biol 4:733–738
Boyer JL, Lazarowski ER, Chen XH, Harden TK (1993) Identification of a P2Y-purinergic receptor that inhibits adenylyl cyclase. J Pharmacol Exp Ther 267:1140–1146
Braganhol E, Huppes D, Bernardi A, Wink MR, Lenz G, Battastini AM (2009) A comparative study of ectonucleotidase and P2 receptor mRNA profiles in C6 cell line cultures and C6 ex vivo glioma model. Cell Tissue Res 335:331–340
Brazil DP, Hemmings BA (2001) Ten years of protein kinase B signalling: a hard Akt to follow. Trends Biochem Sci 26:657–664
Brismar T (1995) Physiology of transformed glial cells. Glia 15:231–243
Burnstock G (1997) The past, present and future of purine nucleotides as signalling molecules. Neuropharmacology 36:1127–1139
Burnstock G (2002) Potential therapeutic targets in the rapidly expanding field of purinergic signalling. Clin Med 2:45–53
Burnstock G (2007) Physiology and pathophysiology of purinergic neurotransmission. Physiol Rev 87:659–797
Cantley LC, Neel BG (1999) New insights into tumor suppression: PTEN suppresses tumor formation by restraining the phosphoinositide 3-kinase/AKT pathway. Proc Natl Acad Sci U S A 96:4240–4245
Carrasquero LM, Delicado EG, Jimenez AI, Perez-Sen R, Miras-Portugal MT (2005) Cerebellar astrocytes co-express several ADP receptors. Presence of functional P2Y(13)-like receptors. Purinergic Signal 1:153–159
Carrasquero LM, Delicado EG, Bustillo D, Gutierrez-Martin Y, Artalejo AR, Miras-Portugal MT (2009) P2X7 and P2Y13 purinergic receptors mediate intracellular calcium responses to BzATP in rat cerebellar astrocytes. J Neurochem 110:879–889
Carter RL, Fricks IP, Barrett MO, Burianek LE, Zhou Y, Ko H, Das A, Jacobson KA, Lazarowski ER, Harden TK (2009) Quantification of Gi-mediated inhibition of adenylyl cyclase activity reveals that UDP is a potent agonist of the human P2Y14 receptor. Mol Pharmacol 76:1341–1348
Chen L, Brosnan CF (2006) Regulation of immune response by P2X7 receptor. Crit Rev Immunol 26:499–513
Cheng X, Ji Z, Tsalkova T, Mei F (2008) Epac and PKA: a tale of two intracellular cAMP receptors. Acta Biochim Biophys Sin (Shanghai) 40:651–662
Chiono M, Mahey R, Tate G, Cooper DM (1995) Capacitative Ca2+ entry exclusively inhibits cAMP synthesis in C6-2B glioma cells. Evidence that physiologically evoked Ca2+ entry regulates Ca(2+)-inhibitable adenylyl cyclase in non-excitable cells. J Biol Chem 270:1149–1155
Chou RC, Langan TJ (2003) In vitro synchronization of mammalian astrocytic cultures by serum deprivation. Brain Res Brain Res Protoc 11:162–167
Claes P, Grobben B, Van Kolen K, Roymans D, Slegers H (2001) P2Y(AC)(−)-receptor agonists enhance the proliferation of rat C6 glioma cells through activation of the p42/44 mitogen-activated protein kinase. Br J Pharmacol 134:402–408
Claes P, Van Kolen K, Roymans D, Blero D, Vissenberg K, Erneux C, Verbelen JP, Esmans EL, Slegers H (2004) Reactive blue 2 inhibition of cyclic AMP-dependent differentiation of rat C6 glioma cells by purinergic receptor-independent inactivation of phosphatidylinositol 3-kinase. Biochem Pharmacol 67:1489–1498
Collins VP (1998) Gliomas. Cancer Surv 32:37–51
Communi D, Govaerts C, Parmentier M, Boeynaems JM (1997) Cloning of a human purinergic P2Y receptor coupled to phospholipase C and adenylyl cyclase. J Biol Chem 272:31969–31973

Communi D, Gonzalez NS, Detheux M, Brezillon S, Lannoy V, Parmentier M, Boeynaems JM (2001) Identification of a novel human ADP receptor coupled to G(i). J Biol Chem 276:41479–41485
Czajkowski R, Barańska J (2002) Cross-talk between the ATP and ADP nucleotide receptor signalling pathways in glioma C6 cells. Acta Biochim Pol 49:877–889
Czajkowski R, Lei L, Sabała P, Barańska J (2002) ADP-evoked phospholipase C stimulation and adenylyl cyclase inhibition in glioma C6 cells occur through two distinct nucleotide receptors, P2Y(1) and P2Y(12). FEBS Lett 513: 179–183
Czajkowski R, Banachewicz W, Ilnytska O, Drobot LB, Barańska J (2004) Differential effects of P2Y1 and P2Y12 nucleotide receptors on ERK1/ERK2 and phosphatidylinositol 3-kinase signalling and cell proliferation in serum-deprived and nonstarved glioma C6 cells. Br J Pharmacol 141:497–507
Dahl D (1981) The vimentin-GFA protein transition in rat neuroglia cytoskeleton occurs at the time of myelination. J Neurosci Res 6:741–748
de Rooij J, Zwartkruis FJ, Verheijen MH, Cool RH, Nijman SM, Wittinghofer A, Bos JL (1998) Epac is a Rap1 guanine-nucleotide-exchange factor directly activated by cyclic AMP. Nature 396:474–477
De Vuyst E, Decrock E, De Bock M, Yamasaki H, Naus CC, Evans WH, Leybaert L (2007) Connexin hemichannels and gap junction channels are differentially influenced by lipopolysaccharide and basic fibroblast growth factor. Mol Biol Cell 18:34–46
Di Virgilio F, Chiozzi P, Ferrari D, Falzoni S, Sanz JM, Morelli A, Torboli M, Bolognesi G, Baricordi OR (2001) Nucleotide receptors: an emerging family of regulatory molecules in blood cells. Blood 97:587–600
Duan S, Anderson CM, Keung EC, Chen Y, Swanson RA (2003) P2X7 receptor-mediated release of excitatory amino acids from astrocytes. J Neurosci 23:1320–1328
Dugan LL, Kim JS, Zhang Y, Bart RD, Sun Y, Holtzman DM, Gutmann DH (1999) Differential effects of cAMP in neurons and astrocytes. Role of B-raf. J Biol Chem 274:25842–25848
Erb L, Liu J, Ockerhausen J, Kong Q, Garrad RC, Griffin K, Neal C, Krugh B, Santiago-Perez LI, Gonzalez FA, Gresham HD, Turner JT, Weisman GA (2001) An RGD sequence in the P2Y(2) receptor interacts with alpha(V) beta(3) integrins and is required for G(o)-mediated signal transduction. J Cell Biol 153:491–501
Ferrari D, Chiozzi P, Falzoni S, Dal Susino M, Collo G, Buell G, Di Virgilio F (1997) ATP-mediated cytotoxicity in microglial cells. Neuropharmacology 36:1295–1301
Fischer W, Appelt K, Grohmann M, Franke H, Norenberg W, Illes P (2009) Increase of intracellular Ca2+ by P2X and P2Y receptor-subtypes in cultured cortical astroglia of the rat. Neuroscience 160:767–783
Fricks IP, Maddileti S, Carter RL, Lazarowski ER, Nicholas RA, Jacobson KA, Harden TK (2008) UDP is a competitive antagonist at the human P2Y14 receptor. J Pharmacol Exp Ther 325:588–594
Fricks IP, Carter RL, Lazarowski ER, Harden TK (2009) Gi-dependent cell signaling responses of the human P2Y14 receptor in model cell systems. J Pharmacol Exp Ther 330:162–168
Fumagalli M, Brambilla R, D'Ambrosi N, Volonte C, Matteoli M, Verderio C, Abbracchio MP (2003) Nucleotide-mediated calcium signaling in rat cortical astrocytes: role of P2X and P2Y receptors. Glia 43:218–230
Fumagalli M, Trincavelli L, Lecca D, Martini C, Ciana P, Abbracchio MP (2004) Cloning, pharmacological characterisation and distribution of the rat G-protein-coupled P2Y(13) receptor. Biochem Pharmacol 68:113–124
Furnari FB, Huang HJ, Cavenee WK (1998) The phosphoinositol phosphatase activity of PTEN mediates a serum-sensitive G1 growth arrest in glioma cells. Cancer Res 58:5002–5008
Gachet C (2006) Regulation of platelet functions by P2 receptors. Annu Rev Pharmacol Toxicol 46:277–300
Grobben B, Claes P, Van Kolen K, Roymans D, Fransen P, Sys SU, Slegers H (2001) Agonists of the P2Y(AC)-receptor activate MAP kinase by a ras-independent pathway in rat C6 glioma. J Neurochem 78:1325–1338
Grobben B, De Deyn PP, Slegers H (2002) Rat C6 glioma as experimental model system for the study of glioblastoma growth and invasion. Cell Tissue Res 310:257–270
Guo LH, Trautmann K, Schluesener HJ (2004) Expression of P2X4 receptor in rat C6 glioma by tumor-associated macrophages and activated microglia. J Neuroimmunol 152:67–72
Hanson MG Jr, Shen S, Wiemelt AP, McMorris FA, Barres BA (1998) Cyclic AMP elevation is sufficient to promote the survival of spinal motor neurons in vitro. J Neurosci 18:7361–7371
Harden K, Barnard EA, Boeynaems JM, Burnstock G, Dubyak G, Hourani SM, Insel PA (1998) P2Y receptors. In: Girdlestone D (ed) The IUPHAR compendium of receptor characterization and classification, 2nd edn. IUPHAR Media Ltd, London
Harden TK, Sesma JI, Fricks IP, Lazarowski ER (2010) Signalling and pharmacological properties of the P2Y14 receptor. Acta Physiol (Oxf) 199:149–160
Hirano Y, Okajima F, Tomura H, Majid MA, Takeuchi T, Kondo Y (1991) Change of intracellular calcium of neural cells induced by extracellular ATP. FEBS Lett 284:235–237
Hollopeter G, Jantzen HM, Vincent D, Li G, England L, Ramakrishnan V, Yang RB, Nurden P, Nurden A, Julius D, Conley PB (2001) Identification of the platelet ADP receptor targeted by antithrombotic drugs. Nature 409:202–207
Illes P, Alexandre Ribeiro J (2004) Molecular physiology of P2 receptors in the central nervous system. Eur J Pharmacol 483:5–17

Ito A, Satoh T, Kaziro Y, Itoh H (1995) G protein beta gamma subunit activates Ras, Raf, and MAP kinase in HEK 293 cells. FEBS Lett 368:183–187
Jalink K, Hordijk PL, Moolenaar WH (1994) Growth factor-like effects of lysophosphatidic acid, a novel lipid mediator. Biochim Biophys Acta 1198:185–196
Jin J, Tomlinson W, Kirk IP, Kim YB, Humphries RG, Kunapuli SP (2001) The C6-2B glioma cell P2Y(AC) receptor is pharmacologically and molecularly identical to the platelet P2Y(12) receptor. Br J Pharmacol 133:521–528
Kim S, Jee K, Kim D, Koh H, Chung J (2001) Cyclic AMP inhibits Akt activity by blocking the membrane localization of PDK1. J Biol Chem 276:12864–12870
Kim SG, Gao ZG, Soltysiak KA, Chang TS, Brodie C, Jacobson KA (2003) P2Y6 nucleotide receptor activates PKC to protect 1321N1 astrocytoma cells against tumor necrosis factor-induced apoptosis. Cell Mol Neurobiol 23:401–418
King BF, Burnstock G, Boyer JL, Boeynaems J-M, Weisman GA, Kennedy C, Jacobson KA, Humphries RG, Abbracchio MP, Miras-Portugal MT (2000) The P2Y receptors. In: Girdlestone D (ed) The IUPHAR compendium of receptor characterization and classification. IUPHAR Media Ltd, London
Kitanaka J, Hashimoto H, Gotoh M, Mayumi T, Baba A (1992) Mechanism of extracellular ATP-stimulated phosphoinositide hydrolysis in rat glioma C6 cells. J Pharmacol Exp Ther 263:1248–1252
Kopperud R, Krakstad C, Selheim F, Doskeland SO (2003) cAMP effector mechanisms. Novel twists for an 'old' signaling system. FEBS Lett 546:121–126
Koschel K, Tas PW (1993) Lysophosphatidic acid reverts the beta-adrenergic agonist-induced morphological response in C6 rat glioma cells. Exp Cell Res 206:162–166
Kranenburg O, Moolenaar WH (2001) Ras-MAP kinase signaling by lysophosphatidic acid and other G protein-coupled receptor agonists. Oncogene 20:1540–1546
Krzemiński P, Supłat D, Czajkowski R, Pomorski P, Barańska J (2007) Expression and functional characterization of P2Y1 and P2Y12 nucleotide receptors in long-term serum-deprived glioma C6 cells. FEBS J 274:1970–1982
Krzemiński P, Pomorski P, Barańska J (2008) The P2Y14 receptor activity in glioma C6 cells. Eur J Pharmacol 594: 49–54
Kubiatowski T, Jang T, Lachyankar MB, Salmonsen R, Nabi RR, Quesenberry PJ, Litofsky NS, Ross AH, Recht LD (2001) Association of increased phosphatidylinositol 3-kinase signaling with increased invasiveness and gelatinase activity in malignant gliomas. J Neurosurg 95:480–488
Kurino M, Fukunaga K, Ushio Y, Miyamoto E (1996) Cyclic AMP inhibits activation of mitogen-activated protein kinase and cell proliferation in response to growth factors in cultured rat cortical astrocytes. J Neurochem 67:2246–2255
Lazarowski E (2006) Regulated release of nucleotides and UDP sugars from astrocytoma cells. Novartis Found Symp 276:73–84, discussion 84–90, 107–112, 275–181
Leon C, Hechler B, Vial C, Leray C, Cazenave JP, Gachet C (1997) The P2Y1 receptor is an ADP receptor antagonized by ATP and expressed in platelets and megakaryoblastic cells. FEBS Lett 403:26–30
Liao Z, Seye CI, Weisman GA, Erb L (2007) The P2Y2 nucleotide receptor requires interaction with alpha v integrins to access and activate G12. J Cell Sci 120:1654–1662
Lin WW, Chuang DM (1993) Extracellular ATP stimulates inositol phospholipid turnover and calcium influx in C6 glioma cells. Neurochem Res 18:681–687
Lin WW, Chuang DM (1994) Different signal transduction pathways are coupled to the nucleotide receptor and the P2Y receptor in C6 glioma cells. J Pharmacol Exp Ther 269:926–931
Locovei S, Scemes E, Qiu F, Spray DC, Dahl G (2007) Pannexin1 is part of the pore forming unit of the P2X(7) receptor death complex. FEBS Lett 581:483–488
McKenzie FR, Pouyssegur J (1996) cAMP-mediated growth inhibition in fibroblasts is not mediated via mitogen-activated protein (MAP) kinase (ERK) inhibition. cAMP-dependent protein kinase induces a temporal shift in growth factor-stimulated MAP kinases. J Biol Chem 271:13476–13483
Messens J, Slegers H (1992) Synthesis of glial fibrillary acidic protein in rat C6 glioma in chemically defined medium: cyclic AMP-dependent transcriptional and translational regulation. J Neurochem 58:2071–2080
Moolenaar WH (1999) Bioactive lysophospholipids and their G protein-coupled receptors. Exp Cell Res 253:230–238
Moore DJ, Murdock PR, Watson JM, Faull RL, Waldvogel HJ, Szekeres PG, Wilson S, Freeman KB, Emson PC (2003) GPR105, a novel Gi/o-coupled UDP-glucose receptor expressed on brain glia and peripheral immune cells, is regulated by immunologic challenge: possible role in neuroimmune function. Brain Res Mol Brain Res 118:10–23
Neary JT, Zimmermann H (2009) Trophic functions of nucleotides in the central nervous system. Trends Neurosci 32:189–198
Nicholas RA, Watt WC, Lazarowski ER, Li Q, Harden K (1996) Uridine nucleotide selectivity of three phospholipase C-activating P2 receptors: identification of a UDP-selective, a UTP-selective, and an ATP- and UTP-specific receptor. Mol Pharmacol 50:224–229
North RA (2002) Molecular physiology of P2X receptors. Physiol Rev 82:1013–1067
North RA, Surprenant A (2000) Pharmacology of cloned P2X receptors. Annu Rev Pharmacol Toxicol 40:563–580
Oey J (1975) Noradrealine induces morphological alterations in nucleated and enucleated rat C6 glioma cells. Nature 257:317–319

Pelegrin P, Surprenant A (2006) Pannexin-1 mediates large pore formation and interleukin-1beta release by the ATP-gated P2X7 receptor. EMBO J 25:5071–5082
Putney JW Jr, Bird GS (1993) The inositol phosphate-calcium signaling system in nonexcitable cells. Endocr Rev 14:610–631
Qiu W, Zhuang S, von Lintig FC, Boss GR, Pilz RB (2000) Cell type-specific regulation of B-Raf kinase by cAMP and 14-3-3 proteins. J Biol Chem 275:31921–31929
Ralevic V, Burnstock G (1998) Receptors for purines and pyrimidines. Pharmacol Rev 50:413–492
Roymans D, Grobben B, Claes P, Slegers H (2001) Protein tyrosine kinase-dependent regulation of adenylate cyclase and phosphatidylinositol 3-kinase activates the expression of glial fibrillary acidic protein upon induction of differentiation in rat c6 glioma. Cell Biol Int 25:467–474
Ryten M, Dunn PM, Neary JT, Burnstock G (2002) ATP regulates the differentiation of mammalian skeletal muscle by activation of a P2X5 receptor on satellite cells. J Cell Biol 158:345–355
Ryu JK, Jantaratnotai N, Serrano-Perez MC, McGeer PL, McLarnon JG (2011) Block of purinergic P2X7R inhibits tumor growth in a C6 glioma brain tumor animal model. J Neuropathol Exp Neurol 70:13–22
Sabała P, Czajkowski R, Przybyłek K, Kalita K, Kaczmarek L, Barańska J (2001) Two subtypes of G protein-coupled nucleotide receptors, P2Y(1) and P2Y(2) are involved in calcium signalling in glioma C6 cells. Br J Pharmacol 132:393–402
Sak K, Illes P (2005) Neuronal and glial cell lines as model systems for studying P2Y receptor pharmacology. Neurochem Int 47:401–412
Salcman M (1995) Glioblastoma and malignant astrocytoma. In: Kaye AH, Laws ER (eds) Brain tumors: an encyclopedic approach. Churchill Livingstone, New York
Savi P, Labouret C, Delesque N, Guette F, Lupker J, Herbert JM (2001) P2y(12), a new platelet ADP receptor, target of clopidogrel. Biochem Biophys Res Commun 283:379–383
Schachter JB, Li Q, Boyer JL, Nicholas RA, Harden TK (1996) Second messenger cascade specificity and pharmacological selectivity of the human P2Y1-purinoceptor. Br J Pharmacol 118:167–173
Schachter JB, Boyer JL, Li Q, Nicholas RA, Harden TK (1997) Fidelity in functional coupling of the rat P2Y1 receptor to phospholipase C. Br J Pharmacol 122:1021–1024
Scrivens M, Dickenson JM (2005) Functional expression of the P2Y14 receptor in murine T-lymphocytes. Br J Pharmacol 146:435–444
Scrivens M, Dickenson JM (2006) Functional expression of the P2Y14 receptor in human neutrophils. Eur J Pharmacol 543:166–173
Skelton L, Cooper M, Murphy M, Platt A (2003) Human immature monocyte-derived dendritic cells express the G protein-coupled receptor GPR105 (KIAA0001, P2Y14) and increase intracellular calcium in response to its agonist, uridine diphosphoglucose. J Immunol 171:1941–1949
Sliwa M, Markovic D, Gabrusiewicz K, Synowitz M, Glass R, Zawadzka M, Wesolowska A, Kettenmann H, Kaminska B (2007) The invasion promoting effect of microglia on glioblastoma cells is inhibited by cyclosporin A. Brain 130:476–489
Soulet C, Sauzeau V, Plantavid M, Herbert JM, Pacaud P, Payrastre B, Savi P (2004) Gi-dependent and -independent mechanisms downstream of the P2Y12 ADP-receptor. J Thromb Haemost 2:135–146
Srinivas M, Calderon DP, Kronengold J, Verselis VK (2006) Regulation of connexin hemichannels by monovalent cations. J Gen Physiol 127:67–75
Stork PJ, Schmitt JM (2002) Crosstalk between cAMP and MAP kinase signaling in the regulation of cell proliferation. Trends Cell Biol 12:258–266
Supłat-Wypych D, Dygas A, Barańska J (2010) 2′, 3′-O-(4-benzoylbenzoyl)-ATP-mediated calcium signaling in rat glioma C6 cells: role of the P2Y(2) nucleotide receptor. Purinergic Signal 6:317–325
Sutherland EW (1972) Studies on the mechanism of hormone action. Science 177:401–408
Tas PW, Koschel K (1998) Sphingosine-1-phosphate induces a Ca2+ signal in primary rat astrocytes and a Ca2+ signal and shape changes in C6 rat glioma cells. J Neurosci Res 52:427–434
Tu MT, Luo SF, Wang CC, Chien CS, Chiu CT, Lin CC, Yang CM (2000) P2Y(2) receptor-mediated proliferation of C(6) glioma cells via activation of Ras/Raf/MEK/MAPK pathway. Br J Pharmacol 129:1481–1489
van Corven EJ, Groenink A, Jalink K, Eichholtz T, Moolenaar WH (1989) Lysophosphatidate-induced cell proliferation: identification and dissection of signaling pathways mediated by G proteins. Cell 59:45–54
Van Kolen K, Slegers H (2004) P2Y12 receptor stimulation inhibits beta-adrenergic receptor-induced differentiation by reversing the cyclic AMP-dependent inhibition of protein kinase B. J Neurochem 89:442–453
Van Kolen K, Slegers H (2006a) Atypical PKCzeta is involved in RhoA-dependent mitogenic signaling by the P2Y(12) receptor in C6 cells. FEBS J 273:1843–1854
Van Kolen K, Slegers H (2006b) Integration of P2Y receptor-activated signal transduction pathways in G protein-dependent signalling networks. Purinergic Signal 2:451–469
van Koppen C, Meyer zu Heringdorf M, Laser KT, Zhang C, Jakobs KH, Bunemann M, Pott L (1996) Activation of a high affinity Gi protein-coupled plasma membrane receptor by sphingosine-1-phosphate. J Biol Chem 271:2082–2087

Vossler MR, Yao H, York RD, Pan MG, Rim CS, Stork PJ (1997) cAMP activates MAP kinase and Elk-1 through a B-Raf- and Rap1-dependent pathway. Cell 89:73–82

Wan G, Zhou L, Lim QE, Wong YH, Too HP (2011) Cyclic AMP signalling through PKA but not Epac is essential for neurturin-induced biphasic ERK1/2 activation and neurite outgrowths through GFRalpha2 isoforms. Cell Signal 23: 1727–1737

Wang L, Liu F, Adamo ML (2001) Cyclic AMP inhibits extracellular signal-regulated kinase and phosphatidylinositol 3-kinase/Akt pathways by inhibiting Rap1. J Biol Chem 276:37242–37249

Wang M, Kong Q, Gonzalez FA, Sun G, Erb L, Seye C, Weisman GA (2005) P2Y nucleotide receptor interaction with alpha integrin mediates astrocyte migration. J Neurochem 95:630–640

Webb TE, Feolde E, Vigne P, Neary JT, Runberg A, Frelin C, Barnard EA (1996) The P2Y purinoceptor in rat brain microvascular endothelial cells couple to inhibition of adenylate cyclase. Br J Pharmacol 119:1385–1392

Weber G (2002) Reciprocal regulation: recognition of pattern of gene expression in cancer cells. Adv Enzyme Regul 42:83–100

Wei W, Ryu JK, Choi HB, McLarnon JG (2008) Expression and function of the P2X(7) receptor in rat C6 glioma cells. Cancer Lett 260:79–87

White N, Burnstock G (2006) P2 receptors and cancer. Trends Pharmacol Sci 27:211–217

Wildman SS, Unwin RJ, King BF (2003) Extended pharmacological profiles of rat P2Y2 and rat P2Y4 receptors and their sensitivity to extracellular H+and Zn2+ ions. Br J Pharmacol 140:1177–1186

Zhang FL, Luo L, Gustafson E, Palmer K, Qiao X, Fan X, Yang S, Laz TM, Bayne M, Monsma F Jr (2002) P2Y(13): identification and characterization of a novel Galphai-coupled ADP receptor from human and mouse. J Pharmacol Exp Ther 301:705–713

Zimmermann H (2000) Extracellular metabolism of ATP and other nucleotides. Naunyn Schmiedebergs Arch Pharmacol 362:299–309

Zwartkruis FJ, Bos JL (1999) Ras and Rap1: two highly related small GTPases with distinct function. Exp Cell Res 253: 157–165

Chapter 4
Calcium Signaling in Glioma Cells – The Role of Nucleotide Receptors

Dorota Wypych and Paweł Pomorski

Abstract Calcium signaling is probably one of the evolutionary oldest and the most common way by which the signal can be transmitted from the cell environment to the cytoplasmic calcium binding effectors. Calcium signal is fast and due to diversity of calcium binding proteins it may have a very broad effect on cell behavior. Being a crucial player in neuronal transmission it is also very important for glia physiology. It is responsible for the cross-talk between neurons and astrocytes, for microglia activation and motility. Changes in calcium signaling are also crucial for the behavior of transformed glioma cells. The present Chapter summarizes molecular mechanisms of calcium signal formation present in glial cells with a strong emphasis on extracellular nucleotide-evoked signaling pathways. Some aspects of glioma C6 signaling such as the cross-talk between $P2Y_1$ and $P2Y_{12}$ nucleotide receptors in calcium signal generation will be discussed in-depth, to show complexity of machinery engaged in formation of this signal. Moreover, possible mechanisms of modulation of the calcium signal in diverse environments there will be presented herein. Finally, the possible role of calcium signal in glioma motility is also discussed. This is a very important issue, since glioma cells, contrary to the vast majority of neoplastic cells, cannot spread in the body with the bloodstream and, at least in early stages of tumor development, may expand only by means of sheer motility.

Keywords Calcium signaling • Nucleotide receptors • Store-operated calcium entry

D. Wypych
Nencki Institute of Experimental Biology, Polish Academy of Sciences, 3 Pasteur St., PL 02-093 Warsaw, Poland

Laboratory of Molecular and Cellular Pharmacology of Purinergic Transmission, Department of Pharmacological and Biomolecular Sciences, Università degli Studi di Milano, 9 via Balzaretti, 20133 Milan, Italy
e-mail: d.wypych@nencki.gov.pl

P. Pomorski (✉)
Nencki Institute of Experimental Biology, Polish Academy of Sciences, 3 Pasteur St., PL 02-093 Warsaw, Poland
e-mail: p.pomorski@nencki.gov.pl

J. Barańska (ed.), *Glioma Signaling*, Advances in Experimental Medicine and Biology 986, DOI 10.1007/978-94-007-4719-7_4, © Springer Science+Business Media Dordrecht 2013

Abbreviations

DAG	Diacylglycerol
ER	Endoplasmic reticulum
GPCR	G-protein coupled receptor
IP_3	Inositol 1,4,5-trisphosphate
IP_3R	IP_3 receptor
2MeSADP	2-Methylthio ADP
MLC	Myosin light chain
NCX	Sodium/calcium exchanger
PI3K	Phosphatidylinositol 3-kinase
PIP_2	Phosphatidylinositol 4,5-biphosphate
PLC	Phospholipase C
PM	Plasma membrane
PMCA	Plasma membrane calcium ATPase
PSF	Point spread function
RyR	Ryanodine receptor
SERCA	Sarco/endoplasmic reticulum calcium ATPase
SOC	Store-operated channel
SOCE	Store-operated calcium entry
STIM1,2	Stromal interaction molecule 1,2
TRP channel	Transient receptor potential channel
TRPA	Ankyrin transient receptor potential channel
TRPC	Canonical transient receptor potential channel
TRPM	Melastatin transient receptor potential channel
TRPV	Vanilloid transient receptor potential channel
VGCC	Voltage-gated calcium channel

4.1 Origin of the Calcium Signaling

The evolutionary origin of the calcium signal is rarely discussed, however it determines how calcium signaling acts, how it is generated and why its pathology is so dangerous for the cell. Life has evolved in seawater, an environment rich in cations, including calcium ions. At the same time, life has become highly dependent on nucleotide chemistry, including the universal energy storage particles: nucleotide triphosphates. However, they have one very unfortunate property: in the seawater context, with plenty of calcium ions, phosphates are able to form insoluble crystals of hydroxyapatite (Krane and Glimcher 1962; McCully 2009). Living organisms started to constantly lose phosphate groups from solution what was balanced by efficient removal of calcium ions from the cytoplasm and keeping tight calcium homeostasis. Calcium removal is facilitated by two types of molecular mechanisms: ATP dependent calcium pumps and calcium exchangers. There are two types of calcium pumps: plasma membrane Ca^{2+} ATPase (PMCA) and sarco/endoplasmic reticulum Ca^{2+} ATPase (SERCA). The former of them transports calcium ions from the cytoplasm to the extracellular environment at the expense of one ATP molecule hydrolyzed for a single ion removal (Carafoli and Stauffer 1994). The later transports two calcium ions into endoplasmic reticulum cisternae at the expense of hydrolysis of one ATP molecule (Lauger 1991). Both pumps transport also protons in exchange of calcium ions with the stoichiometry of one proton per one calcium ion (Hao et al. 1994; Yu et al. 1993). Calcium removal from the cytoplasm is also facilitated by ion exchangers, much faster than ATP dependent pumps but endowed with lower calcium affinity. Sodium/calcium exchanger (NCX) works due to the sodium gradient across the plasma membrane and removes a single calcium ion at the expense of three sodium ions flowing into the cell or a single calcium ion and a single potassium ion at the expense of four sodium ions (Kiedrowski et al. 2004). Calcium may be also efficiently buffered into mitochondria by the calcium uniporter, a transporter driven by electrochemical gradient (Kirichok et al. 2004). Contrary to the previous mechanisms, mitochondrial calcium buffering is not permanent and calcium slowly diffuses back into the cytoplasm as soon as the cytoplasmic concentration of calcium ions falls. All those mechanisms lead to the situation when very low concentration of cytoplasmic calcium, usually around 100 nM, is accompanied by a relatively high level of calcium in the endoplasmic reticulum, (in the high micromolar range) and in the cell environment (in the milimolar range).

In the consequence, an extremely steep calcium gradient across the plasma membrane is formed. This gradient is the main source of the calcium signal – a prerequisite for generating a quick and massive flow of ions into the cytoplasm. In this universal signalization, the important components are duration of the signal, amplitude and frequency of occurrence (Berridge et al. 2000).

4.2 Sources of the Calcium Signal

4.2.1 Extracellular Calcium Signal

The steep gradient of the calcium concentration may be used to form a flow of calcium ions, if the membrane becomes permeable to polar particles. This is facilitated by proteins forming channels permeable to ions through the hydrophobic barrier of the plasma membrane. There are several groups of ion channels and some of them are capable to form calcium currents. From the physiological point of view, they can be divided into three groups, depending on the mechanism of opening. The first one comprises voltage-dependent channels, which open upon changes of electrical polarization of the plasma membrane. They are present in excitable cells such as muscles or neurons (for a wide review of structure and function of voltage dependent calcium channels see: Minor and Findeisen 2010).

The second large group includes channels, which open after binding a chemical ligand. One of the representatives of this group are ATP binding P2X channels. For more information see: (Burnstock and Kennedy 2011) and Chaps. 1 and 3 of this Book. P2X receptors are abundant in the glia and are believed to take part in facilitating information exchange between neurons and astrocytes (Illes et al. 2011).

Finally, there is a big, diverse group of channels, which do not fit to any of the previously mentioned groups. They may be opened by submembranous cytoskeleton tension (stretch activated channels), temperature, osmotic pressure or may depend on the presence of other nearby proteins. To this group belong also the store-operated calcium channels (SOC), which will be described later in more details, as well as the big family of proteins forming transient receptor potential channels (TRP). We know six subfamilies of TRP proteins that exist in mammalian cells: canonical receptor potential channels (TRPC), vanilloid receptor potential channels (TRPV), melastin receptor potential channels (TRPM), polycistein receptor potential channels (TRPP), ankyrin receptor potential channels (TRPA) and mucolipin receptor potential channels (TRPML) localized not on the plasma membrane but on the endosomes inside the cell (Bach 2005; Clapham 2007). Another subfamily, known as no mechanoreceptor potential C – TRPN – is characteristic for lower organisms only (Xiao and Xu 2009). Several of the TRP proteins play a crucial role in glia and glioma signaling. TRPC are believed to be responsible for calcium signaling in astrocytes (Golovina 2005; Vazquez et al. 2004) and will be discussed later. TRPV channels regulate cell volume during changes in the environment osmolarity and serve as mechanoreceptors in astrocytes (Benfenati et al. 2011). Moreover, they are responsible for sensing the hydrostatic pressure by microglia (Sappington and Calkins 2008) and pH sensing by astrocytes (Huang et al. 2010). TRPM channel subfamily can be activated by numerous factors, including pathological conditions such as oxidative stress (Naziroglu 2011) as well as by several ligands, such as derivatives of sphingosine present in oligodendrocytes (Grimm et al. 2005). Channels of the same subfamily are however known to regulate response to osmolarity in other than glial cell types, for example in human embryonic kidney cells (Grimm et al. 2003). There are no data about the role of TRPP or TRPA proteins in glia or glioma physiology. However, the reader can find further details about the vast number of possible roles and partners of TRP proteins in the specialized database "TRIP Database: Mammalian TRansient receptor potential channel-Interacting Protein" (Shin et al. 2011).

4.2.2 Signal Generated by Calcium Stores

Extracellular environment is not the only source of calcium ions that take part in signal formation. Most eucaryotic cells are able to form a calcium signal after receptor activation also in calcium-free medium (Berridge 1993). In this case, the signal may be triggered by one of the two secondary messengers, inositol 1,4,5-trisphosphate (IP_3) or calcium itself (Fabiato 1983), depending on the type of channels present in the endoplasmic reticulum (ER) membrane: IP_3-receptors (IP_3R) (Yoshikawa et al. 1992) or ryanodine receptors (RyR) (Imagawa et al. 1987). From the classification stand point, these receptors belong to the group of ligand-gated calcium channels. They are involved in two calcium-signaling pathways: the store-operated calcium entry (SOCE) and calcium dependent calcium signaling.

4.2.2.1 Store-Operated Calcium Signaling

Store-operated calcium entry occurs when intracellular calcium stores become empty after release of Ca^{2+} from ER. This situation happens after activation of ligand binding receptors, present on the cell surface but not associated with ion channels (Fig. 4.1). Calcium signal evoked by such receptors, including common metabotropic G protein-coupled receptors (GPCR) associated with the α_q subunit of the G protein complex, is preceded by activation of phospholipase C (PLC) type β and Inositol trisphosphate (IP_3) production (Berridge 1995, 2009; Putney and Bird 1993; Vanderheyden et al. 2009). IP_3 acts on its receptor present on the endoplasmic reticulum and causes calcium release from ER cisternae (intracellular calcium stores) *via* IP_3R. The emptying of ER leads in turn to the opening of voltage-independent calcium channels on the plasma membrane, the so-called store-operated calcium channels, and calcium flows in from the cell environment. This influx is also known as capacitative calcium entry (Putney 1990) and was found to be a very important signalization system in many cells (Freichel et al. 2004; Gwack et al. 2008; Lewis 2001; Pinton et al. 2000; Rao et al. 2006; Stiber et al. 2008; Thebault et al. 2006; Vig et al. 2008).

Thus, there are two sources of calcium ions, which participate in cell signalization: intracellular calcium stores and the extracellular space. In the past, many controversies arose over the potential mechanisms of communication between ER and the plasma membrane (PM) during calcium signal generation (for review see: Putney 2009; Venkatachalam et al. 2002). However, since the discovery of the role of the stromal interaction molecule (STIM) proteins in sensing calcium level in ER cisternae, the mechanism of store-operated calcium entry has been, at least partially, unveiled (Fig. 4.1). The STIM protein is considered as the one responsible for signal transduction between calcium stores and the plasma membrane (Liou et al. 2005; Roos et al. 2005; Zhang et al. 2005, 2006, for review see: Hogan et al. 2010). There are two STIM calcium sensors present in the ER membrane: STIM1 and STIM2. It is well documented that the STIM1 protein takes part in signal transmission from ER to the store-operated calcium channels in the plasma membrane. After releasing a calcium ion from its

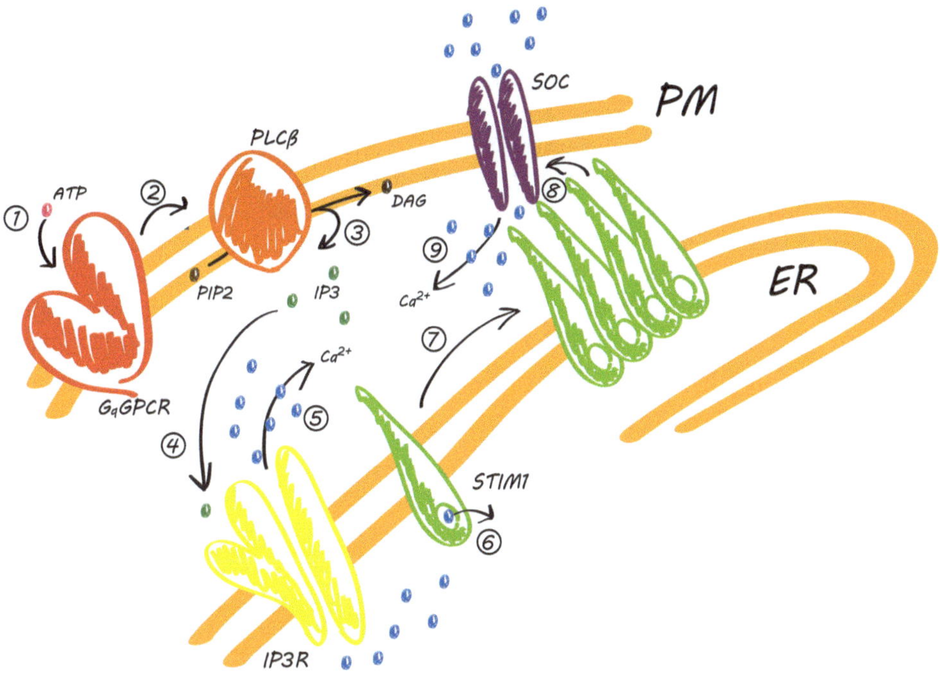

Fig. 4.1 The scheme of the current understanding of store-operated calcium entry mechanism. *1* A ligand (ATP or UTP) binds to the metabotropic GPCR ($P2Y_2$ receptor), *2* G protein subunit α_q released from the receptor activates PLCβ which, *3* catalyses cleavage of phosphatidylinositol 4,5-biphosphate (PIP_2) into inositol (1,4,5) trisphosphate (IP_3) and diacylglycerol (DAG). *4* IP_3 diffuses in cytoplasm and binds to the IP_3 receptor (IP_3R) on the endoplasmic reticulum (ER) membrane. *5* Binding of IP_3 opens IP_3R channel and releases calcium into the cytoplasm. *6* Release of calcium from ER stores lowers the calcium level inside ER which in turn causes dissociation of a calcium ion from the calcium sensor, STIM1. *7* Calcium-free STIM1 rapidly oligomerizes into puncta and binds to the plasma membrane channel protein Orai, *8* This binding opens store-operated channel (SOC) Orai1, and the influx of calcium from cell environment to the cytoplasm starts *9*

EF-hand motif, STIM1 proteins oligomerize and create large complexes, grouped nearby the plasma membrane, visible under the fluorescence microscope (Liou et al. 2005). Such complexes are called puncta and colocalize with SOC channels formed by the Orai protein, however the precise mechanism of this interaction requires further studies (Muik et al. 2011; Potier and Trebak 2008; Wang et al. 2008). Three main hypotheses have been proposed so far (summarized in Potier and Trebak 2008). The first and the most popular model involves a direct binding of the two proteins. The second one suggests that the STIM1-Orai1 interaction occurs through a direct connection between the membrane and / or cytoplasmic proteins, which are constitutively present in these areas or which are recruited into the close space between ER and PM after endoplasmic reticulum emptying. The third theory assumes the existence of a factor produced or released after of ER-PM contact formation, which afterwards activates Orai1. The recent evidence show that electrostatic interaction between STIM1 and Orai1 is required for their coupling (Calloway et al. 2009; Korzeniowski et al. 2010). Moreover, STIM1 is able to form PM junctions (Wang et al. 2010) and recent yeast expression studies show that STIM1 and Orai function independent of other proteins (Zhou et al. 2010).

Thus, the decrease in calcium concentration in ER cisternae results in creation of large-scale protein structures containing the ER calcium sensor, STIM, and the PM calcium channel. This protein-protein interaction needs close proximity between ER and PM. Moreover, in HeLa cells, when ER Ca^{2+} stores were slowly depleted by external EGTA treatment, STIM1 puncta were shown to be formed less efficiently than STIM2 puncta (Brandman et al. 2007). In addition, it has been suggested that STIM1 can create two types of puncta: "resting" oligomers in cells with replenished Ca^{2+} stores or higher-order oligomers in store-depleted cells (Covington et al. 2010). Thus, it seems that STIM1 might respond to EGTA treatment in non-excitable cells, but with a very modest sensitivity.

The role of the STIM2 protein is much more elusive and difficult to interpret. Our results performed on primary culture of cortical neurons showed the relative inability of the STIM2 protein to form puncta after complete depletion of intracellular stores (Gruszczynska-Biegala et al. 2011). STIM2 is believed to be necessary to arouse SOCE (Liou et al. 2005). On the other hand, overexpressed STIM2 can build in into STIM1 puncta (Soboloff et al. 2006), but then it acts as an inhibitor of SOCE, which strongly suggests completely different modes of action of these two proteins. The inhibitory effect of STIM2 on SOCE may depend on its concentration. Putney's group demonstrated that STIM2 overexpression at a lower plasmid concentration did not inhibit thapsigargin-activated Ca^{2+} entry and thus STIM2 no longer behaved like a partial agonist (Bird et al. 2009). Another possible role of the STIM2 protein is that in unstimulated cells, where it regulates resting Ca^{2+} level in ER. The ER Ca^{2+} load is in equilibrium with cytoplasmic free Ca^{2+} level and STIM2 influences the cytoplasmic Ca^{2+} load (Brandman et al. 2007; Zhou et al. 2009). As previously reported, STIM2 has lower ER Ca^{2+} sensitivity than STIM1 (Brandman et al. 2007). Therefore, it translocates to complexes with Orai1 and activates a constitutive calcium influx upon smaller decreases in ER Ca^{2+} concentration in comparison with STIM1. Extended cell incubation in the EGTA-containing medium leads to a decrease in cytoplasmic Ca^{2+} levels and presumably to slowly emptying ER Ca^{2+} stores. The hypothesis of STIM2 acting as an equilibrium regulator is strongly supported by kinetic studies of STIM oligomerization performed by Ikura's group, which showed approximately 70-fold slower aggregation of STIM2 compared with STIM1 (Stathopulos et al. 2009).

4.2.2.2 Calcium-Induced Calcium Signaling

The ryanodine receptors (RyR) were first found on the muscle cell sarcoplasmic network membrane. They take part in another very important calcium signaling pathway namely the calcium-induced calcium release, which strongly depends on intracellular calcium stores. The purpose of this mechanism is to enhance signal after release of very small amount of calcium to the myocyte cytosol. Calcium ions bind to an accessory protein found on the actin filament, which in turn stimulates muscle contraction. Thus, RyR may be considered as the intracellular calcium signal amplifier. The effect of calcium release avalanche from the sarcoplasmic reticulum was described a long time ago (Fabiato 1983) and the calcium-binding calcium channel, known as ryanodine receptor type 1 (RyR1) was described in the striated muscle cells (Inui et al. 1987a). In the same year, another calcium-dependent calcium channel (RyR2) was discovered in the heart muscle cells (Inui et al. 1987b). While those two channels are very specific to the sarcoplasmic reticulum and muscular calcium signaling, the third form of ryanodine receptor (RyR3) is much widely distributed and present in the central nervous system (Hakamata et al. 1992). For review of the calcium-dependent activity of ryanodine receptors see: Hamilton 2005, for extensive review on ryanodine receptors see: Franzini-Armstrong and Protasi 1997.

4.3 Calcium Signaling in Non-Excitable Glial Cells

The term "glia" is broad and far from perfection. Being just a heterogenic class comprising all non-neuronal cells of the central and peripheral nervous system, glia contain cells with separate progenitors (Allen and Barres 2009). While astrocytes and oligodendrocytes originate from ectoderm (Hartline 2011), microglia is a class of immune cells originating from hematopoietic precursors (Gehrmann et al. 1995).

While the differences between gliomas and glia cells are deep and profound, they concern rather the survival and anti-apoptotic aspects and not calcium signaling itself. Proteomic studies reveal changes in some calcium signal-relevant genes, but they encode calcium binding proteins rather than proteins engaged in the actual signal formation (Freije et al. 2004). This is true also for fast moving gliomas (Tatenhorst et al. 2005). Thus, we can assume that gliomas generally inherit calcium signal systems from their healthy progenitors. In our laboratory, we use glioma C6 cell line as the model of astrocyte-derived glioma. More details about this cell line, its origin and the type of nucleotide receptors expressed the reader can find in Chap. 3.

Traditionally, depending on their ability to transmit the signal resulting from plasma membrane depolarization, cells are divided into two large subtypes. Excitable cells, such as neurons, muscles or sensory cells usually express several kinds of Voltage-Gated Calcium Channels (VGCC). The vast majority of eukaryotic cell types, including glial cells, do not develop action potentials and, at least theoretically, should not have VGCC on their surface.

4.3.1 Astrocytes

Contrary to the abovementioned dichotomy of excitable and non-excitable cells, in the 1980sthe evidence was gathered for the presence of VGCC in astrocytes (for review see: Sontheimer 1994). Those results, obtained in electrophysiological experiments on single cell cultures quickly became questioned, as the lack of such activity was observed *in situ*, in brain slice cultures (Carmignoto et al. 1998). The discussion is still open, as up to now several papers have appeared, showing that at least in some situations, VGCC are expressed in astrocytes (Wang et al. 2009) and that the physiological function of glutamate release may be attributed to the hypothetical astrocyte VGCC activity (Yaguchi and Nishizaki 2010). The current hypothesis of astrocyte excitability is, however, based on the primary induction of calcium signaling in a store-dependent manner and its further propagation in RyR-dependent way (Parpura et al. 2011).

It is widely accepted, that the dynamic communication between astrocytes and neurons is secured by the calcium oscillations and calcium waves (Aguado et al. 2002; Parri et al. 2001) and that these oscillations are highly dependent on the SOCE mechanism (Pizzo et al. 2001). The latest research on SOCE show that STIM1 expression is stronger in astrocytes and that of STIM2 in neurons (Gruszczynska-Biegala et al. 2011; Steinbeck et al. 2011). Orai1 was described as the SOCE channel in the immune system cells (Yeromin et al. 2006) and its role in other cells is less evident. In the case of astrocytes, the ability of STIM1 calcium sensor to activate TRPC1-containing channels should be also taken into account (Huang et al. 2006). In the past, it was suggested that TRPC1 could be the main component of astrocyte SOC channels and it was shown the inhibition of TRPC1 by antisense nucleotides (Golovina 2005). However, it was lately shown by using siRNA technique, that SOCE in these cells also highly depends on the Orai1 protein (Linde et al. 2011).

One of the most characteristic properties of calcium signaling in astrocytes is its propagation as a wave in a so called "astrocytic syncytium", which allows signal to spread into large areas (Verkhratsky 2006). This efficient calcium transient transmission between astrocytes was described a long time ago (Cotrina et al. 1998; Finkbeiner 1992; Scemes et al. 2000). Astrocytic syncytium is not a syncytium *sensu stricto*, what means the multinucleate cell. It is rather a broad network of cells connected by multiple gap junctions. It has been shown lately that this vast system of cells bound by gap junctions includes also oligodendrocytes and that bidirectional spreading of calcium waves between these cells occurs (Parys et al. 2010). Astrocytes are connected to each other by homotypic (built from a single connexin type) gap junctions containing connexins 30 or 43 (Theis et al. 2005). Gap junctions between astrocytes and oligodendrocytes are heterotypic, built by connexin 32 on the oligodendrocyte side and connexin 43 or 26 on the astrocyte side.

Connexins may also play an additional role in glioma calcium signaling. It was suggested, that apart from forming gap junctions, connexins might form hemichannels responsible for the release of ATP from astrocytes (Jiang et al. 2011), thus enhancing the spread of the calcium signal.

4.3.2 Microglia

As mentioned before the term "glia" is both old and imprecise. Contrary to other brain cells, which are derived from the neuroectoderm, microglia is a part of the body immune system and arises from the peripheral mesodermal tissue (Chan et al. 2007). Today it is believed that microglia populate the central nervous system during development and stay there, isolated from the rest of the immune system by the brain-blood barrier (Ajami et al. 2007). The morphological differences between microglia and monocytes do not however originate from the long developmental isolation, but from the local environment of the central nervous system since monocytes invading the brain, in case of brain-blood barrier destruction, transform into microglial cells (Mildner et al. 2007).

Since microglia is part of the immune system, the cells have calcium signaling and homeostasis characteristic for immune cells. The classical STIM1/Orai signal coupling was detected in microglial store-operated calcium signaling (Ohana et al. 2009).

It is also known that extracellular nucleotides (see below) play an important role in microglia biology. Beside several GPCR receptors, microglial cells possess $P2X_4$ and $P2X_7$ nucleotide receptors forming ligand activated channels in the plasma membrane (Cavaliere et al. 2005). The level of these receptors strongly depends on the microglia activation (Bianco et al. 2005; Davalos et al. 2005; Gyoneva et al. 2009; Moller et al. 2000; Morigiwa et al. 2000). It is worthy to mention, that nucleotide GPRC receptors coupled to the α_q subunit, notably the $P2Y_2$ receptor, are believed to play an important role in microglia chemotaxis (Davalos et al. 2005; Nimmerjahn et al. 2005).

Moreover, microglia react to the appearance of glioma in the central nervous system. This interesting topic is discussed wider in other Chapters of this Book (see: Chaps. 3 and 6).

4.4 Nucleotide Receptor-Evoked Calcium Signaling in Glial Cells

Nucleotide release in the central nervous system is quite a complex process. ATP, UTP and other nucleotide-derivates are co-released with different neurotransmitters in synaptic vessels as the result of processes of regulated exocytosis (Dean et al. 1984; Gualix et al. 1999, for details see: Chap. 1). The presence of ATP in the extracellular space of *in vitro* cultured astrocytes indicates that the secretion of nucleotides is a common process in these cells, however the mechanism (or mechanisms) as well as the signalling pathways involved in this process are still not clear (Bianco et al. 2005; Coco et al. 2003; Joseph et al. 2003). Coco et. al described the mechanism of ATP release from astrocytes as a process of regulated exocytosis, which strongly depends on calcium ions (Coco et al. 2003). Cortina et. al showed that stimulation of nucleotide receptors by UTP in glioma C6 leads to a calcium-dependent increase in extracellular ATP concentration (Cotrina et al. 1998). ATP release form cultured primary astrocytes and from cancer cells derived from astrocytes is accompanied by an increase of accumulation in extracellular UTP and UDP-glucose. Sugar-derivatives are released from cells in consequence of the vesicular transport from the Golgi apparatus. As other nucleotides (ATP, ADP, UDP) also coexist in this structure, it is possible that vesicular transport is a common way of their release (Lazarowski 2006). On the contrary, UTP is not translocated to either ER or to the Golgi apparatus. However, if UDP can be released *via* the secretion pathway together with UDP-glucose, the alternative source of extracellular UTP can be the vesicular transport of UDP followed by, its extracellular phosphorylation by ectoenzymes. It is also postulated that ATP and UTP existing in the cytosol can be transported to the extracellular space by (not fully identified) plasma membrane transmitters (Lazarowski 2006; Lazarowski et al. 2011).

The concentration of ATP in the extracellular environment at resting conditions (1–10 nM) is the result of dynamic equilibrium between the metabolism of ATP and its continuous extracellular secretion. Unstimulated cells secrete ATP in the amount balancing the level of hydrolyzed ATP. Calculated on this basis, the continuous release of ATP reaches the rate of approximately 10 fmol/min/10^6 in C6 glioma cells and of about 500 fmol/min/10^6 for diverse cultures of primary bronchial epithelial cells (Lazarowski et al. 2003). The biggest source of the sudden increase in the concentration of extracellular ATP is mechanical damage of the tissue or cell death (Burnstock 2006). That is why the purinergic system plays a critical role in the central nervous system both under physiological and pathological conditions.

As mentioned before (see: Chaps. 1, 2, 3, and 4) neural cells express all types of nculeotide receptors and their expression can be dramatically changed upon different conditions. Among them, ionotropic P2X receptors together with G_q-coupled P2Y metabotropic receptors ($P2Y_{1,2,4,6}$) are directly involved in generation of calcium signal. However, the nature of the signal is different, since P2X are ligand-gated ion channels which selectively permeate Ca^{2+} (and Na^+) from the extracellular space while activation of the above mentioned P2Y receptors leads to IP_3-dependent calcium release from the intracellular stores, which is followed by the capacitative calcium entry process.

4.5 Glioma C6: A Case Study

Glioma C6 cells express several nucleotide receptors from both the P2X and the P2Y families (for details see: Chap. 3). It has previously been shown that in these cells ADP, as well as ATP and UTP initiated calcium response is compatible with the capacitative model of Ca^{2+} influx (Baranska et al. 1999; Sabala et al. 2001). Among the large family of metabotropic P2Y nucleotide receptors, the $P2Y_2$ receptor responds to ATP and UTP, whereas $P2Y_1$ and $P2Y_{12}$ receptors respond to ADP and 2MeSADP. $P2Y_1$ and $P2Y_2$ are responsible for Ca^{2+} mobilization from intracellular stores (Baranska et al. 2004; Czajkowski et al. 2004; Sak and Illes 2005), while the $P2Y_{12}$ receptor is negatively coupled to adenylate cyclase (Czajkowski et al. 2002; Hollopeter et al. 2001; Jin et al. 2001). Thus, ADP and unhydrolyzed ADP analogue 2-methylthio ADP (2MeSADP) are agonists of two receptors: $P2Y_1$ and $P2Y_{12}$, which activate different downstream signaling pathways (see: Chap. 3).

4.5.1 The Nature of the Signal – The Role of Nucleotide Receptors in Glioma C6 Cells

In many cells, the cross-talk between $P2Y_1$ and $P2Y_{12}$ receptors is extremely complex (Baranska et al. 2004; Czajkowski and Baranska 2002, see also Chap. 3). In human platelets, Daniel et al. suggested that these two receptors act independently after ADP stimulation (Daniel et al. 1998). On the contrary, Sage et al. (2000) and Fox et al. (2004) suggested that $P2Y_{12}$ might enhance $P2Y_1$-induced cytosolic Ca^{2+} rise. This conflicting evidence was finally explained by Hardy et al. as the different conditions used during platelets preparation (Hardy et al. 2004). Similarly, in glioma C6 cells the differences in the culture conditions may explain the conflicting evidence regarding the role of $P2Y_1$ in ADP-mediated calcium response (Czajkowski and Baranska 2002; Krzeminski et al. 2007). The presence of serum in the culture medium is required for the functional activity of $P2Y_1$ receptors (Baranska et al. 2004; Czajkowski et al. 2004). Otherwise, no IP_3-turnover can be observed after ADP stimulation (Grobben et al. 2001).

Hardy et al. (2004) as well as Sage et al. (2000) suggested the modulatory role of $P2Y_{12}$, which positively regulated the $P2Y_1$-induced Ca^{2+} response. It was known that inhibition of $P2Y_{12}$ receptors reduced the strength of calcium signal by approximately 50 % (Hardy et al. 2004). The stimulatory effect on calcium response was due to the $P2Y_{12}$-induced inhibition of adenylate cyclase and activation of phosphatidylinositol 3-kinase (PI3K). The effect of $P2Y_1$ on PI3K was inhibitory (Hardy et al. 2004). We have previously demonstrated that in glioma C6 cells, the calcium response to ADP or its analogue 2MeSADP, is directly dependent upon activation of the $P2Y_1$ receptor and that the strength of calcium response fully correlates with the $P2Y_1$ protein level. However, inhibition of $P2Y_{12}$ by its specific antagonist, AR-C69931MX, reduced the 2MeSADP-evoked calcium response by 55 % (Suplat et al. 2007). Studies on the cross-talk between nucleotide receptor-induced signalling pathways in glioma C6 cells also demonstrated inhibitory or stimulatory effects on PI3K signaling of $P2Y_1$ and $P2Y_{12}$ respectively (Baranska et al. 2004; Czajkowski et al. 2004; Sabala et al. 2001), see also Chap. 3). Thus, modulatory effect of $P2Y_{12}$ receptors on the $P2Y_1$-induced Ca^{2+} responses in this cell line may occur *via* a mechanism similar to the one suggested for platelets (Hardy et al. 2004).

The modulatory effect of the $P2Y_{12}$ receptor on intracellular free calcium level seems to be a result of the PMCA inhibition (Sage et al. 2000). It is well known that in many cell types the PMCA activity is enhanced by cAMP (Dixon and Haynes 1989; James et al. 1989; Monteith and Roufogalis 1995), thereby limiting the strength of the calcium response. The constant activation of the $P2Y_{12}$ receptor inhibits adenylate cyclase, lowers the level of cAMP and inhibits the PMCA pump. Thus, inhibition of this receptor activity observed in platelets and glioma C6 cells may enhance the pump efficiency and lower the calcium signal (Hardy et al. 2004; Suplat et al. 2007). Moreover, Hardy et al. suggested that PI3K, activated by the $P2Y_{12}$ receptor stimulation, may also activate PLC type γ, leading to the rise in IP_3 and enhancement of the calcium signal. It has been reported that receptors coupled to PI3K may activate PLCγ indirectly in the absence of PLC type γ-tyrosine phosphorylation (Bae et al. 1998; Falasca et al. 1998).

In the 1990s, Sleger's group postulated that the ADP-evoked calcium response is an autonomic action of $P2Y_{12}$ receptors. Since they cultured glioma C6 cells in a serum-free chemically defined medium, they could not evoke $P2Y_1$ receptor-dependent IP_3 turnover. Instead, they suggested that in these cells the $P2Y_{12}$ receptor was not only coupled to the inhibition of adenyl cyclase but also to the activation of PLC-independent Ca^{2+} influx (Grobben et al. 2001). According to the authors, the mechanism of the $P2Y_{12}$-dependent calcium influx remained to be determined. Nonetheless, they proposed an interesting idea that stimulation of the $P2Y_{12}$ receptor might generate, via βγ subunits of the G_i protein, a direct entry of calcium ions from the extracellular space (Van Kolen and Slegers 2006; Van Kolen et al. 2006).

To check this hypothesis we performed a two-phase calcium experiment, in which the ligand was applied in a calcium-free medium to observe calcium release from the calcium stores and then the medium was exchanged to a calcium-containing one to observe SOCE (Fig. 4.2). There was no difference in calcium release form ER between control cells (grey line) and these pretreated with the $P2Y_{12}$ antagonist, AR-C69931MX (green line). This suggests that this component of the signal is $P2Y_1$-dependent and $P2Y_{12}$-independent. Calcium influx from the extracellular space was significantly weaker in presence of the AR-C69931MX when compared to the control. Therefore, if the hypothesis of direct activation of plasmalemmal calcium channels by $P2Y_{12}$ receptors was correct, we should observe inhibition of the calcium entry in cells, in which the $P2Y_{12}$ antagonist was added before stimulation by agonist. At the same time, we should not observe this phenomenon, when the antagonist was added after $P2Y_{12}$ receptor stimulation by an agonist, because the hypothetical channel should be already open. However, the effect of diminished calcium influx was independent of the time the inhibitor was added, which suggests a continuous mechanism of adenyl cyclase inhibition and not a digital in character opening of calcium channels (Bird et al. 2009). Hence, the calcium influx observed in ADP-stimulated glioma C6 cells is consistent with the mechanism of capacitative calcium entry. The strength of the signal depends on the $P2Y_{12}$ and $P2Y_{12}$ cross-talk.

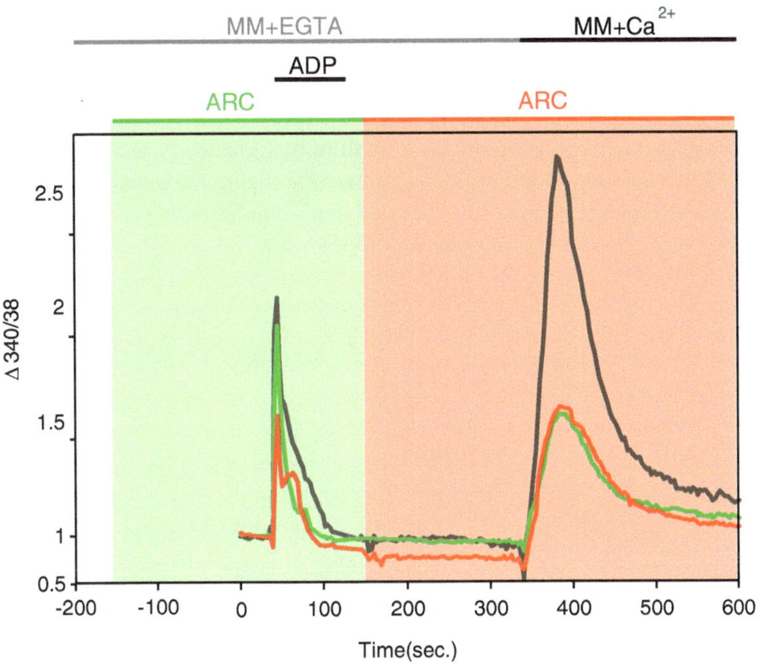

Fig. 4.2 The effect of AR-C69931MX, the $P2Y_{12}$ receptor antagonist, on the first and second phase of calcium response in glioma C6 cells. Cells in Ca^{2+}-free buffer were stimulated by 10 μM ADP. Three cell groups were studied: The control cells were induced with ADP, without antagonist (*grey line*). The second group was treated with 10 μM AR-C69931MX for 3 min before ADP addition and release of calcium from the ER (*green line*). The third group was treated for 3 min with 10 μM AR-C69931MX just after the end of a calcium transient caused by ADP-induced release from the ER (*red line*). Then, in all cell groups, the medium was exchanged for fresh medium containing 2 mM $CaCl_2$ with (the third group) or without (control and the second group) AR-C69931MX (Reproduced from Suplat et al. 2007, changed, with permission from Springer Science)

4.6 Gliomas: The Motile Tumors, Calcium Signaling and Chemotaxis

We, as well as other researchers have previously found that in glioma cells, changes in the actin cytoskeleton architecture strongly interfered with the cell ability to form a calcium signal (Korczynski et al. 2011; Sabala et al. 2002; Suplat et al. 2004). Actually, there are two leading, and somehow contradictory, theories of glia chemotaxis. The first one is based on small GTPase-dependent contractile activity (Liao et al. 2007; Wang et al. 2005) and the second one is based on nucleotide-promoted calcium transients (Striedinger and Scemes 2008; Striedinger et al. 2007). Both theories consider extracellular nucleotides as chemotaxis regulators, which act on GPCR from the P2Y group. The first, GTP-ase dependent theory, postulates that after non-covalent binding of integrins, the $P2Y_2$ receptors change affinity from the G_q to the G_{12} type of α subunit of the heterotrimeric G proteins and thus switch than downstream signaling pathway. This is described in details in another Chapter of this Book (see: Chap. 6). The second theory assumes the crucial role of calcium oscillations, observed after induction of the same metabotropic $P2Y_2$ receptors followed by activation of the classical G_q-dependent calcium signaling pathway. We have recently observed that in glioma C6 cells the UTP-evoked calcium signal was strongly polarized and appeared clearly in the frontal part of the cell body (Wypych and Pomorski, unpublished results). It suggests that both signaling pathways described above may be activated in parallel in the same cell but in different regions. The situation is further complicated by recent revelations from the Margolis group, showing actin-independent motility of glioblastoma cells (Panopoulos et al. 2011).

The ability to transmit the calcium signal from cell to cell is limited in glioma cell cultures. In astrocyte derived gliomas, it was suggested that the expression of connexins and ability to form calcium-permeable gap junctions fall with the growth of malignancy (Caltabiano et al. 2010). It was also shown, that expression of connexin 43 may reverse the malignant phenotype of glioma (Yu et al. 2012). Moreover, expression of connexins 43 and 32 in glioma C6 cells promotes transmission of the calcium signal between cells in culture (Fry et al. 2001).

4.6.1 The Calcium Signal in Glioma C6 Is Strongly Dependent on Actin Cytoskeleton

The observations that in glioma C6 cells SOCE depends on the cytoskeleton state (Sabala et al. 2002; Suplat et al. 2004) are older than our understanding of the molecular mechanism of store-operated calcium entry (Liou et al. 2005; Roos et al. 2005; Vig et al. 2008; Zhang et al. 2006). Apart from the hypothetical nature of all conclusions derived from these old experiments, the additional trouble was their interpretation due to the incoherence of the literature. Depending on the cell type, the influence of cytoskeleton disruption could have positive (Rosado et al. 2000), negative (Targos et al. 2006) or no effect on SOCE (Ribeiro et al. 1997). Now, understanding the mechanism of signal transduction between the calcium depleted ER and the channels in cell plasma membrane, we can fully appreciate how deeply those two processes are entangled. However, our recent experiments on the effect of cross-talk between actomyosin contractility and nucleotide calcium signaling show how complicated the system is (Korczynski et al. 2011).

We have studied the influence of cytoskeleton manipulations on the nucleotide receptor-evoked, as well as on the SERCA pump inhibitor-evoked, calcium responses in glioma C6 cells for several years. The overall result of those studies is that the ability to generate a strong calcium response is dependent on the subtle structure of the actin cytoskeleton and any changes in this structure may reduce the calcium transient (Sabala et al. 2002; Suplat et al. 2004).

The use of Y-27632, a specific inhibitor of ROCK, the main effector kinase dependent on RhoA, a small GTP binding protein (Hirose et al. 1998), provokes to dynamic actin reorganization in glioma C6 cells and subsequent alterations in cell morphology (Korczynski et al. 2011). Contrary to such profound actin cytoskeleton destructing agents like cytochalasine D, that we used previously in our experiments, the action of Y-27632 may be considered as resulting in cytoskeleton modulation. The agent does not change cell behavior deeply and does not prevent spontaneous migration (at least for the time of 30 min used during experiments) even if it inhibits the chemotactic behavior of the cells (Worthylake and Burridge 2003). Our experiments showed that this modulation not only weakens the $P2Y_2$-evoked calcium signal but does it specifically by inhibiting the capacitative calcium influx, leaving calcium release from endoplasmic reticulum unchanged (Korczynski et al. 2011).

Detailed studies of the capacitative calcium signal resulting from $P2Y_2$ receptor activation have shown recently that in motile C6 cells the signal is spatio-temporarily polarized (Fig. 4.3a). Studying fast signal dynamics with the ratiometry method (Roy et al. 2002) we have found, that it originates in the flat, frontal area of the motile cell (Fig. 4.3b, c). Changes in cell morphology, which are observed upon prolonged serum-deprivation, do not influence the expression of $P2Y_2$ receptors in glioma C6 cells (Suplat et al. 2007), but strongly affect the spatio-temporary pattern of calcium wave formation after UTP-stimulation (Wypych and Pomorski, unpublished results). Inhibition of ROCK activity by Y-27632 also leads to the disappearance of the polarity of this signal, what may have influence of Y-27632 incubated cells ability to perform chemotactical behavior (Worthylake and Burridge 2003).

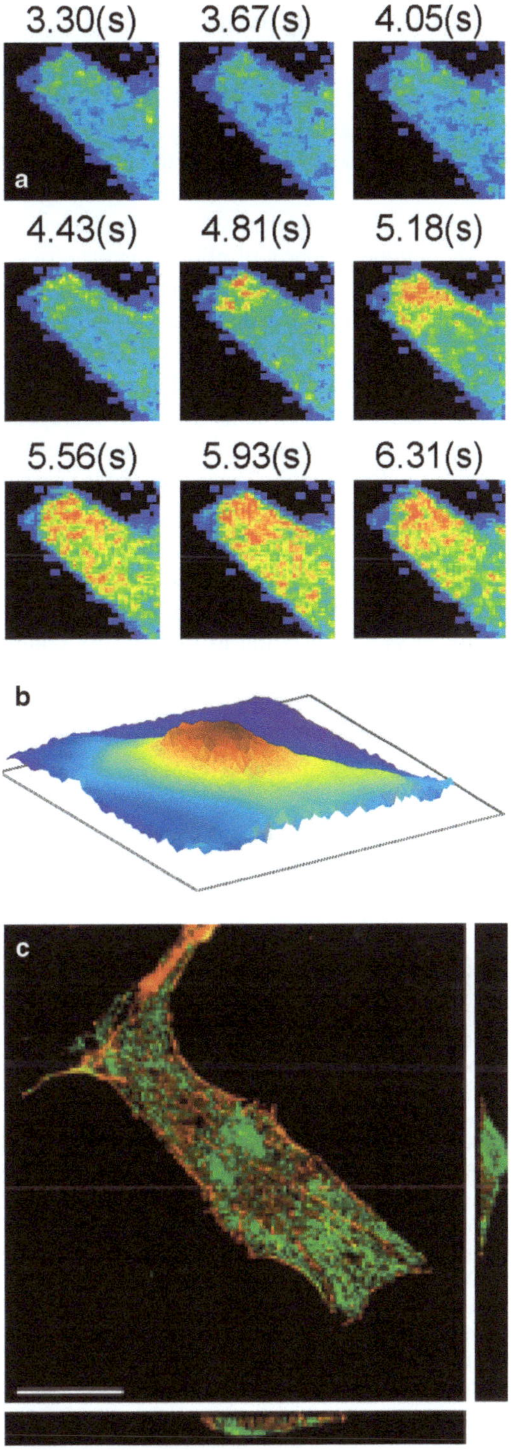

Fig. 4.3 Local nature of the calcium signal in glioma C6 cells. (**a**) Frames from the movie presenting the calcium wave developing after activation of $P2Y_2$ receptor in the flat leading region of the cell. Time counted from the addition of 100 μM UTP. (**b**) Approximation of the cell thickness by the brightness of single channel Fura 2 fluorescence (excitation at 380 nm). (**c**) The real shape of the UTP-induced glioma C6 cell as reconstructed from PSF-deconvolved confocal series. *Red*: Alexa 546 conjugated phalloidin; *green*: staining with antibodies against phosphorylated myosin light chains (Reproduced from Korczynski et al. (2011) with permission from Acta Biochimica Polonica (**c**))

4.7 Concluding Remarks

Concluding, we have to state that glioma cells share most of their calcium signaling machinery with glia cells they have derived from. The point of gravity shifts however strongly with glial cell transformation. Cells loose their ability to spread the calcium signal through wide areas and to synchronize cellular activity. Moreover, calcium signaling becomes important for motility regulation and chemotaxis, which can be interpreted as a mechanism characteristic for undifferentiated cells.

The unique property of gliomas is the strict dependence of tumor spreading on the cell motility. Gliomas are much less dependent on the blood circulation in metastasis formation than most of other human tumors and it makes them more dependent on the calcium signaling during the process of metastasis.

It is also important to notice that calcium signaling is active in all cells in the growing brain tumor and not only in tumor cells. Thus, all therapeutic manipulations of the calcium metabolism in the brain will not act on the tumor cells exclusively, but will affect all cells: neurons, astrocytes, oligodendrocytes and microglia. The activity of the latter cells was shown to have a great influence on tumor development.

Even if the calcium signal in glioma cell is not the crucial event responsible for tumoral transformation and loss of the proliferation control, calcium signaling drives several processes contributing to the malignancy of glia-derived tumors. It appears that the shift in the calcium signaling mechanisms from cell synchronization to the motility control is more pronounced in more malignant types of gliomas. Thus, studies of reciprocal signaling cascades, operating between extracellular nucleotides, cell motility and intercellular communication, all of which are based on calcium signalization, are important for better understanding glia-derived tumors, what in consequence may lead to finding an effective methods of glioma therapy.

Acknowledgements The work in the Author's Laboratory is supported by a grant NN303 070434 from the Ministry of Science and Higher Education.

References

Aguado F, Espinosa-Parrilla JF, Carmona MA, Soriano E (2002) Neuronal activity regulates correlated network properties of spontaneous calcium transients in astrocytes in situ. J Neurosci 22:9430–9444
Ajami B, Bennett JL, Krieger C, Tetzlaff W, Rossi FM (2007) Local self-renewal can sustain CNS microglia maintenance and function throughout adult life. Nat Neurosci 10:1538–1543
Allen NJ, Barres BA (2009) Neuroscience: glia – more than just brain glue. Nature 457:675–677
Bach G (2005) Mucolipin 1: endocytosis and cation channel–a review. Pflugers Arch 451:313–317
Bae YS, Cantley LG, Chen CS, Kim SR, Kwon KS, Rhee SG (1998) Activation of phospholipase C-gamma by phosphatidylinositol 3,4,5-trisphosphate. J Biol Chem 273:4465–4469
Baranska J, Przybylek K, Sabala P (1999) Capacitative calcium entry. Glioma C6 as a model of nonexcitable cells. Pol J Pharmacol 51:153–162
Baranska J, Czajkowski R, Sabala P (2004) Cross-talks between nucleotide receptor-induced signaling pathways in serum-deprived and non-starved glioma C6 cells. Adv Enzyme Regul 44:219–232
Benfenati V, Caprini M, Dovizio M, Mylonakou MN, Ferroni S, Ottersen OP, Amiry-Moghaddam M (2011) An aquaporin-4/transient receptor potential vanilloid 4 (AQP4/TRPV4) complex is essential for cell-volume control in astrocytes. Proc Natl Acad Sci U S A 108:2563–2568
Berridge MJ (1993) Inositol trisphosphate and calcium signalling. Nature 361:315–325
Berridge MJ (1995) Capacitative calcium entry. Biochem J 312(Pt 1):1–11
Berridge MJ (2009) Inositol trisphosphate and calcium signalling mechanisms. Biochim Biophys Acta 1793:933–940
Berridge MJ, Lipp P, Bootman MD (2000) The versatility and universality of calcium signalling. Nat Rev Mol Cell Biol 1:11–21

Bianco F, Fumagalli M, Pravettoni E, D'Ambrosi N, Volonte C, Matteoli M, Abbracchio MP, Verderio C (2005) Pathophysiological roles of extracellular nucleotides in glial cells: differential expression of purinergic receptors in resting and activated microglia. Brain Res Brain Res Rev 48:144–156

Bird GS, Hwang SY, Smyth JT, Fukushima M, Boyles RR, Putney JW Jr (2009) STIM1 is a calcium sensor specialized for digital signaling. Curr Biol 19:1724–1729

Brandman O, Liou J, Park WS, Meyer T (2007) STIM2 is a feedback regulator that stabilizes basal cytosolic and endoplasmic reticulum Ca2+ levels. Cell 131:1327–1339

Burnstock G (2006) Purinergic signalling–an overview. Novartis Found Symp 276:26–48, discussion 48–57, 275–281

Burnstock G, Kennedy C (2011) P2X receptors in health and disease. Adv Pharmacol 61:333–372

Calloway N, Vig M, Kinet JP, Holowka D, Baird B (2009) Molecular clustering of STIM1 with Orai1/CRACM1 at the plasma membrane depends dynamically on depletion of Ca2+ stores and on electrostatic interactions. Mol Biol Cell 20:389–399

Caltabiano R, Torrisi A, Condorelli D, Albanese V, Lanzafame S (2010) High levels of connexin 43 mRNA in high grade astrocytomas. Study of 32 cases with in situ hybridization. Acta Histochem 112:529–535

Carafoli E, Stauffer T (1994) The plasma membrane calcium pump: functional domains, regulation of the activity, and tissue specificity of isoform expression. J Neurobiol 25:312–324

Carmignoto G, Pasti L, Pozzan T (1998) On the role of voltage-dependent calcium channels in calcium signaling of astrocytes in situ. J Neurosci 18:4637–4645

Cavaliere F, Dinkel K, Reymann K (2005) Microglia response and P2 receptor participation in oxygen/glucose deprivation-induced cortical damage. Neuroscience 136:615–623

Chan WY, Kohsaka S, Rezaie P (2007) The origin and cell lineage of microglia: new concepts. Brain Res Rev 53:344–354

Clapham DE (2007) SnapShot: mammalian TRP channels. Cell 129:220

Coco S, Calegari F, Pravettoni E, Pozzi D, Taverna E, Rosa P, Matteoli M, Verderio C (2003) Storage and release of ATP from astrocytes in culture. J Biol Chem 278:1354–1362

Cotrina ML, Kang J, Lin JH, Bueno E, Hansen TW, He L, Liu Y, Nedergaard M (1998) Astrocytic gap junctions remain open during ischemic conditions. J Neurosci 18:2520–2537

Covington ED, Wu MM, Lewis RS (2010) Essential role for the CRAC activation domain in store-dependent oligomerization of STIM1. Mol Biol Cell 21:1897–1907

Czajkowski R, Baranska J (2002) Cross-talk between the ATP and ADP nucleotide receptor signalling pathways in glioma C6 cells. Acta Biochim Pol 49:877–889

Czajkowski R, Lei L, Sabala P, Baranska J (2002) ADP-evoked phospholipase C stimulation and adenylyl cyclase inhibition in glioma C6 cells occur through two distinct nucleotide receptors, P2Y(1) and P2Y(12). FEBS Lett 513: 179–183

Czajkowski R, Banachewicz W, Ilnytska O, Drobot LB, Baranska J (2004) Differential effects of P2Y1 and P2Y12 nucleotide receptors on ERK1/ERK2 and phosphatidylinositol 3-kinase signalling and cell proliferation in serum-deprived and nonstarved glioma C6 cells. Br J Pharmacol 141:497–507

Daniel JL, Dangelmaier C, Jin J, Ashby B, Smith JB, Kunapuli SP (1998) Molecular basis for ADP-induced platelet activation. I. Evidence for three distinct ADP receptors on human platelets. J Biol Chem 273:2024–2029

Davalos D, Grutzendler J, Yang G, Kim JV, Zuo Y, Jung S, Littman DR, Dustin ML, Gan WB (2005) ATP mediates rapid microglial response to local brain injury in vivo. Nat Neurosci 8:752–758

Dean GE, Fishkes H, Nelson PJ, Rudnick G (1984) The hydrogen ion-pumping adenosine triphosphatase of platelet dense granule membrane. Differences from F1F0- and phosphoenzyme-type ATPases. J Biol Chem 259: 9569–9574

Dixon DA, Haynes DH (1989) Kinetic characterization of the Ca2+–pumping ATPase of cardia sarcolemma in four states of activation. J Biol Chem 264:13612–13622

Fabiato A (1983) Calcium-induced release of calcium from the cardiac sarcoplasmic reticulum. Am J Physiol 245: C1–C14

Falasca M, Logan SK, Lehto VP, Baccante G, Lemmon MA, Schlessinger J (1998) Activation of phospholipase C gamma by PI 3-kinase-induced PH domain-mediated membrane targeting. EMBO J 17:414–422

Finkbeiner S (1992) Calcium waves in astrocytes-filling in the gaps. Neuron 8:1101–1108

Fox SC, Behan MW, Heptinstall S (2004) Inhibition of ADP-induced intracellular Ca2+ responses and platelet aggregation by the P2Y12 receptor antagonists AR-C69931MX and clopidogrel is enhanced by prostaglandin E1. Cell Calcium 35:39–46

Franzini-Armstrong C, Protasi F (1997) Ryanodine receptors of striated muscles: a complex channel capable of multiple interactions. Physiol Rev 77:699–729

Freichel M, Vennekens R, Olausson J, Hoffmann M, Muller C, Stolz S, Scheunemann J, Weissgerber P, Flockerzi V (2004) Functional role of TRPC proteins in vivo: lessons from TRPC-deficient mouse models. Biochem Biophys Res Commun 322:1352–1358

Freije WA, Castro-Vargas FE, Fang Z, Horvath S, Cloughesy T, Liau LM, Mischel PS, Nelson SF (2004) Gene expression profiling of gliomas strongly predicts survival. Cancer Res 64:6503–6510
Fry T, Evans JH, Sanderson MJ (2001) Propagation of intercellular calcium waves in C6 glioma cells transfected with connexins 43 or 32. Microsc Res Tech 52:289–300
Gehrmann J, Matsumoto Y, Kreutzberg GW (1995) Microglia: intrinsic immuneffector cell of the brain. Brain Res Brain Res Rev 20:269–287
Golovina VA (2005) Visualization of localized store-operated calcium entry in mouse astrocytes. Close proximity to the endoplasmic reticulum. J Physiol 564:737–749
Grimm C, Kraft R, Sauerbruch S, Schultz G, Harteneck C (2003) Molecular and functional characterization of the melastatin-related cation channel TRPM3. J Biol Chem 278:21493–21501
Grimm C, Kraft R, Schultz G, Harteneck C (2005) Activation of the melastatin-related cation channel TRPM3 by D-erythro-sphingosine [corrected]. Mol Pharmacol 67:798–805
Grobben B, Claes P, Van Kolen K, Roymans D, Fransen P, Sys SU, Slegers H (2001) Agonists of the P2Y(AC)-receptor activate MAP kinase by a ras-independent pathway in rat C6 glioma. J Neurochem 78:1325–1338
Gruszczynska-Biegala J, Pomorski P, Wisniewska MB, Kuznicki J (2011) Differential roles for STIM1 and STIM2 in store-operated calcium entry in rat neurons. PLoS One 6:e19285
Gualix J, Pintor J, Miras-Portugal MT (1999) Characterization of nucleotide transport into rat brain synaptic vesicles. J Neurochem 73:1098–1104
Gwack Y, Srikanth S, Oh-Hora M, Hogan PG, Lamperti ED, Yamashita M, Gelinas C, Neems DS, Sasaki Y, Feske S, Prakriya M, Rajewsky K, Rao A (2008) Hair loss and defective T- and B-cell function in mice lacking ORAI1. Mol Cell Biol 28:5209–5222
Gyoneva S, Orr AG, Traynelis SF (2009) Differential regulation of microglial motility by ATP/ADP and adenosine. Parkinsonism Relat Disord 15(Suppl 3):S195–S199
Hakamata Y, Nakai J, Takeshima H, Imoto K (1992) Primary structure and distribution of a novel ryanodine receptor/calcium release channel from rabbit brain. FEBS Lett 312:229–235
Hamilton SL (2005) Ryanodine receptors. Cell Calcium 38:253–260
Hao L, Rigaud JL, Inesi G (1994) Ca2+/H+countertransport and electrogenicity in proteoliposomes containing erythrocyte plasma membrane Ca-ATPase and exogenous lipids. J Biol Chem 269:14268–14275
Hardy AR, Jones ML, Mundell SJ, Poole AW (2004) Reciprocal cross-talk between P2Y1 and P2Y12 receptors at the level of calcium signaling in human platelets. Blood 104:1745–1752
Hartline DK (2011) The evolutionary origins of glia. Glia 59:1215–1236
Hirose M, Ishizaki T, Watanabe N, Uehata M, Kranenburg O, Moolenaar WH, Matsumura F, Maekawa M, Bito H, Narumiya S (1998) Molecular dissection of the Rho-associated protein kinase (p160ROCK)-regulated neurite remodeling in neuroblastoma N1E-115 cells. J Cell Biol 141:1625–1636
Hogan PG, Lewis RS, Rao A (2010) Molecular basis of calcium signaling in lymphocytes: STIM and ORAI. Annu Rev Immunol 28:491–533
Hollopeter G, Jantzen HM, Vincent D, Li G, England L, Ramakrishnan V, Yang RB, Nurden P, Nurden A, Julius D, Conley PB (2001) Identification of the platelet ADP receptor targeted by antithrombotic drugs. Nature 409:202–207
Huang GN, Zeng W, Kim JY, Yuan JP, Han L, Muallem S, Worley PF (2006) STIM1 carboxyl-terminus activates native SOC, I(crac) and TRPC1 channels. Nat Cell Biol 8:1003–1010
Huang C, Hu ZL, Wu WN, Yu DF, Xiong QJ, Song JR, Shu Q, Fu H, Wang F, Chen JG (2010) Existence and distinction of acid-evoked currents in rat astrocytes. Glia 58:1415–1424
Illes P, Verkhratsky A, Burnstock G, Franke H (2011) P2X receptors and their roles in astroglia in the central and peripheral nervous system. Neuroscientist. doi:10.1177/1073858411418524 [e-pub ahead of print]
Imagawa T, Smith JS, Coronado R, Campbell KP (1987) Purified ryanodine receptor from skeletal muscle sarcoplasmic reticulum is the Ca2+–permeable pore of the calcium release channel. J Biol Chem 262:16636–16643
Inui M, Saito A, Fleischer S (1987a) Purification of the ryanodine receptor and identity with feet structures of junctional terminal cisternae of sarcoplasmic reticulum from fast skeletal muscle. J Biol Chem 262:1740–1747
Inui M, Saito A, Fleischer S (1987b) Isolation of the ryanodine receptor from cardiac sarcoplasmic reticulum and identity with the feet structures. J Biol Chem 262:15637–15642
James PH, Pruschy M, Vorherr TE, Penniston JT, Carafoli E (1989) Primary structure of the cAMP-dependent phosphorylation site of the plasma membrane calcium pump. Biochemistry 28:4253–4258
Jiang S, Yuan H, Duan L, Cao R, Gao B, Xiong YF, Rao ZR (2011) Glutamate release through connexin 43 by cultured astrocytes in a stimulated hypertonicity model. Brain Res 1392:8–15
Jin J, Tomlinson W, Kirk IP, Kim YB, Humphries RG, Kunapuli SP (2001) The C6-2B glioma cell P2Y(AC) receptor is pharmacologically and molecularly identical to the platelet P2Y(12) receptor. Br J Pharmacol 133:521–528
Joseph SM, Buchakjian MR, Dubyak GR (2003) Colocalization of ATP release sites and ecto-ATPase activity at the extracellular surface of human astrocytes. J Biol Chem 278:23331–23342
Kiedrowski L, Czyz A, Baranauskas G, Li XF, Lytton J (2004) Differential contribution of plasmalemmal Na/Ca exchange isoforms to sodium-dependent calcium influx and NMDA excitotoxicity in depolarized neurons. J Neurochem 90:117–128

Kirichok Y, Krapivinsky G, Clapham DE (2004) The mitochondrial calcium uniporter is a highly selective ion channel. Nature 427:360–364
Korczynski J, Sobierajska K, Krzeminski P, Wasik A, Wypych D, Pomorski P, Klopocka W (2011) Is MLC phosphorylation essential for the recovery from ROCK inhibition in glioma C6 cells? Acta Biochim Pol 58:125–130
Korzeniowski MK, Manjarres IM, Varnai P, Balla T (2010) Activation of STIM1-Orai1 involves an intramolecular switching mechanism. Sci Signal 3:ra82
Krane SM, Glimcher MJ (1962) Transphosphorylation from nucleoside Di- and triphosphates by apatite crystals. J Biol Chem 237:2991–2998
Krzeminski P, Suplat D, Czajkowski R, Pomorski P, Baranska J (2007) Expression and functional characterization of P2Y1 and P2Y12 nucleotide receptors in long-term serum-deprived glioma C6 cells. FEBS J 274:1970–1982
Lauger P (1991) Kinetic basis of voltage dependence of the Na, K-pump. Soc Gen Physiol Ser 46:303–315
Lazarowski E (2006) Regulated release of nucleotides and UDP sugars from astrocytoma cells. Novartis Found Symp 276:73–84, discussion 84–90, 107–112, 275–181
Lazarowski ER, Shea DA, Boucher RC, Harden TK (2003) Release of cellular UDP-glucose as a potential extracellular signaling molecule. Mol Pharmacol 63:1190–1197
Lazarowski ER, Sesma JI, Seminario-Vidal L, Kreda SM (2011) Molecular mechanisms of purine and pyrimidine nucleotide release. Adv Pharmacol 61:221–261
Lewis RS (2001) Calcium signaling mechanisms in T lymphocytes. Annu Rev Immunol 19:497–521
Liao Z, Seye CI, Weisman GA, Erb L (2007) The P2Y2 nucleotide receptor requires interaction with alpha v integrins to access and activate G12. J Cell Sci 120:1654–1662
Linde CI, Baryshnikov SG, Mazzocco-Spezzia A, Golovina VA (2011) Dysregulation of Ca2+ signaling in astrocytes from mice lacking amyloid precursor protein. Am J Physiol Cell Physiol 300:C1502–C1512
Liou J, Kim ML, Heo WD, Jones JT, Myers JW, Ferrell JE Jr, Meyer T (2005) STIM is a Ca2+ sensor essential for Ca2+-store-depletion-triggered Ca2+ influx. Curr Biol 15:1235–1241
McCully KS (2009) Chemical pathology of homocysteine. IV. Excitotoxicity, oxidative stress, endothelial dysfunction, and inflammation. Ann Clin Lab Sci 39:219–232
Mildner A, Schmidt H, Nitsche M, Merkler D, Hanisch UK, Mack M, Heikenwalder M, Bruck W, Priller J, Prinz M (2007) Microglia in the adult brain arise from Ly-6ChiCCR2+ monocytes only under defined host conditions. Nat Neurosci 10:1544–1553
Minor DL Jr, Findeisen F (2010) Progress in the structural understanding of voltage-gated calcium channel (CaV) function and modulation. Channels (Austin) 4:459–474
Moller T, Kann O, Verkhratsky A, Kettenmann H (2000) Activation of mouse microglial cells affects P2 receptor signaling. Brain Res 853:49–59
Monteith GR, Roufogalis BD (1995) The plasma membrane calcium pump–a physiological perspective on its regulation. Cell Calcium 18:459–470
Morigiwa K, Quan M, Murakami M, Yamashita M, Fukuda Y (2000) P2 Purinoceptor expression and functional changes of hypoxia-activated cultured rat retinal microglia. Neurosci Lett 282:153–156
Muik M, Fahrner M, Schindl R, Stathopulos P, Frischauf I, Derler I, Plenk P, Lackner B, Groschner K, Ikura M, Romanin C (2011) STIM1 couples to ORAI1 via an intramolecular transition into an extended conformation. EMBO J 30:1678–1689
Naziroglu M (2011) TRPM2 cation channels, oxidative stress and neurological diseases: where are we now? Neurochem Res 36:355–366
Nimmerjahn A, Kirchhoff F, Helmchen F (2005) Resting microglial cells are highly dynamic surveillants of brain parenchyma in vivo. Science 308:1314–1318
Ohana L, Newell EW, Stanley EF, Schlichter LC (2009) The Ca2+ release-activated Ca2+ current (I(CRAC)) mediates store-operated Ca2+ entry in rat microglia. Channels (Austin) 3:129–139
Panopoulos A, Howell M, Fotedar R, Margolis RL (2011) Glioblastoma motility occurs in the absence of actin polymer. Mol Biol Cell 22:2212–2220
Parpura V, Grubisic V, Verkhratsky A (2011) Ca(2+) sources for the exocytotic release of glutamate from astrocytes. Biochim Biophys Acta 1813:984–991
Parri HR, Gould TM, Crunelli V (2001) Spontaneous astrocytic Ca2+ oscillations in situ drive NMDAR-mediated neuronal excitation. Nat Neurosci 4:803–812
Parys B, Cote A, Gallo V, De Koninck P, Sik A (2010) Intercellular calcium signaling between astrocytes and oligodendrocytes via gap junctions in culture. Neuroscience 167:1032–1043
Pinton P, Ferrari D, Magalhaes P, Schulze-Osthoff K, Di Virgilio F, Pozzan T, Rizzuto R (2000) Reduced loading of intracellular Ca(2+) stores and downregulation of capacitative Ca(2+) influx in Bcl-2-overexpressing cells. J Cell Biol 148:857–862
Pizzo P, Burgo A, Pozzan T, Fasolato C (2001) Role of capacitative calcium entry on glutamate-induced calcium influx in type-I rat cortical astrocytes. J Neurochem 79:98–109
Potier M, Trebak M (2008) New developments in the signaling mechanisms of the store-operated calcium entry pathway. Pflugers Arch 457:405–415

Putney JW Jr (1990) Capacitative calcium entry revisited. Cell Calcium 11:611–624

Putney JW (2009) Capacitative calcium entry: from concept to molecules. Immunol Rev 231:10–22

Putney JW Jr, Bird GS (1993) The inositol phosphate-calcium signaling system in nonexcitable cells. Endocr Rev 14:610–631

Rao JN, Platoshyn O, Golovina VA, Liu L, Zou T, Marasa BS, Turner DJ, Yuan JX, Wang JY (2006) TRPC1 functions as a store-operated Ca2+ channel in intestinal epithelial cells and regulates early mucosal restitution after wounding. Am J Physiol Gastrointest Liver Physiol 290:G782–G792

Ribeiro CM, Reece J, Putney JW Jr (1997) Role of the cytoskeleton in calcium signaling in NIH 3T3 cells. An intact cytoskeleton is required for agonist-induced [Ca2+]i signaling, but not for capacitative calcium entry. J Biol Chem 272:26555–26561

Roos J, DiGregorio PJ, Yeromin AV, Ohlsen K, Lioudyno M, Zhang S, Safrina O, Kozak JA, Wagner SL, Cahalan MD, Velicelebi G, Stauderman KA (2005) STIM1, an essential and conserved component of store-operated Ca2+ channel function. J Cell Biol 169:435–445

Rosado JA, Jenner S, Sage SO (2000) A role for the actin cytoskeleton in the initiation and maintenance of store-mediated calcium entry in human platelets. Evidence for conformational coupling. J Biol Chem 275:7527–7533

Roy P, Rajfur Z, Pomorski P, Jacobson K (2002) Microscope-based techniques to study cell adhesion and migration. Nat Cell Biol 4:E91–E96

Sabala P, Czajkowski R, Przybylek K, Kalita K, Kaczmarek L, Baranska J (2001) Two subtypes of G protein-coupled nucleotide receptors, P2Y(1) and P2Y(2) are involved in calcium signalling in glioma C6 cells. Br J Pharmacol 132:393–402

Sabala P, Targos B, Caravelli A, Czajkowski R, Lim D, Gragnaniello G, Santella L, Baranska J (2002) Role of the actin cytoskeleton in store-mediated calcium entry in glioma C6 cells. Biochem Biophys Res Commun 296:484–491

Sage SO, Yamoah EH, Heemskerk JW (2000) The roles of P(2X1)and P(2T AC)receptors in ADP-evoked calcium signalling in human platelets. Cell Calcium 28:119–126

Sak K, Illes P (2005) Neuronal and glial cell lines as model systems for studying P2Y receptor pharmacology. Neurochem Int 47:401–412

Sappington RM, Calkins DJ (2008) Contribution of TRPV1 to microglia-derived IL-6 and NFkappaB translocation with elevated hydrostatic pressure. Invest Ophthalmol Vis Sci 49:3004–3017

Scemes E, Suadicani SO, Spray DC (2000) Intercellular communication in spinal cord astrocytes: fine tuning between gap junctions and P2 nucleotide receptors in calcium wave propagation. J Neurosci 20:1435–1445

Shin YC, Shin SY, So I, Kwon D, Jeon JH (2011) TRIP Database: a manually curated database of protein-protein interactions for mammalian TRP channels. Nucleic Acids Res 39:D356–D361

Soboloff J, Spassova MA, Hewavitharana T, He LP, Xu W, Johnstone LS, Dziadek MA, Gill DL (2006) STIM2 is an inhibitor of STIM1-mediated store-operated Ca2+ Entry. Curr Biol 16:1465–1470

Sontheimer H (1994) Voltage-dependent ion channels in glial cells. Glia 11:156–172

Stathopulos PB, Zheng L, Ikura M (2009) Stromal interaction molecule (STIM) 1 and STIM2 calcium sensing regions exhibit distinct unfolding and oligomerization kinetics. J Biol Chem 284:728–732

Steinbeck JA, Henke N, Opatz J, Gruszczynska-Biegala J, Schneider L, Theiss S, Hamacher N, Steinfarz B, Golz S, Brustle O, Kuznicki J, Methner A (2011) Store-operated calcium entry modulates neuronal network activity in a model of chronic epilepsy. Exp Neurol 232:185–194

Stiber J, Hawkins A, Zhang ZS, Wang S, Burch J, Graham V, Ward CC, Seth M, Finch E, Malouf N, Williams RS, Eu JP, Rosenberg P (2008) STIM1 signalling controls store-operated calcium entry required for development and contractile function in skeletal muscle. Nat Cell Biol 10:688–697

Striedinger K, Scemes E (2008) Interleukin-1beta affects calcium signaling and in vitro cell migration of astrocyte progenitors. J Neuroimmunol 196:116–123

Striedinger K, Meda P, Scemes E (2007) Exocytosis of ATP from astrocyte progenitors modulates spontaneous Ca2+ oscillations and cell migration. Glia 55:652–662

Suplat D, Targos B, Sabala P, Baranska J, Pomorski P (2004) Differentiation of answer of glioma C6 cells to SERCA pump inhibitors by actin disorganization. Biochem Biophys Res Commun 323:870–875

Suplat D, Krzeminski P, Pomorski P, Baranska J (2007) P2Y(1) and P2Y(12) receptor cross-talk in calcium signalling: evidence from nonstarved and long-term serum-deprived glioma C6 cells. Purinergic Signal 3:221–230

Targos B, Pomorski P, Krzeminski P, Baranska J, Redowicz MJ, Klopocka W (2006) Effect of Rho-associated kinase inhibition on actin cytoskeleton structure and calcium response in glioma C6 cells. Acta Biochim Pol 53:825–831

Tatenhorst L, Puttmann S, Senner V, Paulus W (2005) Genes associated with fast glioma cell migration in vitro and in vivo. Brain Pathol 15:46–54

Thebault S, Flourakis M, Vanoverberghe K, Vandermoere F, Roudbaraki M, Lehen'kyi V, Slomianny C, Beck B, Mariot P, Bonnal JL, Mauroy B, Shuba Y, Capiod T, Skryma R, Prevarskaya N (2006) Differential role of transient receptor potential channels in Ca2+ entry and proliferation of prostate cancer epithelial cells. Cancer Res 66:2038–2047

Theis M, Sohl G, Eiberger J, Willecke K (2005) Emerging complexities in identity and function of glial connexins. Trends Neurosci 28:188–195

Van Kolen K, Slegers H (2006) Integration of P2Y receptor-activated signal transduction pathways in G protein-dependent signalling networks. Purinergic Signal 2:451–469
Van Kolen K, Gilany K, Moens L, Esmans EL, Slegers H (2006) P2Y12 receptor signalling towards PKB proceeds through IGF-I receptor cross-talk and requires activation of Src, Pyk2 and Rap1. Cell Signal 18:1169–1181
Vanderheyden V, Devogelaere B, Missiaen L, De Smedt H, Bultynck G, Parys JB (2009) Regulation of inositol 1,4,5-trisphosphate-induced Ca2+ release by reversible phosphorylation and dephosphorylation. Biochim Biophys Acta 1793:959–970
Vazquez G, Wedel BJ, Aziz O, Trebak M, Putney JW Jr (2004) The mammalian TRPC cation channels. Biochim Biophys Acta 1742:21–36
Venkatachalam K, van Rossum DB, Patterson RL, Ma HT, Gill DL (2002) The cellular and molecular basis of store-operated calcium entry. Nat Cell Biol 4:E263–E272
Verkhratsky A (2006) Calcium ions and integration in neural circuits. Acta Physiol (Oxf) 187:357–369
Vig M, DeHaven WI, Bird GS, Billingsley JM, Wang H, Rao PE, Hutchings AB, Jouvin MH, Putney JW, Kinet JP (2008) Defective mast cell effector functions in mice lacking the CRACM1 pore subunit of store-operated calcium release-activated calcium channels. Nat Immunol 9:89–96
Wang M, Kong Q, Gonzalez FA, Sun G, Erb L, Seye C, Weisman GA (2005) P2Y nucleotide receptor interaction with alpha integrin mediates astrocyte migration. J Neurochem 95:630–640
Wang Y, Deng X, Hewavitharana T, Soboloff J, Gill DL (2008) Stim, ORAI and TRPC channels in the control of calcium entry signals in smooth muscle. Clin Exp Pharmacol Physiol 35:1127–1133
Wang D, Yan B, Rajapaksha WR, Fisher TE (2009) The expression of voltage-gated ca2+ channels in pituicytes and the up-regulation of L-type ca2+ channels during water deprivation. J Neuroendocrinol 21:858–866
Wang Y, Deng X, Gill DL (2010) Calcium signaling by STIM and Orai: intimate coupling details revealed. Sci Signal 3:pe42
Worthylake RA, Burridge K (2003) RhoA and ROCK promote migration by limiting membrane protrusions. J Biol Chem 278:13578–13584
Xiao R, Xu XZ (2009) Function and regulation of TRP family channels in *C. elegans*. Pflugers Arch 458:851–860
Yaguchi T, Nishizaki T (2010) Extracellular high K+stimulates vesicular glutamate release from astrocytes by activating voltage-dependent calcium channels. J Cell Physiol 225:512–518
Yeromin AV, Zhang SL, Jiang W, Yu Y, Safrina O, Cahalan MD (2006) Molecular identification of the CRAC channel by altered ion selectivity in a mutant of Orai. Nature 443:226–229
Yoshikawa S, Tanimura T, Miyawaki A, Nakamura M, Yuzaki M, Furuichi T, Mikoshiba K (1992) Molecular cloning and characterization of the inositol 1,4,5-trisphosphate receptor in *Drosophila melanogaster*. J Biol Chem 267:16613–16619
Yu X, Carroll S, Rigaud JL, Inesi G (1993) H+countertransport and electrogenicity of the sarcoplasmic reticulum Ca2+ pump in reconstituted proteoliposomes. Biophys J 64:1232–1242
Yu SC, Xiao HL, Jiang XF, Wang QL, Li Y, Yang XJ, Ping YF, Duan JJ, Jiang JY, Ye XZ, Xu SL, Xin YH, Yao XH, Chen JH, Chu WH, Sun W, Wang B, Wang JM, Zhang X, Bian XW (2012) Connexin 43 reverses malignant phenotypes of glioma stem cells by modulating E-cadherin. Stem Cells 30:108–120
Zhang SL, Yu Y, Roos J, Kozak JA, Deerinck TJ, Ellisman MH, Stauderman KA, Cahalan MD (2005) STIM1 is a Ca2+ sensor that activates CRAC channels and migrates from the Ca2+ store to the plasma membrane. Nature 437:902–905
Zhang SL, Yeromin AV, Zhang XH, Yu Y, Safrina O, Penna A, Roos J, Stauderman KA, Cahalan MD (2006) Genome-wide RNAi screen of Ca(2+) influx identifies genes that regulate Ca(2+) release-activated Ca(2+) channel activity. Proc Natl Acad Sci U S A 103:9357–9362
Zhou Y, Mancarella S, Wang Y, Yue C, Ritchie M, Gill DL, Soboloff J (2009) The short N-terminal domains of STIM1 and STIM2 control the activation kinetics of Orai1 channels. J Biol Chem 284:19164–19168
Zhou Y, Meraner P, Kwon HT, Machnes D, Oh-hora M, Zimmer J, Huang Y, Stura A, Rao A, Hogan PG (2010) STIM1 gates the store-operated calcium channel ORAI1 in vitro. Nat Struct Mol Biol 17:112–116

Chapter 5
Purinergic Signaling in Glioma Progression

Elizandra Braganhol, Márcia Rosângela Wink, Guido Lenz, and Ana Maria Oliveira Battastini

Abstract Among the pathological alterations that give tumor cells invasive potential, purinergic signaling is emerging as an important component. Studies performed in *in vitro*, *in vivo* and *ex vivo* glioma models indicate that alterations in the purinergic signaling are involved in the progression of these tumors. Gliomas have low expression of all E-NTPDases, when compared to astrocytes in culture. Nucleotides induce glioma proliferation and ATP, although potentially neurotoxic, does not evoke cytotoxic action on the majority of glioma cells in culture. The importance of extracellular ATP for glioma pathobiology was confirmed by the reduction in glioma tumor size by apyrase, which degrades extracellular ATP to AMP, and the striking increase in tumor size by over-expression of an ecto-enzyme that degrades ATP to ADP, suggesting the effect of extracellular ATP on the tumor growth depends on the nucleotide produced by its degradation. The participation of purinergic receptors on glioma progression, particularly $P2X_7$, is involved in the resistance to ATP-induced cell death. Although more studies are necessary, the purinergic signaling, including ectonucleotidases and receptors, may be considered as future target for glioma pharmacological or gene therapy.

Keywords ATP • Adenosine • Gliomas • Ectonucleotidases • Animal models of gliomas • E-NTPDases (ectonucleoside triphosphate diphosphohydrolase) • E-NPPs (ectonucleoside pyrophosphatase/phosphodiesterase) • Ecto-5′-nucleotidase/CD73 • $P2X_7$ • Cancer stem cells

E. Braganhol (✉)
Centro de Ciências Químicas, Farmacêuticas e de Alimentos (CCQFA), Universidade Federal de Pelotas (UFPel), Campus Capão do Leão S/N Caixa Postal 354, Pelotas CEP 96010900, RS, Brazil
e-mail: elizandra.braganhol@ufrgs.br

M.R. Wink
Departamento de Ciências Básicas da Saúde, Universidade Federal de Ciências da Saúde de Porto Alegre, 245 Rua Sarmento Leite, Porto Alegre CEP 90050-170, RS, Brazil
e-mail: mwink@ufcspa.edu.br

G. Lenz
Departamento de Biofísica, IB e Centro de Biotecnologia, Universidade Federal do Rio Grande do Sul, 9500 Av. Bento Goncalves, Porto Alegre 61501970, RS, Brazil
e-mail: lenz@ufrgs.br

A.M.O. Battastini
Departamento de Bioquímica, ICBS, Universidade Federal do Rio Grande do Sul, 2600-anexo Rua Ramiro Barcelos, Porto Alegre CEP 90035-003, RS, Brazil
e-mail: abattastini@gmail.com

J. Barańska (ed.), *Glioma Signaling*, Advances in Experimental Medicine and Biology 986,
DOI 10.1007/978-94-007-4719-7_5, © Springer Science+Business Media Dordrecht 2013

Abbreviations

ADA	Adenosine deaminase
ADP	Adenosine diphosphate
Akt	Protein kinase B
ALP	Alkaline phosphatase
AMP	Adenosine monophosphate
APCP	α,β-Methylene ADP
Apyrase	Adenyl-pyrophosphatase
ATP	Adenosine triphosphate
BBB	Brain blood barrier
BBG	Brilliant Blue G
BzATP	2,3-(Benzoyl-4-benzoyl)-ATP
CDK	Cyclin-dependent kinase
CDKN2A (Ink4a/ARF)	Cyclin-dependent kinase inhibitor 2A
CSCs	Cancer stem cell
ECM	Extracellular matrix
Ecto-5′-NT/CD73	Ecto-5′-nucleotidase
EGF	Epidermal growth factor
EGFR	Epidermal growth factor receptor
E-NPP	Ectonucleoside pyrophosphatase/phosphodiesterase
E-NTPDase	Ectonucleoside triphosphate diphosphohydrolase
ERBB2	Human epidermal growth factor receptor 2
ERK	Extracellular signal-regulated kinases
FGF	Fibroblast growth factor
IL-1β	Interleukin 1β
IL-6	Interleukin 6
KO	Knockout
MDM2	Murine double minute 2
MET	NF1, neurofibromatosis 1
MMP-9	Metalloproteinase-9
NPCs	Neural precursor cell
NSC	Neural stem cell
OPC	Oligodendrocyte precursor cell
PDGF	Platelet-derived growth factor
PI3K	Phosphatidylinositol 3-kinase
PTEN	Phosphatase and tensin homolog
Ras/MAPK	Ras/Mitogen activated protein kinase
Ras-GAP	Ras-GTPase activating protein
NF1	Neurofibrimatosis 1
RB1	Retinoblastoma
CNS	Central nervous system
SVZ	Subventricular zone
TNF-α	Tumor necrosis factor

5.1 Introduction

5.1.1 Molecular and Cellular Origins of Gliomas

Two questions are central about the origin of a given cancer: What are the genetic alterations behind formation of this cancer? And in which cells do these alterations find the cellular environment to strive? Both questions are obviously very difficult to answer and, as in all complex illnesses, there is a wide range of possible answers. This is even more true for cancer, in which every new cancer has its own evolutionary history, but which is constrained by the cellular repertoire and the microenvironment of the tissue in which the cancer develops.

Several landmark papers appeared in the last years aiming at these difficult questions regarding gliomagenesis and now there is a good body of evidence for the most frequent molecular alterations observed in gliomas and also some very convincing evidence on the cells that give rise to gliomas.

As suggested by the multitude of familial cancers, even though the mutation in these cases occurs in all cells, only a few cells eventually progress to form a tumor. Therefore, the cell of origin of mutation was in an ancestor of the afflicted patient, but the cell of origin of the tumor is a cell in the patient. Although in non-familial cancers the mutation may occur in the cell of origin, this is not necessarily the case.

Starting at the end of the 1990s, several studies aimed at identifying the glioma cell of origin, which was based on the expression of oncogenes under the control of promoters supposedly specific for a given set of cells (Visvader 2011). Despite the leakiness of these promoters, the picture that emerged was that glioma cell origin is stem cells, but the exact nature of these stem cells was not clear. Recently, using a mosaic analysis with double markers, Liu et al. (2011) pinpointed the cell of origin as being the oligodendrocyte precursor cell (OPC), at least in a murine model in which gliomas were induced by deletion of p53 and activation of the Ras pathway by deletion of neurofibromatosis 1 (NF1). Interestingly, when these alterations were done in neural stem cell (NSC), tumors also arouse, but these were restricted to the OPCs formed from the NSCs. In this case, a multitude of other cells harboring the genetic alterations were formed by the NSCs, but did not produce tumors. This created the concept of cellular environment that allows the development of tumors, clearly differentiating the cell of origin of glioma (OPC), from the cell of mutation, which can potentially be any cell that can give rise to OPC, or OPC itself.

A large analysis of the molecular alterations that drive gliomagenesis revealed that the activation of the Ras/MAPKs and PI3K/Akt pathways, together with the inactivation of p53 are the most common events associated with gliomas.

Oncogenic signals are centered on the activation of Ras/MAPKs and PI3K/Akt, which are mediated by activating genetic alterations in the growth factor receptors EGFR, ERBB2, PDGFRA and MET. Ras activation also occurs through deletion or mutation in the Ras-GAP NF1, while PI3K/Akt activation occurs mainly through deletion and mutation in PTEN, although direct mutations in PI3K was also observed (TCGA 2008).

Among the tumor suppressor pathways inactivated in gliomas, the most prevalent were inactivation of p53, either directly, or through activation of MDM2 or deletion or mutation on CDKN2A (Ink4a/ARF) and inactivation of the CDK inhibitors CDKN2A, 2B and 2C, together with deletion of RB (TCGA 2008). It is important to stress out, however, that these are statistical analysis of the most prominent genetic alterations, and, tumors may arise with genetic alterations other than the one described above.

5.1.2 Glioma Cancer Stem Cells (CSCs)

Since the first histological observation of gliomas, the cytological variability was evident, but only in the last 10 years or so the concept of cancer stem cells was clearly defined and the role of this subtype of cells has been widely studied.

The first identification of glioma CSCs was done by the group of Peter Dirks, which showed that glioma cells expressing CD133 are much more tumorigenic in mice than glioma cell not expressing this cell surface marker (Singh et al. 2003, 2004). Further evidence for the stem like nature of these cells came with the observations that these cells can give rise to functional tumor vasculature, further increasing the complex heterogeneity of glioma tumors (Ricci-Vitiani et al. 2010; Wang et al. 2010).

The importance of CSCs in cancer biology predicts that elimination of this subpopulation of cells, either through induction of cell death or through differentiation should bring a considerable therapeutic benefit. Data from animal models suggest that elimination of CSCs (Vlashi et al. 2009) or the CSC derived endothelial cells of the tumor vasculature (Ricci-Vitiani et al. 2010) is a promising therapeutic strategy. Unfortunately, CSCs are more resistant to radiotherapy due to higher expression of DNA repair enzymes (Bao et al. 2006). On the other hand, chemotherapeutic agents that present the best results in gliomas in humans, such as temozolomide, seems to reduce the proportion of CSCs in the tumors, suggesting that this may be part of the mechanism of action (Beier et al. 2008).

5.1.3 Tumor Microenvironment – Key for Understanding and Targeting Gliomas

A set of genetic alterations in a cell does not guarantee the development of a tumor. Interplay with the tissue microenvironment is a fundamental part of the process of tumorigenesis. Besides being able to form their own blood vessels through differentiation of their CSCs, gliomas also show complex interactions with their neighbor cells and even with the organism. In a breakthrough work, Skog et al. (2008) studied the microvesicles liberated by gliomas in culture and found that these microvesicles contain a set of mRNA enriched in pro-angiogenic genes and that these microvesicles can "transfect" cells. Thus, liberation of microvesicle protected mRNA allows the cancer cells to contribute to the pool of mRNA of their neighbor cells altering their behavior to favor tumor growth. Interestingly, these microvesicles could be observed in the circulations of patients with gliomas, indicating diagnostic potential.

Among the molecules that play important roles in the interplay of gliomas with the microenvironment is glutamate. This molecule, well established as an excitatory aminoacid that is one of the main culprits of neurotoxicity, is secreted by gliomas in rodent models as well as in patients (Buckingham et al. 2011; Marcus et al. 2010; Takano et al. 2001; Ye and Sontheimer 1999).

A recent clinical study with glioma patients using microdialysis technique, found 100-fold higher glutamate concentrations in the tumor resection margin, where the subpopulation of invasive cells is concentrated (Berens and Giese 1999), when compared to the normal peritumoral cortex tissue (Marcus et al. 2010). It has been hypothesized that this peritumoral increasing concentration of glutamate is responsible for the seizures and tumor associated epilepsy, presented by glioma patients as an early and recurrent symptom. The hyper-excitability and epileptic activity of neurons at the peritumoral regions was attributed to glutamate release by gliomas cells via the system x_{c^-}cysteine-glutamate transporter (Buckingham et al. 2011).

Glioma tumors that release glutamate have a substantial growth advantage when compared with those that do not secret this molecule. The role of glutamate is reinforced by the significantly attenuation on tumor growth by the administration of the NMDA receptor antagonist MK801 (dizocilpine), *in vivo*, but not *in vitro*, confirming the microenvironment rather than cell autonomous nature of the action of glutamate (Takano et al. 2001).

Taken together, these findings point to a model where the glutamate released by gliomas affect the peritumoral microenvironment, causing peritumoral seizures and facilitating the tumor progression by damaging normal tissue. Nevertheless the glutamate by itself can not completely explain how the growing tumor mass causes neuronal cell death along the growing tumor margins. Among the signaling pathways that could synergistically act with glutamate, the purinergic system has emerged as an important candidate.

5.2 Purinergic Signaling in Gliomas

As would be expected for cancer cells, different gliomas express different sets of purinergic receptors (Table 5.1), and only for some their role in glioma biology was studied. As commented at the introduction of this chapter, gliomas originate from oligodendroglial precursor cells (OPCs) and a comparison of the receptors expressed by these cells with several gliomas may help pinpointing the involvement of specific receptors in the OPC to glioma transformation.

From the initial studies of our group we noticed that glioma cells respond in diverse ways to ATP and other purinoceptor agonists and antagonists. ATP and adenosine induce proliferation of several human glioma cell lines such as U138MG, U87MG and U251MG (Morrone et al. 2003). In U138MG cells, the proliferation stimulus was also observed with ADP, UTP, inosine and guanosine treatment suggesting the involvement of the subtypes $P2Y_4$ and A_3 in this response. Interestingly, ATP and

Table 5.1 Expression of purinergic receptors in gliomas and glial cells

Receptor	Cells expressing	Receptor	Cells expressing
A_1	OPC (r), Oligo (r), Astrocytes (r), C6 (r), U251MG (h)	$P2Y_{13}$	Astrocytes (r), C6 (r), U138MG (h), U251MG (h)
A_{2A}	OPC (r), Oligo (r), Astrocytes (r), C6 (r), GL261 (m), U87MG (h), U138MG (h), U251MG (h)	$P2Y_{14}$	Astrocytes (r), C6 (r), GL261 (m), U138MG (h)
A_{2B}	OPC (r), Oligo (r), Astrocytes (r), U87MG (h), U138MG (h), U251MG (h)	$P2X_1$	OPC (r), Astrocytes (r), GL261 (m)
A_3	OPC (r), Oligo (r), Astrocytes (r), C6 (r)	$P2X_2$	OPC (r), Astrocytes (r)
$P2Y_1$	OPC (r), Astrocytes (r), C6 (r), GL261 (m)	$P2X_3$	OPC (r), Astrocytes (r), C6 (r), GL261 (m)
$P2Y_2$	OPC (r), Astrocytes (r), C6 (r)	$P2X_4$	OPC (r), Astrocytes (r), C6 (r), GL261 (m), U87MG (h), U251MG (h)
$P2Y_4$	OPC (r), Astrocytes (r), C6 (r), GL261 (m), U138MG (h)	$P2X_5$	Astrocytes (r)
$P2Y_6$	Astrocytes (r), U251MG (h), C6 (r)	$P2X_6$	Astrocytes (r), U87MG (h), U138MG (h)
$P2Y_{12}$	Astrocytes (r) C6 (r), GL261 (m), U138MG (h), U251MG (h)	$P2X_7$	OPC (r), Astrocytes (r), GL261 (m),C6 (r), U87MG (h), U251MG (h)

Designations:

Oligo, Oligodendrocytes; OPC, Oligodendrocyte Precursor Cells; (r), Rat; (m), mouse; (h), human; C6, GL261, U251MG, U87MG, U138MG: Glioma cell lines

Data are from Agresti et al. (2005), Verkhratsky et al. (2009) (OPCs and Oligodendrocytes), Braganhol et al. (2008), Ledur et al. (2011), Morrone et al. (2003), Tamajusuku et al. (2010) (Gliomas), Fumagalli et al. (2003), Jacques-Silva et al. (2004a), Lenz et al. (2001), Verkhratsky et al. (2009) (astrocytes) adenosine also increased the transport of thymidine into the cell, suggesting a much broader effect than the regulation of cell cycle. ATP and adenosine activate the ERK, MAPK and PI3K/Akt pathways, and both pathways are important in mediating the proliferative effects of these purinergic agonists in U138MG human gliomas (Jacques-Silva et al. 2004a). As discussed below, ATP to induce these effects could stem from cell lysis associated with low degradation of ATP in gliomas.

One important aspect of this model is that glioma cells are resistant to ATP induced cell death. ATP, at concentrations above 1 mM is toxic to several cell lines, mostly mediated by the $P2X_7$ receptor subtype. In neuronal cultures and organotypic cultures of the hippocampus, ATP is toxic at 5 mM (Morrone et al. 2005). Glial cells are normally resistant to ATP-induced cell death, and, in astrocytes, high ATP concentrations and activation of $P2X_7$ actually activates the PI3K/Akt pathway, which normally has an anti-death role (Jacques-Silva et al. 2004b).

Notably, mouse neural progenitor cells (NPCs), which are able to differentiate into neurons, astrocytes and oligodendrocytes are sensitive to extracellular ATP with a pharmacological and molecular profile that suggests the involvement of $P2X_7$ and NPCs from $P2X_7$ KO mice were much more resistant to extracellular ATP (Delarasse et al. 2009). On the other hand, OPCs, which originate from NPCs, but are already restricted to the oligodendrocyte lineage and which recent data indicates to be the main cells of origin of gliomas, express functional $P2X_7$ receptors and suffer cytotoxicity when treated with high doses of BzATP, but not by a long lasting challenge with ATP up to 3 mM (Agresti et al. 2005). Interestingly, mature oligodendrocytes are quite sensitive even to low concentrations of ATP or BzATP (Matute 2008; Verkhratsky et al. 2009). The observation that glioma originate from a cell type resistant to ATP, while their precursor and differentiated cell are sensitive may be indicative of the importance of resistance to ATP cell death for glioma development. However, the observations that some glioma cell lines lost this resistance indicate that this is not an absolute requirement for glioma growth (Tamajusuku et al. 2010).

5.3 Ectonucleotidases

The effects of nucleotides and nucleosides on purinergic receptors are regulated by the action of ectonucleotidases, which includes ectonucleoside triphosphate diphosphohydrolases (E-NTPDases), ectonucleoside pyrophosphatase/phosphodiesterases (E-NPPases), ecto-5′-nucleotidase/CD73 (Ecto-5′-NT/CD73) and alkaline phosphatases (ALP) (Robson et al. 2006; Zimmermann 2001, 2006). These enzymes operate in concert for the complete nucleotide hydrolysis (e.g. ATP) to nucleoside (e.g. adenosine) and represent a powerful manner to control the effects mediated by extracellular purines (Knowles 2011; Yegutkin 2008).

5.3.1 Ectonucleoside Triphosphate Diphosphohydrolases (E-NTPDases)

The occurrence of adenyl-pyrophosphatases, which splits the two phosphate groups from ATP, was for the first time demonstrated in muscle cells by Lohmann (Lohmann 1928). Later, Meyerhof proposed the name "apyrase" for enzymes that hydrolyse ATP, ADP and other triphospho- and diphosphonucleosides to their equivalent monophosphonucleosides and inorganic phosphate (Meyerhof 1945). Over the years, apyrases were identified in many different organisms such as plants; insects; parasites and mammals, and they were classified as ATP-diphosphohydrolase (EC 3.6.1.5) to distinguish them from intracellular ATPases. Until 1995, there was an apparent confusion about the molecular and kinetic identity of the enzymes responsible for the hydrolysis of extracellar ATP. In that year, a complete review describing the ecto-ATPases, which were identified as E-ATPases in analogy to other ATPases, was published (Plesner 1995). Ecto-ATPases as well as ecto-apyrases (or ecto-ATP diphosphohydrolases) are ubiquitous enzymes that hydrolyze extracellular nucleoside tri- and/or diphosphate exhibiting E-type ATPase activity. They hydrolyze several purine and pyrimidine nucleoside tri- and diphosphate but not nucleoside monophosphates. They are dependent on millimolar concentration of Ca^{2+} or Mg^{2+}, are insensitive to specific P-type, V-type and F-ATPases and also alkaline phosphatase inhibitors and have an alkaline optimum pH. Ecto-ATPases/ecto-apyrases are glycoproteins anchored by two transmembrane domains, which makes difficult their solubilization and purification (Plesner 1995).

At present, the previously classified as E-type ATPases are identified as NTPDases belonging to the CD39 family. In mammals, at least eight related and homologous enzymes sharing five apyrase-conserved regions (ACRs), named NTDPase1 to 8, have been cloned and characterized: NTPDase1 (CD39, ATPDase, ecto-apyrase or ecto-ATP diphosphohydrolase), NTPDase2 (CD39L1, ecto-ATPase), NTPDase3 (CD39L3, HB6), NTPDase4 (UDPase, LALP70), NTPDase5 (CD39L4, ER-UDPase, PCPH), NTPDase6 (CD39L2), NTPDase7 (LALP1) and NTPDase8 (Robson et al. 2006; Zimmermann 2001).

NTPDase1-3 and 8 share common membrane topography with two transmembrane domains at the N- and C-terminus and a catalytic site facing the extracellular compartment. These E-NTPDase members differ regarding the preferences for nucleotides, while NTPDase1, 3 and 8 hydrolyses nucleoside tri- and diphosphates equally well, NTPDase2 hydrolyses nucleoside triphosphates with a 30-fold preference (Bigonnesse et al. 2004; Chadwick and Frischauf 1998; Heine et al. 1999; Kaczmarek et al. 1996; Kegel et al. 1997; Smith et al. 1998).

NTPDase4-7 present an intracellular localization such as the Golgi apparatus and endoplasmic reticulum (Biederbick et al. 2000; Shi et al. 2001). NTPDase5 and NTPDase6 lack the C-terminal transmembrane domain and are also expressed in the plasma membrane or as secreted enzymes with specificity for the hydrolysis of nucleoside diphosphates (Hicks-Berger et al. 2000; Oses et al. 2004; Mulero et al. 1999, 2000; Murphy-Piedmonte et al. 2005; Yeung et al. 2000).

Several studies have shown the expression of multiple NTPDases in the CNS (Battastini et al. 1991; Braun et al. 2003; Bruno et al. 2002; Nagy et al. 1986; Pinsky et al. 2002; Wang and Guidotti 1999). NTPDase1, the lymphoid cell activation antigen CD39, is associated with endothelium and vascular smooth muscle cells being strongly expressed by microglia (Braun et al. 2000; Zimmermann 2006). In neurons, an apyrase-like enzyme has been characterized (Boeck et al. 2002) and the expression of multiple ectonucleotidase has been described in PC12 cells (Vollmayer et al. 2001). NTPDase2 is the dominant ectonucleotidase expressed by rat astrocytes (Wink et al. 2006) being also identified in immature Schwann cells of the peripheral nervous system, in satellite glia cells of dorsal root ganglia and sympathetic ganglia and in enteric glia (Zimmermann 2006). This enzyme is also expressed in adult mouse hippocampal progenitors (Shukla et al. 2005) and in type B cells of the subventricular zone (SVZ) (Braun et al. 2003), two neurogenic regions of the adult mammalian brain. Chadwick and Frischauf described the tissue distribution of NTPDase3 in brain (Chadwick and Frischauf 1998). In more recent studies, NTPDase3 was identified in the mitochondrial matrix or closely linked to the inner mitochondrial membrane of hypothalamic neurons. It was also found that decrease of NTPDase-activity resulted in significantly decreased mitochondrial respiratory capacity. The authors assumed that hypothalamic neuronal activity especially that of excitatory neurons may be dependent on the activity of mitochondrial NTPDase3 due to the ATPase activity of this enzyme (Belcher et al. 2006; Kiss et al. 2009). The NTPDase8 is very poorly expressed or absent in brain (Bigonnesse et al. 2004).

5.3.2 *Ectonucleotide Pyrophosphatase/Phosphodiesterases (E-NPPs)*

The second family of enzymes involved in the extracellular nucleotide metabolism is the ectonucleotide pyrophosphatase/phosphodiesterase family (E-NPPs; EC 3.1.4.1). NPP1-3 have been detected in almost all tissues, although individual isoforms are usually confined to specific substructures and/

or cell types (for review see Bollen et al. 2000; Goding et al. 2003). They hydrolyze pyrophosphate or phosphodiester bonds of a variety of extracellular compounds with a broad substrate specificity, including nucleotides, lysophospholipids and choline phosphate esters that may reflect their roles in various physiological and biochemical processes. NPP1 was originally discovered on the surface of mouse B-lymphocytes as the plasma cell differentiation antigen (PC-1) (Takahashi et al. 1970). It is highly expressed in cells from bone and cartilage with intermediate expression in other tissues, including brain capillary endothelium (Goding et al. 2003). It is interesting to observe that NPP1 has not been detected in neurons or glial cells, although it has been detected in rat C6 glioma cells (Grobben et al. 1999). The second member of this family, the NPP2 was discovered as an autocrine motility factor (autotaxin, NPP2a) (Stracke et al. 1992). The expression of NPP2a has been correlated with oligodendrocyte differentiation and in the formation of myelin sheet (Fuss et al. 1997). In addition, rat brain NPP2 is expressed in glial cells of the cerebellum (Goding et al. 2003). NPP3 is a glycoprotein present in a specific subset of glial precursor cells, which expression is dependent on the stage of differentiation (Blass-Kampmann et al. 1997; Deissler et al. 1995b).

5.3.3 Ecto-Alkaline Phosphatases (ALP)

The extracellular nucleotides ATP, ADP (and AMP) can also be hydrolyzed by alkaline phosphatases (ALP, EC 3.1.3.1) (Picher et al. 2003). Human ALPs are encoded by four different gene loci and three of the four isozymes are tissue-specific, the intestinal (IALP), placental (PLALP) and germ cell (GCALP). The fourth ALP is the tissue-nonspecific (TNALP) expressed in relatively high amounts in bone, liver and kidney (Martins et al. 2001). This protein co-localize with detergent-resistant and glycolipid-rich membrane subdomains and is important to control signal transduction and membrane trafficking (Bianchi and Spychala 2003; Matsuoka and Ohkubo 2004). Although TNALP is widely distributed in different tissues, information about its physiological function in the brain is limited (Zimmermann 1996). It was established that this ubiquitous enzyme has a specific regional and subcellular localization in the brain of adult animals and could be involved in neurotransmission, as it exhibits a neuronal activity-dependent regulation and it is localized in the synaptic cleft of cortical synapses (Fonta et al. 2004). TNALP is expressed in the neural tube in different brain regions during embryonic development indicating its involvement in developmental steps with a potential role in cortical plasticity and brain disorders (Fonta et al. 2005; Narisawa et al. 1994). In addition, it was shown that TNALP is expressed in capillary endothelium of blood–brain barrier (BBB) and may participate in extracellular phosphorylation/dephosphorylation processes, which are involved in the modulation of organic cation transport at BBB (Calhau et al. 2002). Finally, considering that ATP may be hydrolyzed by E-NPPs generating PPi and that it is known that TNALP is able to hydrolyse PPi into Pi (Goding et al. 2003), it is possible that in addition to E-NTPDases, these two enzymes might cooperate to hydrolyze ATP directly to AMP without transient production of ADP in the extracellular milieu. Therefore, the combination of more than one enzyme may acts as important regulator of purinergic receptor activation by removing efficiently P2 agonists.

5.3.4 Ecto-5′-Nucleotidase (Ecto-5′-NT/CD73)

The intermediary product of ATP hydrolysis, AMP, can be hydrolyzed by the action of ecto-5′-nucleotidase (ecto-5′-NT/CD73, EC 3.1.3.5), which is a widely distributed enzyme anchored on the outer surface of the plasma membrane via a glycosyl phosphatidylinositol linkage and also co-localizes with detergent resistant and glycolipid-rich membrane subdomains called lipid rafts (Bianchi and

Sypchala 2003). It produces nucleosides from non-cyclic nucleoside monophosphates in the extracellular space, being the best-characterized enzymatic source of extracellular adenosine (Zimmermann 1992). Considering that ATP and ADP can exert an inhibitory action on ecto-5′-NT/CD73 (Dornand et al. 1978), this step of extracellular nucleotide metabolism represents a pivotal role by controlling the activation of P2 and P1 receptors (Komoszynski and Wojtczak 1996). Ecto-5′-NT/CD73 is expressed in many different tissues (Zimmermann 1992) being identified in different brain regions either located as a membrane bound ecto-enzyme or as a soluble enzyme with intracellular localization (Bianchi and Spychala 2003, Heymann et al. 1984). In the Central Nervous System (CNS) it was identified in astrocytes, hippocampal mossy fiber terminals, the choroid plexus and vascular endothelium (Zimmermann 2006). The subcellular distribution of ecto-5′-NT/CD73 in the brain cells shows that the bulk of activity is associated with myelin, synaptossomal and microssomal fractions (Heymann et al. 1984). This enzyme has been proposed as a glial marker and indicator for development, regeneration and plasticity (Lie et al. 1999; Zimmermann et al. 1998). Recently, Stanojevic et al. showed a negative correlation between the enzyme activity and the enzyme protein abundance in the synaptic plasma membrane, indicating additional roles than those related to AMP hydrolysis for ecto-5′-NT/CD73 in the synaptic compartment during postnatal brain development (Stanojevic et al. 2011). The results of this study provided direct evidence for the existence of this ecto-enzyme in the presynaptic compartment and suggest that ecto-5′-NT/CD73 may be a part of general scheme of brain development and synapse maturation. In fact, several studies demonstrate that besides its catalytic function, ecto-5′-NT/CD73 is also an important player on cell-cell and cell-extracellular matrix (ECM) interactions, (Spychala 2000; Stochaj et al. 1989) as well as a modulator of T lymphocytes signaling (for review see Resta et al. 1998). The highly variable level of expression of ecto-5′-NT/CD73 in animal tissues and cells suggests that tissue-specific mechanisms control the expression of this enzyme. Indeed, a number of tissue-specific regulatory elements within the ecto-5′-NT/CD73 promoter were identified suggesting that there might be a high level of complexity in the interactions between binding factors of this gene (Hansen et al. 1995; Spychala et al. 1999).

In addition to the action of ecto-5′-NT/CD73, AMP can also be deaminated to inosine monophosphate (IMP) by an AMP deaminase activity. This alternative enzymatic degradation of AMP might constitute a mechanism to control extracellular adenosine formed from ATP breakdown (Cunha and Sebastião 1991; Goldman et al. 2010).

5.4 Ecto-Adenosine Deaminase (Ecto-ADA)

The extracellular adenosine levels are controlled by the uptake of plasma membrane-located adenosine transporters and/or by the action of adenosine deaminase (ADA; EC 3.5.4.4) that produces inosine (Cunha et al. 2000). ADA was originally considered to be cytosolic but it has been found ubiquitously on the surface of many different cell types including brain synaptosomes and, therefore, it can be also considered an ecto-enzyme (ecto-ADA) (Franco et al. 1997).

Ecto-ADA has an extra-enzymatic function via its interaction with CD26 and other cell-surface proteins (Franco et al. 1998). Ecto-ADA was identified in neuronal populations in the brain of different mammals (Yamamoto et al. 1987). It was also shown that ecto-ADA can be associated with A_1 adenosine receptors (A_1Rs) in different cell types including pig brain cortical membranes (for review see Zimmermann 1996). Accordingly, the co-localization and interaction between A_1Rs and ecto-ADA may be the functional basis of the extra-enzymatic role of ecto-ADA in modulating ligand-induced signaling, desensitization and internalization of A_1Rs (Beraudi et al. 2003; Ciruela et al. 1996; Gines et al. 2001; Saura et al. 1998). Inosine, product of adenosine deamination, was originally thought to have no biological effects, but today it is well established that this nucleoside may participate actively in many biological processes including immunomodulation and neuroprotection.

Inosine preserves the viability of glial cells and neuronal cells during hypoxia and stimulates axonal re-growth after injury. The actions of inosine might involve effects on adenosine receptors, but it is also possible that inosine augments extracellular adenosine levels by competing with the nucleoside transporters and thus generating secondary effects to adenosine binding to its receptors (for review see Haskó et al. 2004). Recently, it was shown that inosine exerts anticonvulsant effect against hyperactivity of the glutamatergic system independently of benzodiazepines or adenosine receptors activation (Ganzella et al. 2011). Thus, besides its non-enzymatic functions, the final step of purine metabolizing cascade catalyzed by ecto-ADA is very important for the control of adenosine availability and inosine production with consequent activation of different signaling pathways.

5.5 Other Ecto-Nucleotide Metabolizing Enzymes

ATP can be also consumed by ecto-enzymes belonging to protein kinase family (ecto-PK, EC 2.7.10-11-12-13) (Ehrlich et al. 1990). These enzymes are located on the surface of normal, transformed or malignant cells (Paas et al. 1999; Redegeld et al. 1999; Seehafer et al. 1984) and the coordinated action of ecto-PK and ALP would participate in the phosphorylation/dephosphorylation mechanisms in a variety of extracellular events, such as cell-cell interaction and transduction of external signals (Calhau et al. 2002; Kübler et al. 1989). Moreover, the presence of purine converting ecto-enzymes such as ecto-nucleotide kinase (adenylate kinase, EC 2.7.4.3) (Nagy et al. 1989), ectonucleoside diphosphate kinase (EC 2.7.4.6) (Yegutkin et al. 2002; Zimmermann 2006) and the identification of ectopic FoF1-ATP synthase on plasma membrane of mammalian cells (Ravera et al. 2011), provide new perspectives for the scenario of extracellular nucleotide metabolism. In summary, the combination of these multiple extracellular ATP-generating systems along with ATP-consuming pathways would complete the panel of enzymatic and non-enzymatic systems involved in the control of availability of specific agonists for nucleotide/nucleoside-selective receptors (Yegutkin et al. 2002).

5.6 Ectonucleotidases in Gliomas

Events that trigger malignant transformation of glial cells into gliomas are poorly understood. Genetic alterations that controls cell proliferation and differentiation, including the regulation in oncogenes expression (MDM2, CDK4, EGFR) and tumor suppressor genes (p53, p16, p15 and RB1) (Louis 1994; Maher et al. 2001; Shapiro 2001; Von Deimling et al. 1995) are common features of glial cell malignant transformation. In addition, glioma progression requires specific microenvironment conditions, which affect tumor cell interactions with neurons, glia and vascular cells in the CNS (Demuth and Berens 2004). Among the pathological alterations that give tumor cells invasive potential, purinergic signaling is emerging as an important component. By activating specific purinergic receptors (P2X and P2Y), extracellular ATP has been shown to mediate events related to cell proliferation, cell differentiation and cell death (White and Burnstock 2006). Nucleotides exert a synergist effect on cell proliferation together with growth factors, chemokines or cytokines (Lemoli et al. 2004). ATP has been identified as a mitogen for v-myc immortalized neural progenitor cells (Ryu et al. 2003). In astrocytes, extracellular ATP regulates ERK function by activating $P2Y_1$, $P2Y_2$ or $P2Y_4$ purinoceptors (Lenz et al. 2000; Neary et al. 2003), indicating the potential for cross-talk with FGF, EGF and PDGF driven cell mitogenic pathways. Adenosine is one of factors that can contribute to tumor progression (Spychala 2000). This nucleoside accumulates at high concentrations in solid tumors and it has been previously shown to stimulate tumor growth and angiogenesis through activation of P1 receptors and to inhibit cytokine synthesis, cell spreading, and adhesion of immune cells to the endothelial wall and the function of T-cells, macrophages and natural killer cells (Spychala 2000).

As described above, the biological effects of nucleotides and nucleosides are regulated by the action of ectonucleotidases, which efficiently control the purinergic receptor activation by hydrolyzing these molecules in the extracellular space. Accumulating evidence suggests that alterations in the extracellular nucleotide/nucleoside metabolism are involved in the growth and progression of gliomas. We demonstrated that glioma cell lines have altered extracellular ATP, ADP and AMP catabolism when compared to astrocytes, showing low rates of extracellular ATP hydrolysis and high rates of extracellular AMP hydrolysis (Wink et al. 2003). C6 glioma presents a low mRNA expression of NTPDases1-6 (Morrone et al. 2006) and human glioma cell lines as well (unpublished data). These results were confirmed by applying a C6 *ex vivo* glioma model, a primary glioma culture obtained directly from rat biopsy specimens that were previously implanted, indicating that the disruption of purinergic signaling is a feature shown not only by glioma cell lineages, but also by *ex vivo* glioma cultures which represent a closer model of original tumors (Braganhol et al. 2008).

In addition, it seems also important to consider the NTPDase5 mRNA expression in C6 cells (Morrone et al. 2006) and in other different glioma cell lines (unpublished results). In addition to the participation of NTPDase5 in the process of re-glucosylation involved in glycoprotein folding in the endoplasmatic reticulum (Tombetta and Helenius 1999), an unexpected role of NTPDase5 in oncogenesis was recently revealed (Villar et al. 2007). NTPDase5 was shown to be identical to the PCPH gene, a human proto-oncogene product expressed in human tumor cell lines (Paez et al. 2001, Rouzaut et al. 2001). The neoplastic transforming activity of the NTPDase5/PCPH oncoprotein is mediated by its ability to promote a Ras-independent, sustained activation of ERK (Recio et al. 2000) and/or render cancer cells resistant to a variety of apoptosis-inducing stimuli (Recio et al. 2000; Velasco et al. 1999), including serum deprivation, hyperthermia, ionizing radiation and chemotherapeutic drugs. The resistance to various stress stimuli elicited by NTPDase5/PCPH was mediated by its ability to hydrolyze ATP, decreasing the phosphate donor availability for the kinases involved in the stress-induced phosphorylation cascades with which it interacts (Recio et al. 2002).

As presented before, extracellular degradation of ATP proceeds by a cascade of cell surface-bound enzymes that also includes the E-NPPs and ALP. Catalysis by E-NPPs affects processes as diverse as cell proliferation and motility, angiogenesis, bone mineralization and digestion. In addition, E-NPPs are also implicated in the pathophysiology of cancer, insulin resistance and calcification diseases (Stefan et al. 2006). NPP1 and –3 are expressed in rat C6 glioma cells where they are responsible for the hydrolysis of low extracellular ATP concentration (1–10 μM) (Grobben et al. 1999; Joseph et al. 2004). Moreover, there is emerging evidence for non-catalytic functions of NPPs in cell signaling. For example, a role of NPP3 in tumor transformation was demonstrated by showing that its expression in glioma cells induced morphological changes and enhanced tumor invasive properties (Deissler et al. 1995a). The role of NPP1 expression in C6 glioma cell motility and invasion remains to be determined. Regarding the involvement of ALP in the tumorigenic process, there are a few studies characterizing its participation in glioma biology. ALP activity is restricted to the capillary wall, being stronger in capillaries from glioblastomas than astrocytomas. Moreover, in opposite to normal cells, the ALP activity in glioblastoma is markedly positive on the luminal surface of plasma membrane of endothelial cells. Phosphatase activities in brain tumor appear to change in localization pattern in association with glioma malignancy, which may reflect a higher permeability of the BBB (Maeda et al. 1985). ALP can also contribute to adenosine production from AMP as a substrate, as demonstrated for the neuroblastoma glioma hybrid NG108-15 cells (Ohkubo et al. 2000).

The AMP produced in the extracellular space is hydrolyzed to adenosine by the action of ecto-5′-NT/CD73 (Zimmermann 1992). A large body of evidence suggests that ecto-5′-NT/CD73 has tumor-promotion functions (Spychala 2000). High ecto-5′-NT/CD73 expression levels have been reported in many human solid tumors, such as colon, lung, pancreas, ovary (Su et al. 2001), melanomas (Sadej et al. 2006), breast carcinomas (Zhi et al. 2007) and glioblastoma cells (Ludwig et al. 1999;

Wink et al. 2003), being its expression correlated with shorter patient survival time (Spychala 2000). Adenosine, the product of ecto-5′-NT/CD73 activity, has been reported as mediator of cell proliferation, angiogenesis and may suppress the anticancer immune response (Gessi et al. 2011; Morrone et al. 2003; Spychala 2000). Adenosine concentrations increase within hypoxic regions of solid tumors, including gliomas (Melani et al. 2003), and it has been recognized to interfere with the recognition of tumor cells by cytolytic effectors cells of the immune system (Blay et al. 1997; Merighi et al. 2003). In particular A_{2A} adenosine receptors on T-cell surface may play an immunosuppressive role in large solid tumors, inhibiting incoming antitumor cytotoxic T lymphocytes from destroying the tumor (Koshiba et al. 1997). The immunosuppressive role and the ability to protect against ischemia suggest that A_{2A} activation improves hypoxic tumor cell survival and immune escaping. A_{2B} adenosine receptors seem to contribute to tumor growth, neovascularization and for the release of a subset of cytokines (Gessi et al. 2011). In U87MG human glioma cells, adenosine increases IL-6 and IL-8 expression via stimulation of A_{2B}, while the stimulation of A_3 receptors induced an increase of metalloproteinase-9 (MMP-9) levels, which was responsible for an increase of glioblastoma cells invasion (Gessi et al. 2010). In a parallel investigation, adenosine promoted an increase in U138MG glioma cell adhesion, which was prevented by adenosine receptor antagonists and dipyridamole, indicating the participation of extra- and intracellular signaling pathways in cell adhesion mediated by this nucleoside (Cappellari et al. 2011).

In addition to produce the pro-tumor nucleoside adenosine, ecto-5′-NT/CD73 has a role in the control of cell growth, maturation, differentiation, cell-cell and cell-matrix interactions (Navaro et al. 1998; Turnay et al. 1989; Vogel et al. 1991; Zhou et al. 2007). Notably, increasing cell confluency and culture times led to an increase in ecto-5′-NT/CD73 activity and expression in different glioma cell lines (Bavaresco et al. 2008). The ECM laminin and chondroitin sulfate modulated the ecto-5′-NT/CD73 activity and glioma adhesion in a parallel manner, suggesting the involvement of purinergic signaling in the effects mediated by ECM components (Cappellari et al. 2011). In line to the role of ecto-5′-NT/CD73 and its product adenosine on tumor-promotion actions, the treatment with APCP, a synthetic inhibitor of this enzyme, and AMP significantly reduced glioma cell proliferation (Bavaresco et al. 2008). Additionally, it was shown that the inhibitory effect of quercetin, dexamethasone and indomethacin on glioma cell proliferation was related to the modulation of ecto-5′-NT/CD73 activity (Bavaresco et al. 2007; Braganhol et al. 2007; Bernardi et al. 2007) and the expression of A_3 adenosine receptor mediates the adenosine cell death actions (Gessi et al. 2011). In conclusion, ecto-5′-NT/CD73 expression is important in the glioma development for different reasons: (a) providing the major source of extracellular adenosine and/or (b) interacting directly with extracellular proteins acting as an adhesion molecule.

5.7 The Purinergic Hypothesis of Glioma Invasion

Previous studies from our group indicate a strong involvement of purinergic signaling in the growth and progression of gliomas. The majority of glioma cell lines present an extracellular nucleotide metabolism that has low ATPase and high AMPase activity, with is strikingly different from normal astrocytes, which have high ATPase and low AMPase activity (Wink et al. 2003).

Adenine nucleotides induce cell proliferation in diverse human glioma cell lines (Morrone et al. 2003) and the majority of glioma cell lines are resistant to cell death induced by cytotoxic ATP concentrations (Morrone et al. 2005). As ATP is poorly hydrolyzed by glioma cells and, in addition, can be ectopically produced in C6 glioma plasma membrane (Ravera et al. 2011), this nucleotide could potentially accumulate within tumor, resulting in glioma cell proliferation and neuronal toxicity. We have proposed that besides the glutamatergic system (Takano et al. 2001), the purinergic signaling could also be involved in this process (Morrone et al. 2003, 2005; Wink et al. 2003). In this model,

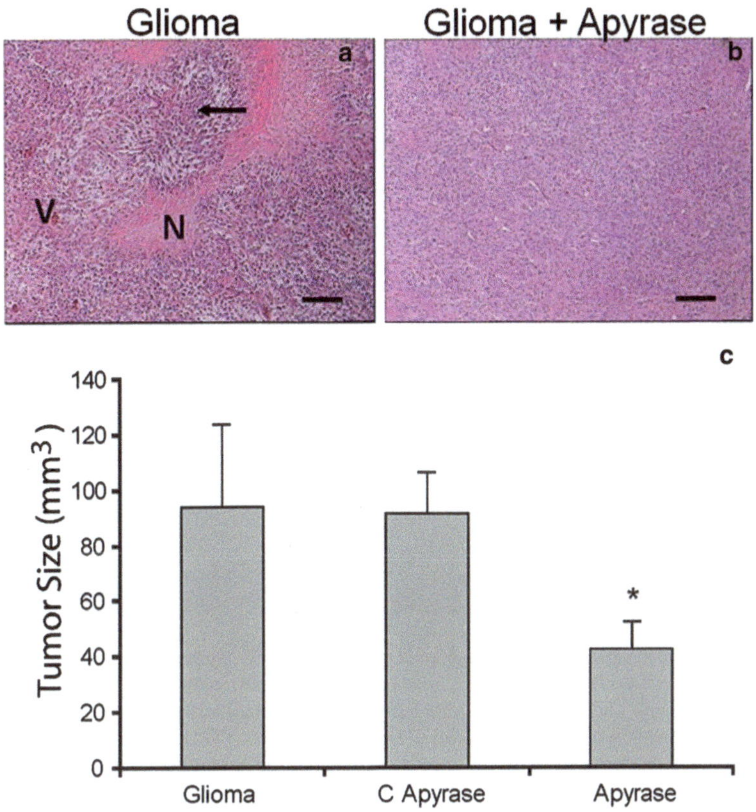

Fig. 5.1 *In vivo* glioblastoma growth is reduced by apyrase co-injection. To determine the effect of apyrase co-injection on *in vivo* glioma growth, C6 cells (1×10^6) were implanted in the right striatum of Wistar rat by stereotaxical surgery in presence (Glioma + Apyrase) or absence of apyrase (Glioma). (**a, b**) Histological characteristics that define glioblastoma multiforme as seen in rats implanted with gliomas and in rats co-injected with apyrase. Scale bars = 100 μm. (**c**) Tumor size quantification of implanted gliomas. Tumor size was evaluated 20 days following glioma implantation. Data represent the mean ± SD of at least six animals per group. Mean ± S.E.M. *$p < 0.05$ for comparison versus control, as determined by ANOVA, followed by Tukey-Krammer test. Glioma, rats implanted with C6 cells; CApyrase, apyrase denatured by boiling and co-injected with C6 cells; Apyrase, apyrase (2 U) co-injected with C6 cells; N, necrosis and microvascular proliferation; V, giant cell formation and nuclear pleomorphism (*arrow*) (Adapted from Morrone et al. (2006). With permission from BioMed Central)

we hypothesize that neuronal death induced by glutamate released from gliomas results in the liberation of ATP and glutamate (normally present in high concentrations in the intracellular milieu) to the extracellular space leading to more neuronal cell death and glioma proliferation in a positive feedback cycle. Because gliomas, contrary to astrocytes in culture, exhibit low NTPDase expression and activity (Morrone et al. 2006) this feedback is not blocked.

To test this hypothesis, we examined the effect of co-injection of apyrase, an ATP/ADP scavenger, in a C6 rat glioma experimental model, which has been extensively used to test antitumor interventions (Takano et al. 2001). The implanted glioma co-injected with apyrase produced smaller tumors when compared with the rats injected only with gliomas or with gliomas plus inactivated apyrase. According to the pathological analysis, the malignant gliomas induced by C6 co-injected with apyrase exhibited a significant reduction in the mitotic index, necrosis and vascular proliferation, pathological characteristics that indicate a less invasive/proliferative tumor (Fig. 5.1). Considering that the injection of apyrase was done only at the moment of implantation, the effect of ATP/ADP depletion is

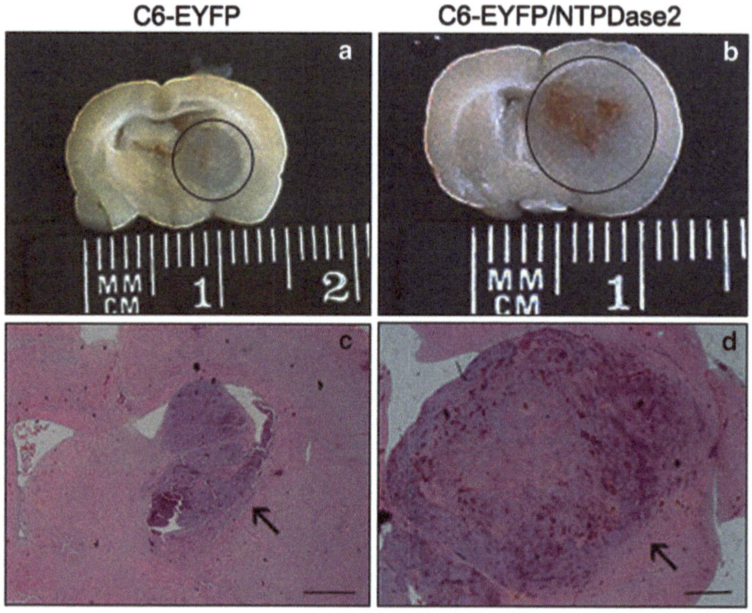

Fig. 5.2 NTPDase2 expression stimulates *in vivo* glioblastoma growth. To determine the glioma growth *in vivo*, equal amounts of C6, C6-EYFP, or C6-EYFP/NTPDase2 cells (1×10^6 cells) were implanted in the right striatum of Wistar rat brains by stereotaxical surgery. The animals were killed 20 days later and glioma sections were dissected and analyzed for tumor growth. (**a**, **b**) Photographs of rat brain slices of C6-EYFP and C6-NTPDase2-implanted gliomas. The gliomas are marked with a circle. (**c**, **d**) Representative sections of C6-EYFP and C6-EYFP/NTPDase2-implanted gliomas stained with HE. Scale bars = 0.5 mm (Adapted from Braganhol et al. 2009. With permission from John Wiley and Sons)

important probably at the implantation and initial growth of the glioma. This could be of therapeutic interest, because the application of apyrase in the surgical resection cavity could be helpful in reducing the initial growth of invaded tumor (Morrone et al. 2006). Considering that NTPDase2 is the dominant E-NTPDase member expressed by astrocytes in culture (Wink et al. 2006) in a further study we better characterized the participation of nucleotides in glioma progression by restoring NTPDase2 expression and activity in rat C6 glioma cells. Surprisingly, NTPDase2 overexpression promoted a dramatic increase in the *in vivo* glioma growth and in the malignant characteristics (Fig. 5.2). A sizable platelet sequestration in the tumor area and an increase in angiogenesis and inflammatory response were observed (Braganhol et al. 2009). The opposite biological outcomes obtained by using NTPDase2 (high ATPase/ADPase ratio) and apyrase (low ATPase/ADPase ratio) as ATP/ADP scavengers, reveal a complex interactions between tumor, immune cells and purinergic mediators. This inverse effect could be due to the fact that whereas NTPDase1 hydrolyses ATP and ADP approximately equally well, the preferred degradation of ATP over ADP of NTPDase2 favors extracellular ADP accumulation (Robson et al. 2006; Zimmermann 2001). Considering that platelets express P2 receptors, which are activated by ADP ($P2Y_1$ and $P2Y_{12}$), we hypothesized that ADP produced by the NTPDase2 overexpressed in the implanted glioma cells could activate these receptors, leading to increased platelet recruitment and activation. The latter processes promote angiogenesis as well as the recruitment of other inflammatory cells (Sierko and Wojtukiewicz 2007). The treatment with clopidogrel, a $P2Y_{12}$ antagonist that prevents the platelet activation by ADP, decreased these parameters to control levels. These data suggest that the ADP derived from NTPDase2 activity stimulates platelet migration to the tumor area and that NTPDase2, by regulating angiogenesis and inflammation, seems to play an important role in tumor progression (Braganhol et al. 2009).

Interestingly, the NTPDase2 reconstitution in gliomas altered not only the local tumor development, but also modulated systemic inflammatory responses. NTPDase2 overexpression promoted an increase in IL-1β, TNF-α and IL-6 pro-inflammatory cytokine production and regulated platelet function *in vivo*. Additionally, pathological alterations in the lungs were observed in rats bearing NTPDase2-gliomas (Braganhol et al. 2011). These data suggest that disruption of purinergic signaling creates an inflammatory microenvironment that dictates tumor cell progression and local invasiveness. Moreover, our findings reveal a previously underestimated role for ADP in tumor promotion and reinforce the important roles carried out by the different E-NTPDase members, which, by working in a highly coordinated enzymatic chain, maintain the extracellular nucleotide equilibrium and control the effects mediated by purinergic receptors.

As presented in the begging of this chapter, in concert with alterations in the extracellular ATP metabolism, modulations of P2 receptors, mainly $P2Y_1$, $P2Y_2$, $P2Y_{12}$ and $P2X_7$, may be important participants in glioma pathology. In C6 glioma cells $P2Y_2$ respond to ATP and UTP while $P2Y_1$ and $P2Y_{12}$ both respond to ADP. Baranska et al. have shown that agonists of these receptors modulate activities of ERK1/2 and PI3K, which are central pathways of cell survival (Baranska et al. 2004). These effects depend on physiological conditions of the cells. Under serum starvation culture conditions, UTP and ADP modulate positively ERK1/2 on C6 rat glioma. In non-starved cells, ADP markedly decreases the PI3K activity, whereas in serum-starved cells it causes an opposite effect. The differential P2Y expression under different culture conditions suggests a cross-talk between $P2Y_1$ and $P2Y_{12}$ receptors in order to favor the glioma cell growth (Baranska et al. 2004).

$P2X_7$ receptor subtype possesses unique biological properties such as the opening of a pore through which molecules up to 900 Da can pass. $P2X_7$ is responsible for ATP-induced cell death in various cell types through mechanisms that involve necrotic features such as swelling, loss of membrane integrity and apoptosis (Taylor et al. 2008; Tsukimoto et al. 2005). In the CNS, ATP is highly toxic to neurons *in vitro* and *ex vivo* (Morrone et al. 2005) and extracellular ATP plays an important role in neuronal death in pathological conditions such multiple sclerosis, Alzheimer and brain ischemia (Burnstock 2008). To date, the involvement and roles of $P2X_7R$ in modulating the growth of brain tumors have been little explored. The $P2X_7$ activation in glioma cells mediates multiple effects that include cell death, survival and modulation of immune system. For example, in rat C6 cell line, C6 *ex vivo* glioma model and human U138MG, the involvement of $P2X_7R$ in the resistance to citotoxicity of ATP has been reported (Braganhol et al. 2008; Morrone et al. 2005). Consistent with the $P2X_7R$ role in tumor progression, immunohistochemical staining of human glioblastoma tissue samples demonstrated greater expression of $P2X_7R$ compared to control non-tumor samples and the pharmacological blockage of $P2X_7R$ inhibits *in vivo* C6 glioma growth (Ryu et al. 2011). On the other hand, GL261 a murine glioma cell line sensitive to high concentrations of ATP, presents a higher ATPase activity, suggesting that the sensitivity may be compensated by a higher degradation rate of ATP for protective reasons. The sensitivity of GL261 to ATP and BzATP was blocked by silencing of $P2X_7R$ (Tamajusuku et al. 2010). The cell death induced by ATP is mainly necrotic of nature, but their involvement in actual tumor growth is not clear. Blocking of $P2X_7R$ with BBG, known $P2X_7$ receptor antagonist, produced a marked reduction in C6 glioma growth in rats, suggesting that cell death of glioma cells is not an important feature of this model (Ryu et al. 2010).

In conclusion, the information presented here supports the idea that alterations in the activity and expression of ectonucleotidases are involved in the glioma progression. The alterations in the extracellular ATP metabolism associated with P1/P2 receptor disruption may have important consequences in events related to tumor advance (Fig. 5.3). Although additional studies are needed to determine whether the altered nucleotide hydrolysis is the cause or consequence of malignant transformation, the ectonucleotidases may be considered as new molecular markers of gliomas and future target for pharmacological or gene therapy.

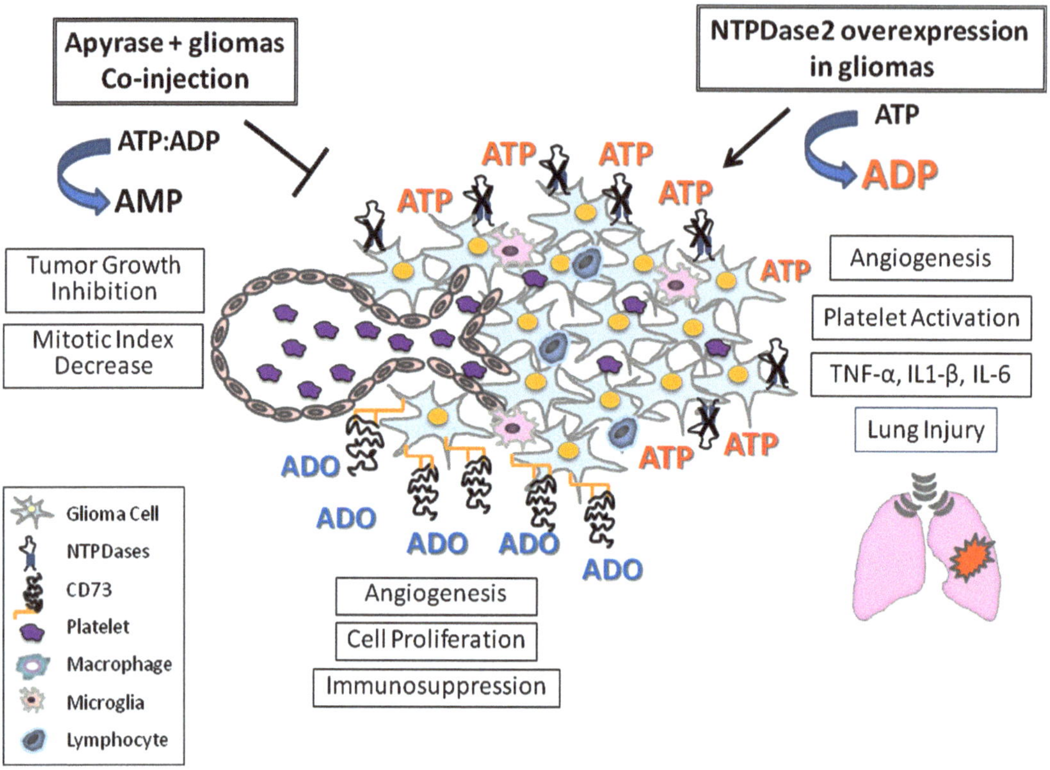

Fig. 5.3 Possible pathways connecting alterations in purinergic signaling and glioma progression**.** Gliomas exhibit an inversion in the extracellular nucleotide metabolism when compared to astrocytes in culture, hydrolyzing poorly the ATP and highly the AMP. This pattern of enzymatic activity may favor the accumulation of extracellular ATP and adenosine in the tumor milieu, inducing cell proliferation, angiogenesis and immunesuppression. Conversely, ATP could induce neuronal cell death, increasing the extracellular ATP pool and further improving the tumor invasion. In accordance with the participation of ATP and ADP in the glioma pathology, the co-injection of apyrase with gliomas decreased the tumor growth and the malignant characteristics, while NTPDase2 produced the opposite effect. We suggest that the ADP formation may be a component of platelet activation, induction of inflammatory cytokine release and the consequent tumor growth observed. The high expression of ecto-5′-NT/CD73, an adhesion and migration cell related protein and main source of enzymatic extracellular adenosine might mediate the interactions between tumor cells and tumor cell-ECM, essential for glioma progression

Acknowledgments We thank Dr. Maria Isabel A. Edelweiss, Departamento de Patologia, HCPA, UFRGS; Fernanda B. Morrone, Faculdade de Farmácia, PUCRS, Porto Alegre, RS, Brasil; Dr. Simon C. Robson, Beth Israel Deaconess Medical Center, Harvard University, Boston, MA, USA and Dr. Jean Sevigny, Centre de Recherche en Rheumatologie et Immunologie, Centre Hospitalier Universitaire de Québec (Pavillon CHUL) and Département de Microbiologie-Infectiologie et d'Immunologie, Faculté de Médecine, Université Laval, Québec, QC, Canada, for their collaboration. We thank the BioMed Central and Willey Editors for the Copyright permission of Figs. 5.1 and 5.2, published in the papers Morrone et al. (2006) (BMC Cancer 23:226) and Braganhol et al. (2009) (Cancer Sci 100(8): 1434–1442), respectively. We also acknowledge the Brazilian Funding Agencies: CNPq, CAPES, FIPE-HCPA and FAPERGS for the financial support.

References

Agresti C, Meomartini ME, Amadio S, Ambrosini E, Serafini B, Franchini L, Volonte C, Aloisi F, Visentin S (2005) Metabotropic P2 receptor activation regulates oligodendrocyte progenitor migration and development. Glia 50:132–144

Bao S, Wu Q, McLendon RE, Hao Y, Shi Q, Hjelmeland AB, Dewhirst MW, Bigner DD, Rich JN (2006) Glioma stem cells promote radioresistance by preferential activation of the DNA damage response. Nature 444:756–760
Baranska J, Czajkowski R, Sabala P (2004) Croos-talks between nucleotide receptor-induced signaling pathways in serum-deprived and non-starved glioma C6 cells. Adv Enzyme Regul 44:219–232
Battastini AMO, da Rocha JBT, Kuplich CK, Dias RD, Sarkis JJF (1991) Characterization of an ATP diphosphohydrolase (EC 3.6.1.5) in synaptosomes from cerebral cortex of adult rats. Neurochem Res 16(12):1303–1310
Bavaresco L, Bernardi A, Braganhol E, Wink MR, Battastini AMO (2007) Dexamethasone inhibits proliferation and stimulates ecto-5′-nucleotidase/CD73 activity in C6 rat glioma cell line. J Neurooncol 84:1–8
Bavaresco L, Bernardi A, Braganhol E, Cappellari AR, Rockenbach L, Farias PF, Wink MR, Delgado-Cañedo A, Battastini AMO (2008) The role of ecto-5′-nucleotidase/CD73 in glioma cell line proliferation. Mol Cell Biochem 319(1–2):61–68
Beier D, Rohrl S, Pillai DR, Schwarz S, Kunz-Schughart LA, Leukel P, Proescholdt M, Brawanski A, Bogdahn U, Trampe-Kieslich A, Giebel B, Wischhusen J, Reifenberger G, Hau P, Beier CP (2008) Temozolomide preferentially depletes cancer stem cells in glioblastoma. Cancer Res 68:5706–5715
Belcher SM, Zsarnovszky A, Crawford PA, Hemani H, Spurling L, Kirley TL (2006) Immunolocalization of ecto-nucleoside triphosphate diphosphohydrolase 3 in rat brain: implications for modulation of multiple homeostatic systems including feeding and sleep-wake behavior. Neuroscience 137:1331–1346
Beraudi A, Traversa U, Villani L, Sekino Y, Nagy JI, Poli A (2003) Distribution and expression of A_1 adenosine receptors, adenosine deaminase and adenosine deaminase-binding protein (CD26) in goldfish brain. Neurochem Int 42(6): 455–464
Berens ME, Giese A (1999) "...those left behind." Biology and oncology of invasive glioma cells. Neoplasia 1:208–219
Bernardi A, Bavaresco L, Wink MR, Jacques-Silva MC, Delgado-Cañedo A, Lenz G, Battastini AMO (2007) Indomethacin stimulates activity and expression of ecto-5′-nucleotidase/CD73 in glioma cell lines. Eur J Pharmacol 569:8–15
Bianchi V, Spychala J (2003) Mammalian 5′-nucleotidases. J Biol Chem 278(47):46195–46198
Biederbick A, Kosan C, Kunz J, Elsasser HP (2000) First apyrase splice variants have different enzymatic properties. J Biol Chem 275:19018–19024
Bigonnesse F, Lévesque SA, Kukulski F, Lecka J, Robson SC, Fernandes MJ, Sévigny J (2004) Cloning and characterization of mouse nucleoside triphosphate diphosphohydrolase-8. Biochemistry 43(18):5511–5519
Blass-Kampmann S, Kindler-Rohrborn A, Deissler H, D'Urso D, Rajewsky MF (1997) In vitro differentiation of neural progenitor cells from prenatal rat brain: common cell surface glycoprotein on three glial cell subsets. J Neurosci Res 48:95–111
Blay J, White TD, Hoskin DW (1997) The extracellular fluid of solid carcinomas contains immunosuppressive concentrations of adenosine. Cancer Res 57:2602–2605
Boeck CR, Sarkis JJF, Vendite D (2002) Kinetic characterization and immunodetection of ecto-ATP diphosphohydrolase (EC 3.6.1.5) in cultured hippocampal neurons. Neurochem Int 40:449–453
Bollen M, Gijsbers R, Ceulemans H, Stalmans W, Stefan C (2000) Nucleotide pyrophosphatases/phosphodiesterases on the move. Crit Rev Biochem Mol Biol 35:393–432
Braganhol E, Tamajusuku ASK, Bernardi A, Wink MR, Battastini AMO (2007) Ecto-5′-nucleotidase/CD73 inhibition by quercetin in the human U138MG glioma cell line. Biochim Biophys Acta 1770:1352–1359
Braganhol E, Huppes D, Bernardi A, Wink MR, Lenz G, Battastini AMO (2008) A comparative study of ectonucleotidase and P2 receptor mRNA profiles in C6 cell line cultures and C6 ex vivo glioma model. Cell Tissue Res 335(2):331–340
Braganhol E, Morrone FB, Bernardi A, Huppes D, Meurer L, Edelweiss MI, Lenz G, Wink MR, Robson SC, Battastini AMO (2009) NTPDase2 expression modulates in vivo rat glioma growth. Cancer Sci 100(8):1434–1442
Braganhol E, Zanin RF, Bernardi A, Bergamin LS, Cappellari AR, Campesato LF, Morrone FB, Campos MM, Calixto JB, Edelweiss MI, Wink MR, Sévigny J, Robson SC, Battastini AM (2011) Overexpression of NTPDase2 in gliomas promotes systemic inflammation and pulmonary injury. Purinergic Signal. doi:10.1007/s11302-011-9276-1
Braun N, Sevigny J, Robson SC, Enjyoji K, Guckelberger O, Hammer K, Di Virgilio F, Zimmermann H (2000) Assignment of ecto-nucleoside triphosphate diphosphohydrolase-1/CD39 expression to microglia and vasculature of the brain. Eur J Neurosci 12:4357–4366
Braun N, Sevigny J, Mishra SK, Robson SC, Barth SW, Gerstberger R, Hammer K, Zimmermann H (2003) Expression of the ecto-ATPase NTPDase2 in the germinal zones of the developing and adult rat brain. Eur J Neurosci 17:1355–1364
Bruno AN, Bonan CD, Wofchuk ST, Sarkis JJ, Battastini AM (2002) ATP diphosphohydrolase (NTPDase 1) in rat hippocampal slices and effect of glutamate on the enzyme activity in different phases of development. Life Sci 71(2):215–225
Buckingham SC, Campbell SL, Haas BR, Montana V, Robel S, Ogunrinu T, Sontheimer H (2011) Glutamate release by primary brain tumors induces epileptic activity. Nat Med 17(10):1269–1274

Burnstock G (2008) Purinergic signalling and disorders of the central nervous system. Nat Rev Drug Discov 7(7):575–590

Calhau C, Martel F, Soares-da-Silva P, Hipólito-Reis C, Azevedo I (2002) Regulation of [(3)H]MPP(+) transport by phosphorylation/dephosphorylation pathways in RBE4 cells: role of ecto-alkaline phosphatase. Naunyn Schmierdebergs Arch Pharmacol 365(5):349–356

Cappellari AR, Vasques GJ, Bavaresco L, Braganhol E, Battastini AMO (2011) Involvement of ecto-5′-nucleotidase/CD73 in U138MG cell adhesion. Mol Cell Biochem 359:315–322

Chadwick BP, Frischauf AM (1998) The CD39-like gene family: identification of three new human members (CD39L2, CD39L3, and CD39L4), their murine homologues, and a member of the gene family from *Drosophila melanogaster*. Genomics 50(3):357–367

Ciruela F, Saura C, Canela EI, Mallol J, Lluis C, Franco R (1996) Adenosine deaminase affects ligand-induced signalling by interacting with cell surface adenosine receptors. FEBS Lett 380:219–223

Cunha RA, Sebastião AM (1991) Extracellular metabolism of adenine nucleotides and adenosine in the innervated skeletal muscle of the frog. Eur J Pharmacol 197:83–92

Cunha RA, Almeida T, Ribeiro JA (2000) Modification by arachidonic acid of extracellular adenosine metabolism and neuromodulatory action in the rat hippocampus. J Biol Chem 275(48):37572–37581

Deissler H, Blass-Kampmann S, Bruyneel E, Mareel M, Rajewsky MF (1995a) Neural cell surface differentiation antigen gp130RB13-6induces fibroblasts and glioma cells to express astroglial proteins and invasive properties. FASEB J 13:657–666

Deissler H, Lottspeich F, Rajewsky MF (1995b) Affinity purification and cDNA cloning of rat neural differentiation and tumor cell surface antigen gp130RB13-6 reveals relationship to human and murine PC-1. J Biol Chem 270:9849–9855

Delarasse C, Gonnord P, Galante M, Auger R, Daniel H, Motta I, Kanellopoulos JM (2009) Neural progenitor cell death is induced by extracellular ATP via ligation of P2X7 receptor. J Neurochem 109:846–857

Demuth T, Berens ME (2004) Molecular mechanisms of glioma cell migration and invasion. J Neurooncol 70:217–228

Dornand J, Bonnafous JC, Mani JC (1978) Purification and properties of 5′-nucleotidase from lymphocyte plasma membranes. Eur J Biochem 87:459–465

Ehrlich YH, Hogan MV, Pawlowska Z, Naik U, Kornecki E (1990) Ectoprotein kinase in the regulation of cellular responsiveness to extracellular ATP. Ann N Y Acad Sci 603:401–416

Fonta C, Negyessy L, Renaud L, Barone P (2004) Areal and subcellular localization of the ubiquitous alkaline phosphatase in the primate cerebral cortex: evidence for a role in neurotransmission. Cereb Cortex 14:595–609

Fonta C, Negyessy L, Renaud L, Barone P (2005) Postnatal development of alkaline phosphatase activity correlates with the maturation of neurotransmission in the cerebral cortex. J Comp Neurol 486(2):179–196

Franco R, Casado V, Ciruela F, Saura C, Mallol J, Canela EI, Lluis C (1997) Cell surface adenosine deaminase: much more than an ecto-enzyme. Prog Neurobiol 52:283–294

Franco R, Valenzuela A, Lluis C, Blanco J (1998) Enzymatic and extraenzymatic role of ecto-adenosine deaminase in lymphocytes. Immunol Rev 161:27–42

Fumagalli M, Brambilla R, D'Ambrosi N, Volonte C, Matteoli M, Verderio C, Abbracchio MP (2003) Nucleotide-mediated calcium signaling in rat cortical astrocytes: role of P2X and P2Y receptors. Glia 43:218–303

Fuss B, Baba H, Phan T, Tuohy VK, Macklin WB (1997) Phosphodiesterase I, a novel adhesion molecule and/or cytokine involved in oligodendrocyte function. J Neurosci 17:9095–9103

Ganzella MG, Faraco RB, Almeida RF, Fernandes VF, Souza DO (2011) Intracerebroventricular administration of inosine is anticonvulsant against quinolinic acid-induced seizures in mice: an effect independent of benzodiazepine and adenosine receptors. Pharmacol Biochem Behav 100:271–274

Gessi S, Sacchetto V, Fogli E, Merighi S, Varani K, Baraldi PG, Tabrizi MA, Leung E, MacLennan S, Borea PA (2010) Modulation of metalloproteinase-9 in U87MG glioblastoma cells by A3 adenosine receptors. Biochem Pharmacol 79:1483–1495

Gessi S, Merighi S, Sacchetto V, Simioni C, Borea PA (2011) Adenosine receptors and cancer. Biochim Biophys Acta 1808:1400–1412

Gines S, Ciruela F, Burgueño J, Casado V, Canela EI, Mallol J, Lluis C, Franco R (2001) Involvement of caveolin in ligand-induced recruitment and internalization of A1 adenosine receptor and adenosine deaminase in an epithelial cell line. Mol Pharmacol 59:1314–1323

Goding JW, Grobben B, Slegers H (2003) Physiological and pathophysiological functions of the ecto-nucleotide pyrophosphatase/phosphodiesterase family. Biochim Biophys Acta 1638:1–19

Goldman N, Chen M, Fujita T, Xu Q, Peng W, Liu W, Jensen TK, Pei Y, Wang F, Han X, Chen JF, Schnermann J, Takano T, Bekar L, Tieu K, Nedergaard M (2010) Nat Neurosci 13(7):883–888

Grobben B, Anciaux K, Roymans D, Stefan C, Bollen M, Esmans EL, Slegers H (1999) An ecto-nucleotide pyrophosphatase is one of the main enzymes involved in the extracellular metabolism of ATP in rat C6 glioma. J Neurochem 72:826–834

Hansen KR, Resta R, Webb CF, Thompson LF (1995) Isolation and characterization of the promoter of the human 5′-nucleotidase 9CD730-encoding gene. Gene 167:307–312

Haskó G, Sitkovsky MV, Szabó C (2004) Immunomodulatory and neuroprotective effects of inosine. Trends Pharmacol Sci 25(3):152–157

Heine P, Braun N, Heilbronn A, Zimmermann H (1999) Functional characterization of rat ecto-ATPase and ecto-ATP diphosphohydrolase after heterologous expression in CHO cells. Eur J Biochem 262:102–107

Heymann D, Reddington M, Kreutzberg GW (1984) Subcellular localization of 5′-nucleotidase in rat brain. J Neurochem 43(4):971–978

Hicks-Berger CA, Chadwick BP, Frischauf AM, Kirley TL (2000) Expression and characterization of soluble and membrane-bound human nucleoside triphosphate diphosphohydrolase 6 (CD39L2). J Biol Chem 275: 34041–34045

Jacques-Silva MC, Bernardi A, Rodnight R, Lenz G (2004a) ERK, PKC and PI3K/Akt pathways mediate extracellular ATP and adenosine-induced proliferation of U138-MG human glioma cell line. Oncology 67:450–459

Jacques-Silva MC, Rodnight R, Lenz G, Liao Z, Kong Q, Tran M, Kang Y, Gonzalez FA, Weisman GA, Neary JT (2004b) P2X7 receptors stimulate AKT phosphorylation in astrocytes. Br J Pharmacol 141:1106–1117

Joseph SM, Pifer MA, Przybylski RJ, Dubyak GR (2004) Methylene ATP analogs as modulators of extracellular ATP metabolism and accumulation. Br J Pharmacol 142:1002–1014

Kaczmarek E, Koziak K, Sevigny J, Siegel JB, Anrather J, Beaudoin AR, Bach FH, Robson SC (1996) Identification and characterization of CD39/vascular ATP diphosphohydrolase. J Biol Chem 271:33116–33122

Kegel B, Braun N, Heine P, Maliszewski CR, Zimmermann H (1997) An ecto-ATPase and an ecto-ATP diphosphohydrolase are expressed in rat brain. Neuropharmacology 36(9):1189–1200

Kiss DS, Zsarnovszky A, Horvath K, Gyorffy A, Bartha T, Hazai D, Sotonyi P, Somogyi V, Frenyo LV, Diano S (2009) Ecto-nucleoside triphosphate diphosphohydrolase 3 in the ventral and lateral hypothalamic area of female rats: morphological characterization and functional implications. Reprod Biol Endocrinol 7:31–43

Knowles AF (2011) The GDA1_CD39 superfamily: NTPDases with diverse functions. Purinergic Signal 7:21–45

Komoszynski M, Wojtczak A (1996) Apyrases (ATP diphosphohydrolases, EC 3.6.1.5): function and relationship to ATPases. Biochim Biophys Acta 1310:233–241

Koshiba M, Kojima H, Huang S, Apasov S, Sitkovsky MV (1997) Memory of extracellular adenosine A2A purinergic receptor-mediated signalling in murine T cells. J Biol Chem 272:25881–25889

Kubler D, Pyerin W, Bill O, Hotz A, Sonka J, Kinzel V (1989) Evidence for ecto-protein kinase activity that phosphorylates kemptide in a cyclic AMP-dependent model. J Biol Chem 264(24):14549–14555

Ledur PF, Villodre ES, Paulus R, Cruz LA, Flores DG, Lenz G (2011) Extracellular ATP reduces tumorsphere growth and cancer stem cell population in U87 human glioblastoma cells. J Cell Biochem. doi:10.1007/s11302-011-9252-9

Lemoli RM, Ferrari D, Fogli M, Rossi L, Pizzirani C, Forchap S, Chiozzi P, Vaselli D, Bertoli F, Foutz T, Aluigi M, Baccarani M, Di Virgilio F (2004) Extracellular nucleotides are potent stimulators of human hematopoietic stem cells in vitro and in vivo. Blood 104:1662–1670

Lenz G, Gottfried C, Luo Z, Avruch J, Rodnight R, Nie WJ, Kang Y, Neary JT (2000) P2Y purinergic receptor subunits recruit different Mek activators in astrocytes. Br J Pharmacol 129:927–936

Lenz G, Goncalves D, Luo Z, Avruch J, Rodnight R, Neary JT (2001) Extracellular ATP stimulates an inhibitory pathway towards growth factor-induced cRaf-1 and MEKK activation in astrocyte cultures. J Neurochem 77: 1001–1009

Lie AA, Blümcke I, Beck H, Wiestler OD, Elger CE, Schoen SW (1999) 5′-Nucleotidase activity indicates sites of synaptic plasticity and reactive synaptogenesis in the human brain. J Neuropathol Exp Neurol 58(5):451–458

Liu C, Sage JC, Miller MR, Verhaak RG, Hippenmeyer S, Vogel H, Foreman O, Bronson RT, Nishiyama A, Luo L, Zong H (2011) Mosaic analysis with double markers reveals tumor cell of origin in glioma. Cell 146:209–221

Lohmann K (1928) The isolation of various natural phosphoric acid compounds and their identity. Biochem Z 194: 306–327

Louis DN (1994) The p53 gene and protein in human brain tumors. J Neuropathol Exp Neurol 53:11–21

Ludwig HC, Rausch S, Schallock K, Markakis E (1999) Expression of CD73 (ecto-5′-nucleotidase) in 165 glioblastomas by immunohistochemistry and electromicroscopic histochemistry. Anticancer Res 19:1747–1752

Maeda T, Nishiyama F, Ogashiwa M, Takeuchi K, Hirano H (1985) Phosphatase activities in human glioma cells as revealed by light and electron microscopy-a preliminary study. J Neurooncol 3(3):211–216

Maher EA, Furnari FB, Bachoo RM, Rowitch DH, Louis DN, Cavenee WK, DePinho RA (2001) Malignant glioma: genetics and biology of a grave matter. Genes Dev 15:1311–1333

Marcus HJ, Carpenter KL, Price SJ, Hutchinson PJ (2010) In vivo assessment of high-grade glioma biochemistry using microdialysis: a study of energy-related molecules, growth factors and cytokines. J Neurooncol 97:11–23

Martins MJ, Negrão MR, Hipólito-Reis C (2001) Alkaline phosphatase from rat liver and kidney is differentially modulated. Clin Biochem 34:463–468

Matsuoka I, Ohkubo S (2004) ATP- and adenosine-mediated signaling in the central nervous system: adenosine receptor activation by ATP through rapid and localized generation of adenosine by ecto-nucleotidases. J Pharmacol Sci 94:95–99

Matute C (2008) P2X7 receptors in oligodendrocytes: a novel target for neuroprotection. Mol Neurobiol 38:123–128

Melani A, De Micheli E, Pinna G, Alfieri A, Corte LD, Pedata F (2003) Adenosine extracellular levels in human brain gliomas: an intraoperative microdialysis study. Neurosci Lett 346(1–2):93–96

Merighi S, Mirandola P, Varani K, Gessi S, Leung E, Baraldi PG, Tabrizi MA, Borea PA (2003) A glance at adenosine receptors: novel target for antitumor therapy. Pharmacol Ther 100:31–48

Meyerhof O (1945) The origin of reaction of Harden and Young in cell-free alcoholic fermentation. J Biol Chem 157:105–109

Morrone FB, Jacques-Silva MC, Horn AP, Bernardi A, Schwartsmann G, Rodnight R, Lenz G (2003) Extracellular nucleotides and nucleosides induce proliferation and increase nucleoside transport in human glioma cell line. J Neurooncol 64:211–218

Morrone FB, Horn AP, Stella J, Spiller F, Sarkis JJ, Salbego CG, Lenz G, Battastini AM (2005) Increased resistance of glioma cell lines to extracellular ATP cytotoxicity. J Neurooncol 71:135–140

Morrone FB, Oliveira DL, Gamermann P, Stella J, Wofchuk S, Wink MR, Meurer L, Edelweiss MIA, Lenz G, Battastini AMO (2006) In vivo glioblastoma growth is reduced by apyrase activity in a rat glioma model. BMC Cancer 23:226

Mulero JJ, Yeung G, Nelken ST, Ford JE (1999) CD39-L4 is a secreted human apyrase, specific for the hydrolysis of nucleoside diphosphates. J Biol Chem 274:20064–20067

Mulero JJ, Yeung G, Nelken ST, Bright JM, McGowan DW, Ford JE (2000) Biochemical characterization of CD39L4. Biochemistry 39(42):12924–12928

Murphy-Piedmonte DM, Crawford PA, Kirley TL (2005) Bacterial expression, folding, purification and characterization of soluble NTPDase5 (CD39L4) ecto-nucleotidase. Biochim Biophys Acta 1747:251–259

Nagy AK, Shuster TA, Delgado-Escueta AV (1986) Ecto-ATPase of mammalian synaptosomes:identification and enzymic characterization. J Neurochem 47:976–986

Nagy A, Shuster TA, Delgado-Escueta AV (1989) Rat brain synaptosomal ATP:AMP-phosphotransferase acitvity. J Neurochem 53:1166–1172

Narisawa S, Hasegawa H, Watanabe K, Millán JL (1994) Stage-specific expression of alkaline phosphatase during neural development of the mouse. Dev Dyn 201:227–235

Navaro JM, Olmo N, Turnay J, López-Conejo MT, Lizarbe MA (1998) Ecto-5′-nucleotidase from a human colon adenocarcinoma cell line. Correlation between enzyme activity and levels in intact cells. Mol Cell Biochem 187:121–131

Neary JT, Kang Y, Willoughby KA, Ellis EF (2003) Activation of extracellular signal-regulated kinase by stretch-induced injury in astrocytes involves extracellular ATP and P2 purinergic receptors. J Neurosci 23:2348–2356

Ohkubo S, Kimura J, Matsuoka I (2000) Ecto-alkaline phosphatase in NG108-15 cells: a key enzyme mediating P1 antagonist-sensitive ATP response. Br J Pharmacol 131(8):1667–1672

Oses JP, Cardoso CM, Germano RA, Kirst IB, Rücker B, Fürstenau CR, Wink MR, Bonan CD, Battastini AM, Sarkis JJ (2004) Soluble NTPDase: an additional system of nucleotide hydrolysis in rat blood serum. Life Sci 74(26): 3275–3284

Paas Y, Bohana-Kashtan O, Fishelson Z (1999) Phophorylation of the complement component, C9, by an ecto-protein kinase of human leukemic cells. Immunopharmacology 42:175–185

Paez JG, Recio JÁ, Rouzaut A, Notario V (2001) Identity between the PCPH proto-oncogene and the CD39L4 (NTPDase5) ectonucleoside triphosphate diphosphohydrolase gene. Int J Oncol 19:1249–1254

Picher M, Burch LH, Hirsh AJ, Spychala J, Boucher RC (2003) Ecto 5′-nucleotidase and nonspecific alkaline phosphatase. Two AMP-hydrolyzing ecto-enzymes with distinct roles in human airways. J Biol Chem 278:13468–13479

Pinsky DJ, Broekman MJ, Peschon JJ, Stocking KL, Fujita T, Ramasamy R, Connolly ES Jr, Huang J, Kiss S, Zhang Y, Choudhri TF, McTaggart RA, Liao H, Drosopoulos JH, Price VL, Marcus AJ, Maliszewski CR (2002) Elucidation of the thromboregulatory role of CD39/ectoapyrase in the ischemic brain. J Clin Invest 109:1031–1040

Plesner L (1995) Ecto-ATPases: identities and functions. Int Rev Cytol 158:141–214

Ravera S, Aluigi MG, Calzia D, Ramoino P, Morelli A, Panfoli I (2011) Evidence for ectopic aerobic ATP production on C6 glioma cell plasma membrane. Cell Mol Neurobiol 31:313–321

Recio JA, Paez JG, Maskeri B, Loveland M, Velasco JA, Notario V (2000) Both normal and transforming PCPH proteins have guanosine diphosphatase activity but only the oncoprotein cooperates with Ras in activating extracellular signal-regulated kinase ERK1. Cancer Res 60:1720–1728

Recio JA, Paez JG, Sanders S, Kawakami T, Notario V (2002) Partial depletion of intracellular ATP mediates the stress-survival function of the PCPH oncoprotein. Cancer Res 62:2690–2694

Redegeld FA, Caldwell CC, Sitkovsky MV (1999) Ecto-protein kinases: ecto-domain phosphorylation as a novel target for pharmacological manipulation? Trends Pharmacol Sci 20:453–459

Resta R, Yamashita Y, Thompson LF (1998) Ectoenzyme and signaling functions of lymphocyte. Immunol Rev 161:95–109

Ricci-Vitiani L, Pallini R, Biffoni M, Todaro M, Invernici G, Cenci T, Maira G, Parati EA, Stassi G, Larocca LM, De Maria R (2010) Tumour vascularization via endothelial differentiation of glioblastoma stem-like cells. Nature 468:824–828
Robson SC, Sévigny J, Zimmermann H (2006) The E-NTPDase family of ectonucleotidases: structure function relationship and pathophysiological significance. Purinergic Signal 2:409–430
Rouzaut A, Recio JA, Notario V (2001) Expression of the protein product of the PCHP proto-conogene in human tumor cell lines. Radiat Res 155:181–187
Ryu JK, Choi HB, Hatori K, Heisel RL, Pelech SL, McLarnon JG, Kim SU (2003) Adenosine triphosphate induces proliferation of human neural stem cells: role of calcium and p70 ribosomal protein S6 kinase. J Neurosci Res 72: 352–362
Ryu JK, Jantaratnotai N, Serrano-Perez MC, McGeer PL, McLarnon JG (2010) Block of purinergic P2X7R inhibits tumor growth in a C6 glioma brain tumor animal model. J Neuropathol Exp Neurol 70:13–22
Ryu JK, Jantaratnotai N, Serrano-Perez MC, McGeer PL, McLarnon JG (2011) Block of purinergic P2X7R inhibits tumor growth in a C6 glioma brain tumor animal model. J Neuropathol Exp Neurol 70(1):13–22
Sadej R, Spychala J, Skladanowski AC (2006) Expression of ecto-5′-nucleotidase (eN, CD73) in cell lines from various stages of human melanoma. Melanoma Res 16(3):213–222
Saura C, Mallol J, Canela E, Lluis C, Franco R (1998) Adenosine deaminase and A1 adenosine receptors internalize together followingagonist-induced receptor desensitization. J Biol Chem 273:17610–17617
Seehafer J, Longenecker BM, Shaw AR (1984) Biochemical characterization of human carcinoma surface antigen associated with protein kinase activity. Int J Cancer 34(6):821–829
Shapiro JR (2001) Genetics of brain neoplasms. Curr Neurol Neurosci Rep 1:217–224
Shi JD, Kukar T, Wang CY, Li QZ, Cruz PE, Davoodi-Semiromi A, Yang P, Gu Y, Lian W, Wu DH, She JX (2001) Molecular cloning and characterization of a novel mammalian endo-apyrase (LALP1). J Biol Chem 276: 17474–17478
Shukla V, Zimmermann H, Wanf L, Kettenmann H, Raab S, Hammer K, Sévigny J, Robson SC, Braun N (2005) Functional expression of the ecto-ATPase NTPDase2 and of nucleotide receptors by neuronal progenitor cells in the adult murine hippocampus. Neurosci Res 80(5):600–610
Sierko E, Wojtukiewicz MZ (2007) Inhibition of platelet function: does it offer a chance of better cancer progression control? Semin Thromb Hemost 33:712–721
Singh SK, Clarke ID, Terasaki M, Bonn VE, Hawkins C, Squire J, Dirks PB (2003) Identification of a cancer stem cell in human brain tumors. Cancer Res 63:5821–5828
Singh SK, Hawkins C, Clarke ID, Squire JA, Bayani J, Hide T, Henkelman RM, Cusimano MD, Dirks PB (2004) Identification of human brain tumour initiating cells. Nature 432:396–401
Skog J, Wurdinger T, van Rijn S, Meijer DH, Gainche L, Sena-Esteves M, Curry WT Jr, Carter BS, Krichevsky AM, Breakefield XO (2008) Glioblastoma microvesicles transport RNA and proteins that promote tumour growth and provide diagnostic biomarkers. Nat Cell Biol 10:1470–1476
Smith TM, Carl SA, Kirley TL (1998) Immunological detection of ecto-ATPase in chicken and rat tissues: characterization, distribution, and a cautionary note. Biochem Mol Biol Int 45(5):1057–1066
Spychala J (2000) Tumor-promoting functions of adenosine. Pharmacol Ther 87(2–3):161–173
Spychala J, Zimmermann AG, Mitchell BS (1999) Tissue-specific regulation of the Ecto-5′-nucleotidase promoter. J Biol Chem 274:22705–22712
Stanojević I, Bjelobaba I, Nedeljković N, Drakulić D, Petrović S, Stojiljković M, Horvat A (2011) (2011) Ontogenetic profile of ecto-5′-nucleotidase in rat brain synaptic plasma membranes. Int J Dev Neurosci 29(4):397–403
Stefan C, Jansen S, Bollen M (2006) Modulation of purinergic signaling by NPP-type ectophosphodiesterases. Purinergic Signal 2:361–370
Stochaj U, Dieckhoff J, Mollenhauer J, Cramer M, Mannherz HG (1989) Evidence for the direct interaction of chicken gizzard 5′-nucleotidase with laminin and fibronectin. Biochim Biophys Acta 992(3):385–392
Stracke ML, Krutzsch HC, Unsworth EJ, Arestad A, Cioce V, Schiffmann E, Liotta LA (1992) Identification, purification, and partial sequence analysis of autotaxin, a novel motility-stimulation protein. J Biol Chem 267:2524–2529
Su AI, Welsh JB, Sapinoso LM, Kern SG, Dimitrov P, Lapp H, Schultz PG, Powell SM, Moskaluk CA, Frierson HF Jr, Hampton GM (2001) Molecular classification of human carcinomas by use of gene expression signatures. Cancer Res 61(20):7388–7393
Takahashi T, Old LJ, Boyse EA (1970) Surface alloantigens of plasma cells. J Exp Med 131:1325–1341
Takano T, Lin JH, Arcuino G, Gao K, Yang J, Nedergaard M (2001) Glutamato release promotes growth of malignant gliomas. Nat Med 7:1010–1015
Tamajusuku AS, Villodre ES, Paulus R, Coutinho-Silva R, Battasstini AM, Wink MR, Lenz G (2010) Characterization of ATP-induced cell death in the GL261 mouse glioma. J Cell Biochem 109:983–991
Taylor SR, Gonzalez-Begne M, Dewhurst S, Chimini G, Higgins CF, Melvin JE, Elliott JI (2008) Sequential shrinkage and swelling underlie P2X7-stimulated lymphocyte phosphatidylserine exposure and death. J Immunol 180: 300–308

TCGA (2008) Comprehensive genomic characterization defines human glioblastoma genes and core pathways. Nature 455:1061–1068
Tombetta ES, Helenius Q (1999) Glycoprotein reglucosylation and nucleotide sugar utilization in the secretory pathway: identification of a nucleoside diphosphatase in the endoplasmic reticulum. EMBO J 18:3282–3292
Tsukimoto M, Harada H, Ikari A, Takagi K (2005) Involvement of chloride in apoptotic cell death induced by activation of ATP-sensitive P2X7 purinoceptor. J Biol Chem 280:2653–2658
Turnay J, Olmo N, Reence G, Von der Mark K, Lizarbe MA (1989) 5′-Nucleotidase activity in cultured cell lines: effect of different assay conditions and correlation with cell proliferation. In Vitro Cell Dev Biol 25:1055–1061
Velasco JA, Avila MA, Notario V (1999) The product of the cph oncogene is a truncated, nucleotide-binding protein that enhances cellular survival to stress. Oncogene 18:689–701
Verkhratsky A, Krishtal OA, Burnstock G (2009) Purinoceptors on neuroglia. Mol Neurobiol 39:190–208
Villar J, Arenas MI, MacCarthy CM, Blanquez MJ, Tirado OM, Notario V (2007) PCPH/ENTPD5 expression enhances the invasiveness of human prostate cancer cells by a protein kinase CD–dependent mechanism. Cancer Res 67(22):10859–10868
Visvader JE (2011) Cells of origin in cancer. Nature 469:314–322
Vlashi E, Kim K, Lagadec C, Donna LD, McDonald JT, Eghbali M, Sayre JW, Stefani E, McBride W, Pajonk F (2009) In vivo imaging, tracking, and targeting of cancer stem cells. J Natl Cancer Inst 101:350–359
Vogel M, Kowalewski HJ, Zimmermann H, Janetzko A, Margolis RU, Wollny HE (1991) Association of the HNK-1 epitope with 5′-nucleotidase from *Torpedo marmorata*. Biochem J 278:199–202
Vollmayer P, Koch M, Braun N, Heine P, Servos J, Israr E, Kegel B, Zimmermann H (2001) Multiple ecto-nucleotidase in PC12 cells: identification and cellular distribution after heterologous expression. J Neurochem 78:1019–1028
Von Deimling A, Louis DN, Wiestler OD (1995) Molecular pathways in the formation of gliomas. Glia 15:328–338
Wang T-F, Guidotti G (1999) Widespread expression of ecto-apyrase (CD39) in the central nervous system. Brain Res 790:318–322
Wang R, Chadalavada K, Wilshire J, Kowalik U, Hovinga KE, Geber A, Fligelman B, Leversha M, Brennan C, Tabar V (2010) Glioblastoma stem-like cells give rise to tumour endothelium. Nature 468:829–833
White N, Burnstock G (2006) P2 receptors and cancer. Trends Pharmacol Sci 27:211–217
Wink MR, Lenz G, Braganhol E, Tamajusuku AS, Schwartsmann G, Sarkis JJ, Battastini AM (2003) Altered extracellular ATP, ADP and AMP catabolism in glioma cell lines. Cancer Lett 198(2):211–218
Wink MR, Braganhol E, Tamajusuku ASK, Lenz G, Zerbini LF, Libermann TA, Sévigny J, Battastini AM, Robson SC (2006) Nucleoside triphosphate diphosphohydrolase-(NTPDase2/CD39L1) is the dominant ectonucleotidase expressed by rat astrocytes. Neuroscience 138:421–432
Yamamoto T, Geiger JD, Daddona PE, Nagy JI (1987) Subcellular, regional and immunohistochemical localization of adenosine deaminase in various species. Brain Res Bull 19:473–484
Ye ZC, Sontheimer H (1999) Glioma cells release excitotoxic concentrations of glutamate. Cancer Res 59:4383–4391
Yegutkin GG (2008) Nucleotide- and nucleoside-converting ectoenzymes: important modulators of purinergic signalling cascade. Biochim Biophys Acta 1783(5):673–694
Yegutkin GG, Henttinen T, Samburski SS, Spychala J, Jalkanen S (2002) The evidence for two opposite, ATP-generating and ATP-consuming, extracellular pathways on endothelial and lymphoid cells. Biochem J 367:121–128
Yeung G, Mulero JJ, McGowan DW, Bajwa SS, Ford JE (2000) CD39L2, a gene encoding a human nucleoside diphosphatase, predominantly expressed in the heart. Biochemistry 39(42):12916–12923
Zhi X, Chen S, Zhou P, Shao Z, Wang L, Ou Z, Yin L (2007) RNA interference of ecto-5′-nucleotidase (CD73) inhibits human breast cancer cell growth and invasion. Clin Exp Metastasis 24(6):439–448
Zhou P, Zhi X, Zhou T, Chen S, Li X, Wang L, Yin L, Shao Z, Ou Z (2007) Overexpression of Ecto-5′-nucleotidase (CD73) promotes T-47D human breast cancer cells invasion and adhesion to extracellular matrix. Cancer Biol Ther 6(3):426–431
Zimmermann H (1992) 5′-Nucleotidase: molecular structure and functional aspects. Biochem J 285:345–365
Zimmermann H (1996) Biochemistry, localization and functional roles of ecto-5′-nucleotidase in the nervous system. Prog Neurobiol 49(6):589–618
Zimmermann H (2001) Ectonucleotidases: some recent developments and a note on nomenclature. Drug Dev Res 52:44–56
Zimmermann H (2006) Purinergic signalling in neuron-glia interactions. In: Novartis foundation symposium 276, Wiley, Chichester, p 113–130
Zimmermann H (2006) Ectonucleotidases in the nervous system. Novartis Found Symp 276:113–128
Zimmermann H, Braun N, Kegel B, Heine P (1998) New insights into molecular structure and function of ectonucleotidases in the nervous system. Neurochem Res 32:421–425

Chapter 6
Cytoskeleton and Nucleotide Signaling in Glioma C6 Cells

Wanda Kłopocka, Jarosław Korczyński, and Paweł Pomorski

Abstract This chapter describes signaling pathways stimulated by the $P2Y_2$ nucleotide receptor ($P2Y_2R$), that regulate cellular processes dependent on actin cytoskeleton dynamics in glioma C6 cells. $P2Y_2R$ coupled with G-proteins, in response to ATP or UTP, regulates the level of phosphatidylinositol-4,5-bisphosphate (PIP_2) which modulates a variety of actin binding proteins and is involved in calcium response and activates Rac1 and RhoA proteins. The RhoA/ROCK signaling pathway plays an important role in contractile force generation needed for the assembly of stress fibers, focal adhesions and for tail retraction during cell migration. Blocking of this pathway by a specific Rho-kinase inhibitor induces changes in F-actin organization and cell shape and decreases the level of phosphorylated myosin II and cofilin. In glioma C6 cells these changes are reversed after UTP stimulation of $P2Y_2R$. Signaling pathways responsible for this compensation are connected with calcium signaling. Stimulation of the Rac1 mediated pathway *via* G_0 proteins needs additional interaction between $\alpha_v\beta_5$ integrins and $P2Y_2Rs$. Rac1 activation is necessary for cofilin phosphorylation as well as integrin activation needed for focal complexes formation and stabilization of lamellipodium. Inhibition of positive Rac1 regulation prevents glioma C6 cells from recovery of control cell like morphology.

Keywords $P2Y_2$ • PIP_2 • Rho • Rac • Small G-proteins • Cofilin • Myosin II • Actin dynamics • Integrins • Cell migration • Glioma C6 cells

Abbreviations

DAG	Diacylglycerol
EGFRs	Epidermal growth factor receptors
FAK	Focal Adhesion Kinase
GEFs	Guanine nucleotide exchange factors
GPCRs	G protein-coupled receptors
IP_3	Inositol-1,4,5-trisphosphate

W. Kłopocka (✉) • J. Korczyński • P. Pomorski
Nencki Institute of Experimental Biology, Polish Academy of Sciences, 3 Pasteur St, PL 02-093 Warsaw, Poland
e-mail: w.klopocka@nencki.gov.pl; j.korczynski@nencki.gov.pl; p.pomorski@nencki.gov.pl

J. Barańska (ed.), *Glioma Signaling*, Advances in Experimental Medicine and Biology 986,
DOI 10.1007/978-94-007-4719-7_6, © Springer Science+Business Media Dordrecht 2013

LIMK	LIM kinase
MARCKS	Myristoylated alanine rich C-kinase substrate
MHC	Myosin heavy chains
MLC	Myosin light chain
MLCK	Myosin light chain kinase
MLCP	Myosin light chain phosphatase
$P2Y_2R$	$P2Y_2$ receptor
PAK	p21 Activated kinase
PDGFRs	Platelet-derived growth factor receptors
PI3K	Phosphatidylinositol 3-kinase
PIP_2	Phosphatidylinositol-4,5-bisphosphate
PIP 5-kinase	Phosphatidylinositol-4-phosphate 5-kinase
PIP	Phosphoinositol phosphate
PKC	Protein kinase C
PLC	Phospholipase C
P-MLC	Phosphorylated myosin light chain
PTX	Pertussis toxin
ROCK	Rho-associated protein kinase
SSH	Slingshot homolog
TESK	Testis-specific kinase

6.1 Introduction

The way in which simple, chemical reactions of the signaling pathways translate itself into the cell behavior is crucial for understanding of the living world. This chapter describes how the signal from one nucleotide receptor can influence and regulate complex phenomena of cell structure and motility.

This is the ability to actively change shape and travel from one location to another what we intuitively assume as one of the most fundamental attributes of being alive. Being motile is the common feature of multiple free living protista (Chung et al. 2001; Kłopocka and Pomorski 1996; Pomorski et al. 2007; Stockem and Kłopocka 1988) as well as tissue cells in higher organisms (Pomorski et al. 2004). The motility is crucial for organism development and everyday functions. While most of cells in human body can use bloodstream as a fast way of traveling long distance, the existence of blood-brain-barrier isolates cells of central nervous system from the passive way to spread around and force to use active crawling to change position. This is as true for the microglia cell looking for the inflammation as for the spreading cells of malignant glioma. The existence of the barrier between blood and brain creates very special environment, where every cell at least on some period in the life have to be actively motile.

The $P2Y_2$ nucleotide receptor is located in the very centre of signaling network regulating cell motility and cytoskeleton shape. There are two basic systems regulating cytoskeleton and motility: Rho-family pathways (Kłopocka and Rędowicz 2003, 2004; Kłopocka et al. 2005; Schmitz et al. 2000; Worthylake et al. 2001) and calcium signaling (Pletjushkina et al. 2001; Pomorski et al. 2004; Somlyo and Somlyo 2003). In this chapter we will show how $P2Y_2$ receptor signaling can influence both those systems. While the Chap. 4 focus on the calcium signaling in glia and glioma cells, here we will discuss the complex relations between this two signaling systems and show why nucleotide receptors and nucleotide signaling are so important for them.

In the chapter we will describe crucial pathways regulating glioma C6 cells cytoskeleton as well as the peculiar role of $P2Y_2$ in cross-talk between them. Finally we will show how experimental induction of this receptor may influence inhibition of classical RhoA cytoskeleton and motility-regulating pathway.

6.2 The Role of the P2Y$_2$ Receptor in Actin Cytoskeleton Organization

Signaling cascades evoked by the P2Y$_2$ receptor (P2Y$_2$R) stimulation operate by the sequential activation and deactivation of small G proteins, phospholipases and protein kinases that regulate cellular processes dependent on actin cytoskeleton dynamics: endo- and exocytosis, cytokinesis, intracellular transport, cell to cell adhesion and cell migration. Actin cytoskeleton reorganization in response to extra-cellular stimuli is crucial in these processes. The dynamic transformation of actin filaments that involves actin polymerization and depolymerization, formation and degradation of filamentous networks and bundles is regulated by integrated signaling cascades. Important factors in actin cytoskeletal reorganization are small GTPases of the Rho family, Rho, Rac, and Cdc42 (Bishop and Hall 2000; Hall 1998) and phosphoinositides, especially phosphatidylinositol-4,5-bisphosphate (PIP$_2$) that binds directly to a variety of actin regulatory proteins and modulates their functions (Sechi and Wehland 2000; Takenawa and Itoh 2001). It means that changes in PIP$_2$ levels influence the remodeling of the actin cytoskeleton. The P2Y$_2$ receptor is coupled to G$_q$, G$_o$ and G$_{12}$ proteins. G$_q$-coupled receptors stimulate phospholipase C (PLC) activity that hydrolyzes PIP$_2$ and generates two second messengers, inositol 1,4,5-trisphosphate (IP$_3$) and diacylglycerol (DAG). IP$_3$ mobilizes Ca^{2+} from the endoplasmic reticulum, whereas DAG activates protein kinase C (PKC). Calcium transient promotes actin cytoskeleton reorganization by activation of Ca^{2+}-calmodulin-dependent myosin light chain kinase (MLCK). MLCK phosphorylates myosine light chains (MLCs) at Ser19 and/or Thr18 residues (Katoh et al. 2001; Somlyo and Somlyo 2003) and increases acto-myosin contractility at the cell periphery (Totsukawa et al. 2000).

The P2Y$_2$ nucleotide receptor contains the integrin-binding domain, arginine-glycine-aspartic acid (RGD), in its first extracellular loop (Lustig et al. 1996; van Rhee et al. 1998). The integrin binding domain in P2Y$_2$R enables it to interact selectively with $\alpha_v\beta_3$ and $\alpha_v\beta_5$ integrins (Erb et al. 2001) and is required for G$_o$-mediated Rac1 activation (Bagchi et al. 2005; Erb et al. 2001). Interaction with $\alpha_v\beta_5$ is also necessary for coupling the P2Y$_2$ receptor to G$_{12}$ (Liao et al. 2007).

The cross-talk between integrins and P2Y$_2$ receptors has been shown to enable receptor coupling to specific G proteins involved in cell migration. It was demonstrated that α_v integrin expression is required for P2Y$_2$R to induce cell migration in many cell types (Bagchi et al. 2005; Wang et al. 2005). P2Y$_2$ receptor has also been suggested to play a role in the wound-healing process (Greig et al. 2003).

In addition to modulation of ion channel activity (Chap. 3), P2Y receptors can modulate the activity of receptor tyrosine kinases. SH3-binding domains, that are necessary to bind and activate the non-receptor tyrosine kinase Src, have been identified in C-terminal tail of the P2Y$_2$. UTP stimulation causes Src activation and subsequent recruitment of the epidermal growth factor receptors (EGFRs) into a protein complex with the P2Y$_2$R that allows Src to phosphorylate EGFR (Liu et al. 2004). Responses to P2Y$_2$R stimulation depend on cell types and the cross-talk between various signaling pathways (Erb et al. 2006).

6.2.1 *Regulation by PIP$_2$*

In response to external stimuli G protein-coupled receptors (GPCRs) and tyrosine kinase receptors (EGFRs and PDGFRs) regulate the level of PIP$_2$ by activating Rac and/or PLC. GTP-bound Rac activates phosphatidylinositol-4-phosphate 5-kinase (PIP 5-kinase) responsible for PIP$_2$ synthesis (Hartwig et al. 1995; Tolias et al. 2000), whereas activation of PLC leads to hydrolysis of PIP$_2$ and to formation of DAG and IP$_3$ (Boarder and Hourani 1998).

PIP$_2$ plays a critical role in controlling actin cytoskeleton dynamics by modulating a variety of actin regulatory proteins (Cooper and Schafer 2000). PIP$_2$ inhibits the activities of profilin, severing proteins ADF/cofilin and capping proteins such as CapZ and gelsolin, whereas it activates the cross linking activity of α-actinin (Fukami et al. 1992). PIP_2 also activates ERM (ezrin/radixin/moesin) and vinculin that link actin filaments to the plasma membrane (Takenawa and Itoh 2001). WASP proteins, MARCKS and pleckstrin regulate actin cytoskeleton also in a PIP_2-dependent manner (Toker 2002). PIP_2 promotes actin nucleation and polymerization by interacting directly with profilin. Profilin sequesters actin monomers, thus preventing polymerization, and PIP_2 disrupts this association (Sohn et al. 1995). Actin filaments are also prevented from spontaneously polymerizing by capping proteins that cap the barbed ends. Interaction of PIP_2 with capping proteins dissociates them from the barbed ends of actin filaments. Uncapping of the barbed ends leads to elongation of the existing microfilaments (Takenawa and Itoh 2001; Toker 2002). The activity of cofilin is regulated by several mechanisms, but the most important are phosphorylation (Pak et al. 2008) and the binding of PIP_2 (van Rheenen et al. 2007). By binding unphosphorylated cofilin PIP_2 inhibits its actin-depolymerizing and actin-severing activity. Stimulation of $P2Y_2R$ by ATP or UTP results in activation of PLC and decrease in the level of PIP_2 with the subsequent release and activation of cofilin. Upon activation, cofilin diffuses from the plasma membrane to the F-actin compartment, where it binds to and severs actin filaments, which results in filaments with more free barbed ends and the formation of cofilin-G-actin complexes (van Rheenen et al. 2007). These complexes diffuse to the cytosol, where cofilin is phosphorylated and released from the cofilin-G-actin complex by LIM kinase (LIMK) (Song et al. 2006; van Rheenen et al. 2007). LIMK is activated nearly simultaneously with PLC (Song et al. 2006) through Rho/Rho associated kinase (ROCK) and Rac/p21-activated kinase (PAK)-dependent pathway (Dan et al. 2001). Thus the translocation of cofilin from the plasma membrane to the F-actin compartment is induced by PLC-mediated decrease in PIP_2 levels, whereas the translocation from the cytosol to the plasma membrane compartment is induced by dephosphorylation of cofilin (van Rheenen et al. 2009). In mammary-tumor cells the reduction in PIP_2 levels by PLC, with the subsequent release of cofilin, is the key to regulating cofilin activity after receptor stimulation (Mouneimne et al. 2004). By contrast, in leukocytes, the primary activation of cofilin is through dephosphorylation (Boldt et al. 2006; Sun et al. 2007) although the binding of PIP_2 might be the key to regulating the activity locally at the cell periphery.

PIP_2 plays also an important role in focal adhesion formation. It binds and regulates the function of proteins that link the extracellular matrix to actin filaments via integrins. PIP_2 binds to vinculin and promotes its binding to actin filaments and to talin (Gilmore and Burridge 1996; Huttelmaier et al. 1998). The interaction between actin cytoskeleton and plasma membrane not only modulates local adhesion energy but also influences cell shape and migration. The ERM family members link actin filaments and surface glycoproteins, and this association is enhanced by PIP_2. ERM proteins are regulated by PIP_2 in synergy with Rac since they are effectors of this small GTP-binding protein (Hirao et al. 1996; Mackay et al. 1997). Similarly, an effector of Cdc42, WASP, binds PIP_2 (Miki et al. 1996). PIP_2 also interacts with the PKC substrates, plekstrin and MARCKS, and promotes their association with actin cytoskeleton (Glaser et al. 1996; Harlan et al. 1995).

6.2.2 Regulation by the Small GTP-binding Proteins: Rho, Rac and Cdc42

Small GTP-binding proteins of the Rho family belong to the Ras superfamily and are intracellular signaling proteins known to act as molecular switches to control actin cytoskeleton dynamics. The best characterized members: RhoA, Rac1, and Cdc42 were first identified in the early 1990s (Ridley and Hall 1992). They play an essential role in the regulation of various cellular processes, including cell migration, smooth muscle contraction, endocytosis, cytokinesis, neurite outgrowth and retraction. They are conformationally regulated by the binding of GTP and GDP. The active form of Rho is GTP-bound (Hall 1998; Van Aelst and D'Souza-Schorey 1997). The exchange of GDP by GTP, catalyzed by guanine nucleotide exchange factors (GEFs), allows Rho proteins to interact with and activate their downstream effectors and thereby transmit signals (Schmitz et al. 2000) (see Chap. 7). It is established that activation of Rho GTPases is controlled by several mechanisms including the stimulation of receptor tyrosine kinases and that many among the G protein-coupled receptors (GPCRs), particularly those coupled to the G_{12}, G_o and G_q type of heterotrimeric G proteins, are upstream regulators of the Rho proteins. Rac activation by both tyrosine kinases and G-coupled receptors is dependent on phosphatidylinositol 3-kinase (PI3K) activity (Royal et al. 2000). During lamellipodia extension phosphoinositol phosphates also bind and activate GEFs that regulate the activity of Rac (Kaibuchi et al. 1999). Rho GTPase downstream target proteins include protein kinases, lipid-modifying enzymes and activators of the Arp2/3 complex (Etienne-Manneville and Hall 2002).

The ability of a cell to move requires an asymmetrical localization of cellular activities. The front of the migrating cell generates a protrusive force, generally associated with the extension of a lamellipodium and development of new cell adhesions to the extracellular substrate. Cell contractility is required to allow cell body to follow the extending front (Ridley 2001; Worthylake et al. 2001). Through its ability to promote both protrusion and cell contraction, the actin cytoskeleton provides the driving force for cell migration. RhoA, Rac1 and Cdc42 proteins are the key regulators of actin cytoskeleton dynamics involved in cell shape changes, protrusion, adhesion and contractility (Clark et al. 1998; Nobes and Hall 1995, 1999). Distinctive effects of Rho, Rac, and Cdc42 activation on the organization of the actin cytoskeleton have been observed in many cell types, including fibroblasts, macrophages, epithelial cells, endothelial cells, astrocytes, neurons and in cancer cells as well as in circulating cells such as lymphocytes, mast cells and platelets (Allen et al. 1997; Hartwig et al. 1995; Ridley et al. 1995).

Cdc42 is the principal determinant in establishing cell polarity and stabilization of the directional movement (chemotaxis) (Itoh et al. 2002). Rac and Cdc42 regulate also small integrin clusters at the leading edge, known as focal complexes, that stabilize the lamellipodium (Ridley et al. 2003). Rac and Cdc42 induce Arp2/3 complex- mediated actin polymerization in an area limited to the cell protrusion and thus enable directional migration. When the Arp2/3 complex is bound with proteins of the WASP or WAVE family, it is activated and catalyzes nucleation of actin polymerization and the formation of new filament branches (Condeelis 2001; Pollard et al. 2000). The product of the human Wiskott-Aldrich syndrome gene, WASP, has been identified as a Cdc42-specific target (Aspenstrom 1997; Welch and Mullins 2000). WAVE is itself regulated by Rac (Cory and Ridley 2002; Ridley 2001). Rac interacts with the insulin receptor tyrosine kinase substrate p53 (IRSp53) (Miki et al. 2000), which in turn interacts through an Src-homologous domain 3 (SH3) with a number of the WAVE family proteins, which then bind to and activate the Arp2/3 complex. An active Arp2/3 complex binds to proteins called nucleation promoting factors (Campellone and Welch 2010; Chesarone and Goode 2009). Another target of Rac that may be involved in actin polymerization is PIP 5-kinase, the enzyme that converts phosphoinositol phosphate (PIP) to PIP_2 (Hartwig et al. 1995). Rac can also control the activity of a crucial regulator of actin assembly, cofilin. Through its effector, PAK, which phosphorylates and activates LIM kinase, Rac regulates LIMK activity (Dan et al. 2001; Edwards et al. 1999; Maekawa et al. 1999; Yang et al. 1998). RhoA-mediated reorganization of the filamentous actin network involves RhoA downstream effectors, among them, the Rho family of serine-threonine protein kinases (ROCKs) that, in turn, affect their specific substrate proteins (Schmitz et al. 2000; Tsuji et al. 2002). ROCK acts upon LIM kinase and testis-specific kinase (TESK), which phosphorylate cofilin at Ser 3 (Arber et al. 1998; Bernard 2007; Maekawa et al. 1999; Pak et al. 2008). LIMK phosphorylates and inactivates cofilin in the cytosol (Delorme et al. 2007). Cofilin ability to bind actin, and therefore its actin severing and depolymerizing activity is inhibited upon phosphorylation (Bamburg et al. 1999; DesMarais et al. 2005). Dephosphorylation of cofilin by the phosphatase Slingshot homolog (SSH) and chronophin (CIN) (Huang et al. 2006; Niwa et al. 2002; Ohta et al. 2003) circumvents this inhibition. Cofilin can increase the number of barbed ends by severing the existing microfilaments (Ichetovkin et al. 2002) and by its actin-nucleation activity (Andrianantoandro and

Pollard 2006). The actin-depolymerization activity of cofilin increases the pool of actin monomers available for polymerization (van Rheenen et al. 2009).

A number of myosins have been implicated in cell migration (Mermall et al. 1998). Myosin light chain phosphorylation is enhanced in the lamellipodial region of cells (Matsumura et al. 1998), which suggests a role for myosins in lamellipodium extension. Rac can affect the phosphorylation of both myosin heavy chains (MHC) (van Leeuwen et al. 1999) and myosin light chains via activation of its downstream kinase, PAK (Brzeska et al. 2004; Klosses et al. 2001). In polarized cells Rac activity is constrained to the cell front. FRET (fluorescence resonance energy transfer) microscopy has revealed that Rac GTP levels are highest at the leading edge of a migrating cell (Itoh et al. 2002). Rho GTP is restricted to the cell body and excluded from the leading edge (Worthylake et al. 2001). Coordination of localized actin polymerization at the front of the cell with contraction at the rear can propel cells forward.

Cell body contraction is dependent on acto-myosin contractility. RhoA plays an important role in contractile force generation needed for the assembly of stress fibers (Fukata et al. 2001), and large integrin clusters – focal adhesions (Ridley et al. 2003) as well as tail retraction, which is a myosin dependent process (Worthylake et al. 2001). ROCK affects the acto-myosin II contractility by phosphorylation and inhibition of the myosin-binding subunit of myosin light chain phosphatase (MLCP) at Thr853 and Thr696 (Hartshorne et al. 1998; Kawano et al. 1999; Kimura et al. 1996; Ramachandran et al. 2011). The actin-activated ATPase activity of myosin II is also increased by direct phosphorylation of MLC at Ser19 (Amano et al. 1996; Matsumura 2005; Moussavi et al. 1993; Riento and Ridley 2003). The ROCK-mediated activation of myosin II is suggested to cause assembly of stress fibers and focal adhesions in nonmuscle cells (Chrzanowska-Wodnicka and Burridge 1996; Kawano et al. 1999; Kimura et al. 1996). The RhoA/ROCK-mediated increase in the level of phosphorylated MLC and contraction occurs in the absence of Ca^{2+} (Kimura et al. 1996) while myosin light chain kinase phosphorylates MLC at Ser19 and/or Thr18 residues in a Ca^{2+}-dependent way (Katoh et al. 2001; Somlyo and Somlyo 2003). Accordingly, G protein-coupled receptors (GPCRs) that mobilize intracellular-free Ca^{2+} activate MLCK and induce MLC phosphorylation.

These two kinases, ROCK and MLCK, may play distinct roles in the spatial regulation of myosin II activity. The RhoA/ROCK signaling pathway plays an important role in generating myosin II-based contractility and stress fiber assembly in the center of cells (Chrzanowska-Wodnicka and Burridge 1996; Kawano et al. 1999; Kimura et al. 1996; Totsukawa et al. 2000). Ca^{2+}/calmodulin MLCK regulation is primarily responsible for phosphorylating MLC at the cell periphery (Totsukawa et al. 2000). A reduction in Rho or Rho-kinase activity affects morphology, motility and invasion of all cell types examined until now.

Another RhoA effector, mDia1, induces actin polymerization and cooperates with ROCK in stress fiber formation (Watanabe et al. 1999). Although not direct targets of Rho, the ERM proteins are essential for Rho- and Rac-induced cytoskeletal effects. Their interaction with a transmembrane protein, CD44, and filamentous actin can be regulated by Rho (Fukata et al. 1998; Hirao et al. 1996; Mackay et al. 1997; Matsui et al. 1998).

Axon guidance can be thought of as a specialized form of directed migration: the cell body does not move, but the growth cone at the axon tip responds to attractive and repulsive extracellular cues (Pak et al. 2008). Cdc42 and Rac are positive regulators of neurite outgrowth, whereas RhoA inhibits neurite extension (Luo 2000). Rac activation induces growth cone advance and increases the number of dendritic spines, RhoA has the opposite effect (Wong et al. 2000). Axonal extension is driven by actin polymerization within the growth cone consisting of filopodial and lamellipodial protrusions. Activation of Rac and Cdc42 in neuroblastoma cells has been shown to promote the formation of lamellipodia and filopodia, respectively, along the neurite extensions (Luo 2000).

The assembly and disassembly of peripheral actin filaments can be utilized to promote localized changes in the structure of the plasma membrane which subsequently affect membrane-controlled processes such as phagocytosis and pinocytosis. Two distinct signaling pathways regulate phagocytosis.

When induced by immunoglobulin receptor stimulation, phagocytosis requires Rac and Cdc42, and when induced by the complement receptor it requires Rho (Caron and Hall 1998).

In addition to their effect on migration and endocytosis, Rho GTPases contribute to the regulation of the mitotic phase of the cell cycle. Mitotic cell rounding is the process of cell shape change. In many types of cells the rearrangement of the actin cytoskeleton and an increase in cortical rigidity that accompany mitotic cell rounding require RhoA and Rac1 (Glotzer 2001; Maddox and Burridge 2003). The cell cycle is completed with cytokinesis, and in animal cells this is driven by an actin and myosin contractile ring, which constricts to form the two daughter cells. Inhibition of Rho as well as expression of constitutively activated Rho prevents the assembly of the contractile ring in a variety of mammalian cells (Kosako et al. 2000). Activated Rho recruits ROCK and citron kinase and probably also the Diaphanous-related proteins, to assemble the contractile ring, mediate cleavage furrow ingression and promote abscission (Eda et al. 2001; Gruneberg et al. 2006; Kosako et al. 2000; Madaule et al. 1998; Piekny et al. 2005). ROCK is responsible for phosphorylation of MLC in the cleavage furrow. Citron kinase phosphorylates MLC at both Ser19 and Thr18 and increases diphosphorylation of MLC during cytokinesis. Both kinases accumulate at the cleavage furrow in several types of cultured cells (Eda et al. 2001; Yamashiro et al. 2003).

6.3 Compensation of ROCK Inhibition by P2Y$_2$R Activated Signaling Pathways

Blocking of the RhoA/ROCK signaling pathway in glioma C6 cells causes serious changes in the organization of actin cytoskeleton, cell shape and migration. These changes can be compensated by UTP stimulation of the P2Y$_2$R. Recovery of the cells from ROCK inhibition requires P2Y$_2$R coupling with both G$_q$ and G$_o$ proteins. Induction of G$_o$-mediated signal transduction depends on $\alpha_v\beta_5$ integrin activation.

6.3.1 ROCK Inhibition in Glioma C6 Cells

Addition of a specific inhibitor of ROCK – Y-27632 inhibits stress fibers and focal adhesions in several cell lines. In glioma C6 cells ROCK inhibition induces changes in F-actin organization and cell shape similar to those observed in other types of cells with blocked RhoA/ROCK signaling pathway.

Our results have shown that actin reorganization and subsequent alterations in glioma C6 cell morphology and migration correlate with the level of phosphorylated MLC (P-MLC) and phosphorylated cofilin (P-cofilin) – two actin binding proteins downstream of RhoA/ROCK. ROCK inhibition by Y-27632 reduces the level of P-MLC (Fig. 6.1a) (Korczyński et al. 2011), the level of P-cofilin and simultaneously increases the level of Rac1 GTP (Fig. 6.1b, unpublished results) (Korczyński et al. in preparation). The observed activation of Rac1 following ROCK inhibition may result from the influence of Rho on Rac, as it has been observed in neurons (Kozma et al. 1997; van Leeuwen et al. 1997), Swiss 3T3 cells (Tsuji et al. 2002) and in malignant astrocytoma (Salhia et al. 2005). Tsuji and coworkers have revealed that the Rho-dependent Rac activation is mediated by mDia1 through Cas phosphorylation (Tsuji et al. 2002).

Glioma C6 cells normally exhibit a polygonal shape and are well spread on the substrate. They are usually characterized by a fibroblast-like morphology (Fig. 6.2a) (see Chap. 3). Actin filaments form a cortical network just beneath the plasma membrane and stress fibers in the cytoplasm. Phospho-myosin II is distributed along the stress fibers and under the cell membrane (Fig. 6.2b). Cells are polarized with clearly defined lamellipodium. Decreasing the level of phosphorylated MLC and

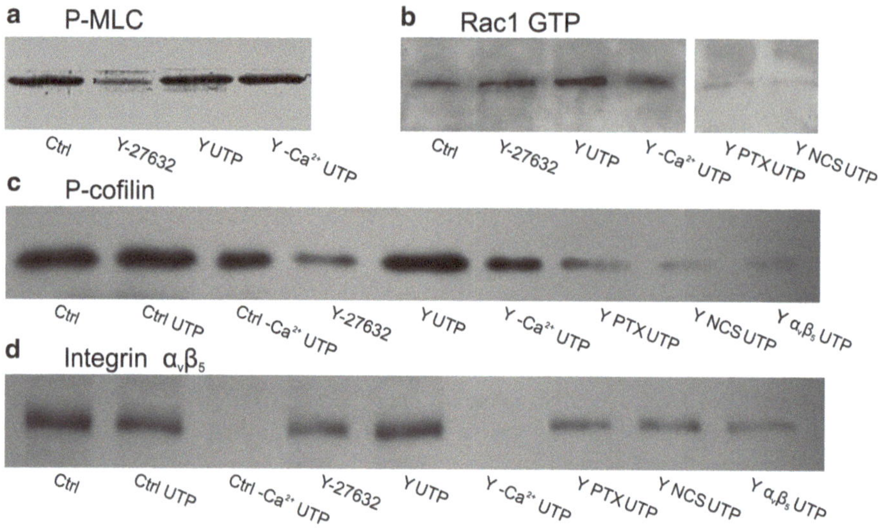

Fig. 6.1 Changes in protein phosphorylation level in glioma C6 cells after ROCK inhibition with Y-27632. Western blot analysis of (**a**) P-MLC, (**b**) Rac1 GTP, (**c**) P-cofilin and (**d**) Integrin $\alpha_v\beta_5$ level in cells treated with different reagents before material collection. Abbreviations for experimental models: **Ctrl –** control cells, **Ctrl UTP –** cells stimulated with UTP, **Ctrl -Ca^{2+} UTP –** control cells stimulated with UTP in calcium-free medium, **Y-27632** – cells pretreated with Y-27632 inhibitor, **Y UTP** – UTP stimulated Y-27632-pretreated cells, **Y -Ca^{2+} UTP** – Y-27632 pretreated cells stimulated with UTP in calcium-free medium, **Y PTX UTP** – cells pretreated with Y-27632 and pertussis toxin (PTX) stimulated with UTP, **Y ML-7 UTP** – Y-27632 and ML-7-pretreated cells followed by UTP stimulation, **Y NCS UTP** – Y-27632 and NCS 23766-pretreated cells stimulated with UTP, **Y $\alpha_v\beta_5$ UTP** – cells pretreated with Y-27632 and $\alpha_v\beta_5$ integrin antibody followed by UTP stimulation

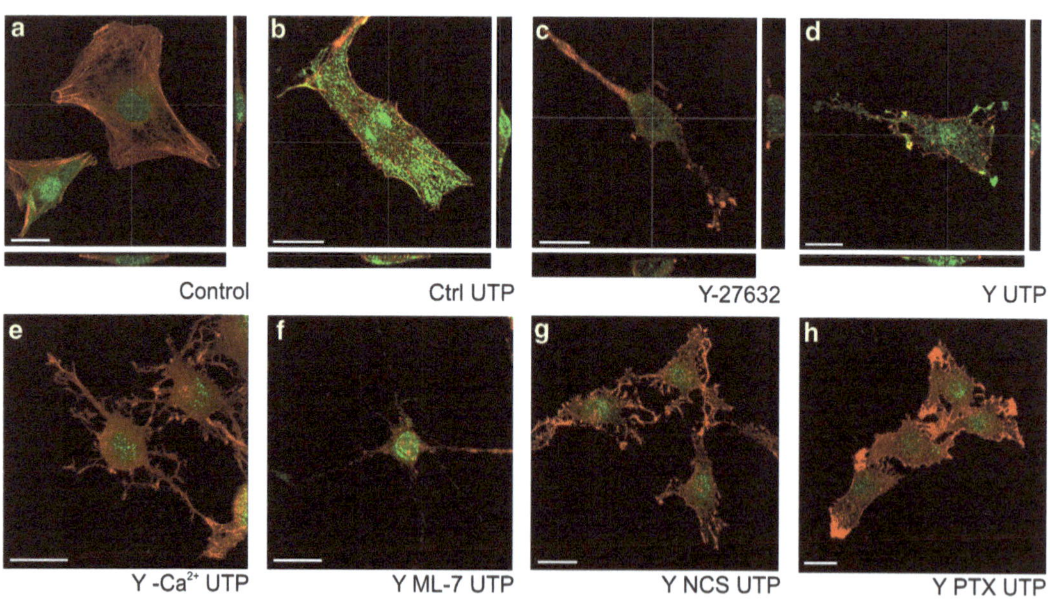

Fig. 6.2 Changes in morphology of glioma C6 cells in different experimental models (for abbreviations see Fig. 6.1). 3D orthogonal sections (**a–d**) and single images (**e–h**) from confocal microscopy. Immunofluorescence staining for F-actin (*red*) and P-cofilin (images **a**, **c**, **e**, **g**, **h**; *green*) or P-MLC (images **b**, **d**, **f**; *green*). Scale bars 20 μm

cofilin results in the loss of cell polarization, in stress fibers decomposition and promotes membrane ruffling and multiple lamellipodia formation. The cells round up and develop long outgrowths. Rounded cells exhibit a high volume to surface ratio as the 3D images show (Fig. 6.2c). F-actin is concentrated mainly under the cell surface and in the elongated processes.

Myosin II inactivation in the presence of Y-27632 contributes to destabilization of the actin cytoskeleton. Since the F- to G-actin ratio is not affected under such conditions (Targos et al. 2006) changes observed in actin cytoskeleton organization (stress fiber disappearance, relocation of F-actin) and in cell morphology seem to be caused by disorganization of the functional state of acto-myosin II system by inhibition of Rho-kinase. Dephosphorylation of MLC would favor myosin remaining in the low-affinity state for actin and subsequently dissociating from actin (Sutton et al. 2001). Phosphorylated myosin II diffuses all over the rounded cell body and does not co-localize with actin filaments. Temporal separation between myosin II inactivation and a complete disruption of stress fibers (Sutton et al. 2001) suggests that other processes, such as actin severing, may be involved in stress fiber disassembly. In glioma C6 cells suppression of LIMK-mediated phosphorylation of cofilin might be responsible for microfilament destabilization in the cell center. However, the ability of cofilin to affect actin dynamics is restricted at the leading edge because in this region microfilaments are not saturated with tropomyosin and not resistant to cofilin severing (DesMarais et al. 2002; Gupton et al. 2005). Changes in the actin cytoskeleton organization described above, together with the dysfunction of cell adhesion regulation, impair directional cell migration. Human TM cells treated with ROCK inhibitors showed a decrease in the electrical cell-substrate resistance and a decrease in tyrosine phosphorylation of paxilin and focal adhesion kinase (FAK) (Ramachandran et al. 2011).

6.3.2 *Effect of P2Y$_2$ Receptor Stimulation on MLC Phosphorylation – The Role of MLCK*

In glioma C6 cells UTP stimulation induces an about 15 % increase in the pool of P-MLC. However, it does not cause visible changes in F-actin distribution and cell morphology even though P2Y$_2$R promotes the formation of stress fibers (Fig. 6.2b) and cell migration (Korczyński et al. 2011) similarly as in astrocytoma cells (Peterson et al. 2010). Moreover stimulated glioma C6 cells are more spread than the control ones. In cells pretreated with ROCK inhibitor, stimulation of P2Y$_2$R induces almost a 65 % increase in P-MLC level (Fig. 6.1a) (Korczyński et al. 2011). The relatively lower level of P-MLC after UTP stimulation in cells with inhibited ROCK, as compared to control cells, seems to be the result of constitutively active myosin II phosphatase which is inhibited by ROCK under normal conditions.

UTP stimulation of Y-27632 pretreated glioma C6 cells correlates with a dynamic actin cytoskeleton reorganization and subsequent recovery of control cell-like morphology (Fig. 6.2d). Myosin II phosphorylation occurs rapidly, and temporally correlates with the assembly of short, weak stress fibers at the cell periphery (Fig. 6.2d). Stress fiber localization and the fact that myosin II phosphorylation follows G$_q$-mediated calcium mobilization indicates that Ca^{2+}/calmodulin dependent myosin light chain kinase may be responsible for the phosphorylation of MLC under such experimental conditions. It was shown that in cells with blocked ROCK and activated MLCK stress fibers are not assembled in the cell center (Katoh et al. 2001; Totsukawa et al. 2000). MLCK inhibition by a specific inhibitor, ML-7, almost completely prevented MLC phosphorylation in Y-27632 pretreated glioma C6 cells and in consequence made actin cytoskeleton reorganization and cell recovery impossible (Fig. 6.2f) (Korczyński et al. 2011). Similar results concerning MLCK inhibition were obtained also for non-transformed astrocytes (Baorto et al. 1992).

Apart from calcium, other upstream messengers may participate in the regulation of MLCK after $P2Y_2$ stimulation. Active Rac, for example, may inhibit MLCK at the leading edge (Brzeska et al. 2004; Sanders et al. 1999). The regulatory mechanism could be complex and the activity of MLCK probably vary depending on the cell type. The Rac/PAK signaling pathway is also involved in direct phosphorylation of myosin light chains (Brzeska et al. 2004).

6.3.3 Effect of $P2Y_2$ Receptor Stimulation on Cofilin Phosphorylation – The Role of Rac1 Protein

Our results have also shown the effect of UTP on cofilin phosphorylation in glioma C6 cells. The $P2Y_2$ receptor activation significantly increases the level of phosphorylated cofilin as well as that of Rac1 GTP in cells with blocked ROCK (Fig. 6.1b). It has been established that activated Rac increases association of PAK with LIMK (Edwards et al. 1999) and in consequence enhances the activity of LIMK (Yang et al. 1998). Signaling pathways which negatively regulate cofilin in glioma C6 cells stimulated with UTP are unknown. We suggest that activation of LIMK by Rac1/PAK signaling pathway may be responsible for cofilin phosphorylation in glioma C6 cells with blocked ROCK, and that activation of Rac1 protein is mediated by G_o protein. The level of phosphorylated cofilin is not increased in cells pretreated with NCS 23766 (Rac1 inhibitor) or with pertussis toxin (PTX) – $G_{i/o}$ inhibitor (Fig. 6.1c). Under such experimental conditions the recovery from ROCK inhibition is not completed (Fig. 6.2g, h) although G_q-mediated intracellular calcium mobilization and myosin II activation is not affected. The complete reversal of the effects of ROCK inhibition in glioma C6 cells requires induction of G_q and G_o mediated signaling pathways and subsequent increasing the level of P-MLC as well as phosphorylated cofilin (Korczyński et al. in preparation). However, in mammary-tumor cells the phosphorylation status of cofilin is not directly related to cofilin activity, and actin depolymerization depends on PLC activity but not on cofilin dephosphorylation. (Mouneimne et al. 2004; van Rheenen et al. 2009; Song et al. 2006). In carcinoma cells P-cofilin level rises upon EGF stimulation while the actin-severing activity of cofilin is at its highest level (Song et al. 2006). If so, cofilin dephosphorylation in glioma C6 cells with blocked ROCK should not be identify with its ability to actin binding. Dephosphorylated cofilin is immediately bound by PIP_2 and inactivated (Rheenen et al. 2009). The morphological effect of decreasing the level of phosphorylated cofilin in glioma C6 cells with blocked ROCK may be compared to suppression of cofilin in MTC cells, a non-metastatic mesenchymal-type tumor cells with high polarity and movement (Shestakova et al. 1999), that caused changes in the cell shape, the formation of numerous lamellipodia and subsequent decrease in the directionality of migration (Sidani et al. 2007). This may suggests that the activity status of cofilin in glioma C6 cells does not only depend on phosphorylation-dephosphorylation processes, and that in cells with blocked ROCK the level of cofilin activity is suppressed.

Stimulation of $P2Y_2$ receptor, *via* G_q protein, activates PLC and a subsequent decrease in the level of PIP_2. Cofilin molecules are released and diffuse to the cortical layer, where cofilin binds and severs actin filaments to force actin polymerization (van Rheenen et al. 2007). However, cofilin activation by PLC is not sufficient for glioma C6 cells recovery from ROCK inhibition since blocking the Rac1 activity prevents from this process. Simultaneously with PLC activation of LIMK through Rac1/PAK-dependent pathway phosphorylates and inactivates cofilin molecules that diffuse away from the plasma membrane (Delorme et al. 2007). Cofilin inactivation together with myosin II activation through calcium dependent MLCK and Rac/PAK signaling pathways (Brzeska et al. 2004) cause actin cytoskeleton reorganization leading to stress fiber assembly in the cell periphery, lamellipodium formation, alterations in cell morphology and migration. It is also possible that UTP stimulation of glioma C6 cells induces G_{12}-mediated RhoA activation similar to that observed in human astrocytoma cells (Liao et al. 2007). This in turn activates a downstream effector mDia1 by disrupting its intermolecular interactions. Active mDia1 induces actin polymerization and the formation of thin actin stress fibers, which however are disorganized when ROCK is inhibited (Watanabe et al. 1999).

6.3.4 *Effect of Integrin on Glioma C6 Cell Recovery from ROCK – Essential Role of Rac1 Protein*

The $\alpha_v\beta_3$ and $\alpha_v\beta_5$ integrins play essential roles in migration of numerous cell types (see Chap. 7). In astrocytes $P2Y_2R$ interaction with $\alpha_v\beta_5$ integrin increases chemotactic and chemokinetic cell response. UTP stimulation also increases the expression of $\alpha_v\beta_{3/5}$ integrins in astrocytes (Wang et al. 2005). Activation of $\alpha_v\beta_5$ integrin in UTP stimulated glioma C6 cells (Fig. 6.1d, unpublished results) is important for cell spreading and for coupling $P2Y_2R$ to G_o. G_o-mediated Rac1 activation thus requires interaction of $P2Y_2R$ with $\alpha_v\beta_5$ integrin and is crucial for morphology remodeling of cells with blocked ROCK (Korczyński et al. in preparation). In calcium-free medium, in detached cells and in cells pretreated with anti-$\alpha_v\beta_5$ antibody integrins are not activated (Fig. 6.1d), and their interaction with $P2Y_2Rs$ and in consequence coupling the $P2Y_2R$ to G_o protein is blocked. Under such experimental conditions Rac1 activation and subsequent Rac1-mediated cofilin phosphorylation is inhibited. The morphological effect of blocking $\alpha_v\beta_5$ integrin is similar to that of ROCK inhibition through Y-27632 (Korczyński et al. in preparation).

Under UTP stimulation of glioma C6 cells, the $P2Y_2$ receptor initiates a typical store operated calcium signal (Suplat-Wypych et al. 2010). This signal pathway enhances MLC phosphorylation and, in cells with blocked ROCK, temporally correlates with renormalization (Fig. 6.2d). However, if UTP stimulation is performed in calcium-free medium, and the calcium response is limited to Ca^{2+} release from intracellular stores, actin cytoskeleton reorganization and the morphology recovery do not take place even though MLC phosphorylation facilitates cell contractility as manifested by cell spreading measured by confocal imaging (Fig. 6.2e). MLC phosphorylation is not sufficient for the reversal of cells with blocked ROCK to their normal phenotype. For cell recovery from ROCK inhibition two signaling pathways must be triggered: G_q-mediated PLC activation for mobilization of intracellular calcium and cofilin release from PIP_2 and G_o-mediated Rac1 activation necessary for cofilin phosphorylation as well as integrin activation (Fig. 6.3). Rac and Cdc42 regulate small integrin clusters at the leading edge known as focal complexes that stabilize the lamellipodium (Ridley et al. 2003). In UTP stimulated glioma C6 cells with blocked ROCK, $\alpha_v\beta_5$ integrins are activated by Rac1. Inhibition of positive Rac1 regulation by NCS or PTX block integrins activation (Fig. 6.1d) and cell adhesion as well as cofilin phosphorylation and cell polarization. Together with suppression of cofilin phosphorylation prevents glioma C6 cells from recovery of control cell like morphology.

6.4 Summary

Central nervous system glial cells release and respond to nucleotides under both physiological and pathological conditions, suggesting that these molecules play key roles in both normal brain function and in repair after damage. Signaling cascades evoked by $P2Y_2$ receptor plays an important role in the phenomena dependent on the actin cytoskeleton dynamics endocytosis, cell division, adhesion, intracellular transport and migration. In astrocytes and microglial cells the agonists activate $P2Y_2Rs$ to induce integrin-dependent activation of Rho and Rac, promoting glial cells migration. Thus, signaling pathways related to $P2Y_2Rs$ are of special interest in glioma tumor cells, where cytoskeleton driven cell motility is the main way of cancer invasion in brain.

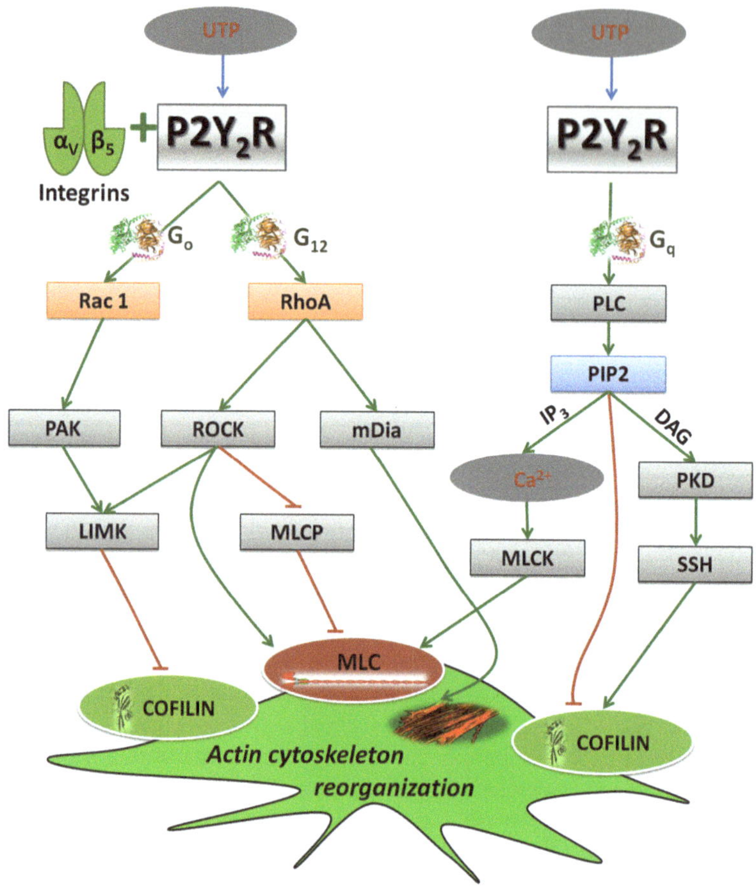

Fig. 6.3 Schematic diagram of signaling pathways activated by external UTP stimulation of P2Y$_2$ receptor in glioma C6 cells. *Green arrows* on the scheme indicate activation of downstream substrates, *red arrows* show inactivation of cofilin or MLC. For details, see text

UTP stimulation of P2Y$_2$R in glioma C6 cells compensates changes in the organization of actin cytoskeleton, cell shape and migration caused by blocking of the RhoA/ROCK signaling pathway. The complete renormalization of glioma C6 cells with blocked ROCK requires: (1) induction of G$_q$-mediated PLC activation and mobilization of intracellular calcium for increasing the level of P-MLC, (2) cofilin release from PIP$_2$, (3) G$_o$-mediated Rac1 activation necessary for both cofilin phosphorylation and integrin activation. Induction of G$_o$-mediated signal transduction depends on $\alpha_v\beta_5$ integrins activation (Korczyński et al. in preparation). In calcium-free medium, as well as in detached cells, UTP stimulation of glioma C6 cells with blocked ROCK does not cause $\alpha_v\beta_5$ integrins activation, blocking the interaction between P2Y$_2$Rs and $\alpha_v\beta_5$ integrins and in consequence coupling the P2Y$_2$R to G$_o$ protein. For cell recovery from ROCK inhibition $\alpha_v\beta_5$ integrins activation is necessary for signal transduction and focal complexes formation that stabilize new formed lamellipodium, fix cell polarization and directional migration. In UTP stimulated cells with blocked ROCK, $\alpha_v\beta_5$ integrins are activated by Rac1. Inhibition of positive Rac1 regulation by NCS or PTX decreases integrins activation and cell adhesion and together with suppression of cofilin phosphorylation prevents glioma C6 cells to complete recovery of control cell like morphology. Inhibition of this signaling pathway can also affect directional migration of the cell as cofilin activity is necessary to induce protrusions and to define the direction of cell motility (Ghosh et al. 2004).

Since the RhoA/ROCK signaling pathway is crucial for migration, infiltrative growth and expansion of brain tumors, the attempts are made to inhibit the development of this type of cancer by inhibition of RhoA or ROCK. However numerous publications mentioned in this chapter indicate a high possibility of compensation of this inhibition by alternative signaling pathways especially related to cell stimulation by nucleotides. This fact can significantly impede the use of currently known inhibitors in medical therapies of brain tumors.

References

Allen WE, Jones GE, Pollard JW, Ridley AJ (1997) Rho, Rac and Cdc42 regulate actin organization and cell adhesion in macrophages. J Cell Sci 110:707–720

Amano M, Ito M, Kimura K, Fukata Y, Chihara K, Nakano T, Matsuura Y, Kaibuchi K (1996) Pohsphorylation and activation of myosin by Rho-associated kinase (Rho-kinase). J Biol Chem 271:20246–20249

Andrianantoandro E, Pollard TD (2006) Mechanism of actin filament turnover by severing and nucleation at different concentrations of ADF/cofilin. Mol Cell 24:13–23

Arber S, Barbayannis FA, Hanser H, Schneider C, Stanyon CA, Bernard O, Caroni P (1998) Regulation of actin dynamics through phosphorylation of cofilin by LIM-kinase. Nature 393:805–809

Aspenstrom P (1997) A Cdc42 target protein with homology to the non-kinase domain of FER has a potential role in regulating the actin cytoskeleton. Curr Biol 7:479–487

Bagchi S, Liao Z, Gonzales FA, Chorna NE, Seye CI, Weisman GA, Erb L (2005) The P2Y2 nucleotide receptor interacts with alpha integrins to activate Go and induce cell migration. J Biol Chem 280:39050–39057

Bamburg JR, McGough A, Ono S (1999) Putting a new twist on actin: ADF/cofilin modulate actin dynamics. Trends Cell Biol 9:364–370

Baorto DM, Mellado W, Shelanski ML (1992) Astrocyte process growth induction by actin breakdown. J Cell Biol 117:357–367

Bernard O (2007) Lim kinases, regulators of actin dynamics. Int J Biochem Cell Biol 39:1071–1076

Bishop AL, Hall A (2000) Rho GTPases and their effector proteins. Biochem J 348:241–255

Boarder MR, Hourani SM (1998) The regulation of vascular function by P2 receptors: multiple sites and multiple receptors. Trends Pharmacol Sci 19:99–107

Boldt K, Rist W, Weiss SM, Weith A, Lenter MC (2006) FPRL-1 induces modifications of migration-associated proteins in human neutrophils. Proteomics 6:4790–4799

Brzeska H, Szczepanowska J, Matsumura F, Korn ED (2004) Rac-induced increase of phosphorylation of myosin regulatory light chain in HeLa cells. Cell Motil Cytoskel 58:186–199

Campellone KG, Welch MD (2010) A nucleator arms race: cellular control of actin assembly. Nat Rev Mol Cel Biol 11:237–251

Caron E, Hall A (1998) Identification of two distinct mechanisms of phagocytosis controlled by different Rho GTPases. Science 282:1717–1721

Chesarone MA, Goode BL (2009) Acti nucleation and elongation factors: mechanisms and interplay. Curr Opin Cell Biol 21:28–37

Chrzanowska-Wodnicka M, Burridge K (1996) Rho-stimulated contractility drives the formation of stress fibers and focal adhesions. J Cell Biol 133:1403–1415

Chung CY, Funamoto S, Firtel RA (2001) Signaling pathways controlling cell polarity and chemotaxis. Trends Biochem Sci 26:557–566

Clark EA, King WG, Brugge JS, Symons M, Hynes RO (1998) Integrin-mediated signals regulated by members of the Rho family of GTPases. J Cell Biol 142:573–586

Condeelis J (2001) How is actin polymerization nucleated in vivo? Trends Cell Biol 11:288–293

Cooper JA, Schafer DA (2000) Control of actin assembly and disassembly at filament ends. Curr Opin Cell Biol 12:97–103

Cory GO, Ridley AJ (2002) Cell motility: braking WAVEs. Nature 418:732–733

Dan C, Kelly A, Bernard O, Minden A (2001) Cytoskeletal changes regulated by the PAK4 serine/threonine kinase are mediated by LIM kinase 1 and cofilin. J Biol Chem 276:32115–32121

Delorme V, Machacek M, DerMardirossian C, Anderson KL, Wittmann T, Hanein D, Waterman-Storer C, Danuser G, Bokoch GM (2007) Cofilin activity downstream of Pak1 regulates cell protrusion efficiency by organizing lamellipodium and lamella actin network. Dev Cell 13:646–662

DesMarais V, Ichetovkin I, Condeelis J, Hitchcock-DeGregori SE (2002) Spatial regulation of actin dynamics: a tropomyosin-free, actin-rich compartment at the leading edge. J Cell Sci 115:4649–4660

DesMarais V, Ghosh M, Eddy R, Condeelis J (2005) Cofilin takes the lead. J Cell Sci 118:19–26
Eda M, Yonemura S, Kato T, Watanabe N, Ishizaki T, Madaule P, Narumiya S (2001) Rho-dependent transfer of Citron-kinase to the cleavage furrow of dividing cells. J Cell Sci 114:3273–3284
Edwards DC, Sanders LC, Bokoch GM, Gill GN (1999) Activation of LIM-kinase by Pak1 couples Rac/Cdc42 GTPase signaling to actin cytoskeleton dynamics. Nat Cell Biol 1:253–259
Erb L, Liu J, Ockerhausen J, Kong Q, Garrad RC, Griffin K, Neal C, Krugh B, Santiago-Pérez LI, González FA, Gresham HD, Turner JT, Weisman GA (2001) An RGD sequence in the P2Y$_2$ receptor interacts with $\alpha_v\beta_3$ ntegrins and is required for G$_0$-mediated signal transduction. J Cell Biol 153:491–501
Erb L, Liao Z, Seye CI, Weismann GA (2006) P2 receptors: intracellular signaling. Eur J Physiol 452:552–562
Etienne-Manneville S, Hall A (2002) RhoGTPases in cell biology. Nature 420:629–635
Fukami K, Furuhashi M, Inagaki M, Endo T, Hatano S, Tekanawa T (1992) Requirement of phosphatidylinositol 4,5-biphosphate for alpha-actinin function. Nature 359:150–152
Fukata Y, Kimura K, Oshiro N, Saya H, Matsumura Y, Kaibuchi K (1998) Association of the myosin-binding subunit of myosin phosphatase and moesin: dual regulation of moesin phosphorylation by Rho-associated kinase and myosin phosphatase. J Cell Biol 141:409–418
Fukata Y, Amano M, Kaibuchi K (2001) Rho-Rho-kinase pathway in smooth muscle contraction and cytoskeletal reorganization of non-muscle cells. Trends Pharmacol 22:32–39
Ghosh M, Song X, Mouneimne G, Sidani M, Lawrence DS, Condeelis JS (2004) Cofilin promotes actin polymerization and defines the direction of cell motility. Science 304:743–746
Glaser M, Wanaski S, Buser CA, Boguslavsky V, Rashidzada W, Morris A, Rebecchi M, Scarlata SF, Runnels LW, Prestwich GD, Chen J, Aderem A, Ahn J, McLaughlin S (1996) Myristoylated alanine-rich C kinase substrate (MARCKS) produces reversible inhibition of phospholipase C by sequestering phosphatidylinositol 4,5-bisphosphate in lateral domains. J Biol Chem 271:26187–26193
Gilmore AP, Burridge K (1996) Regulation of vinculin binding to talin and actin by phosphatidylinositol 4,5-biphosphate. Nature 381:531–535
Glotzer M (2001) Animal cell cytokinesis. Annu Rev Cell Dev Biol 17:351–386
Greig AV, James SE, McGrouther DA, Terenghi G, Burnstock G (2003) Purinergic receptor expression in the regeneration epidermis in a rat model of normal and delayed wound healing. Exp Dermatol 12:860–871
Gruneberg U, Neef R, Li X, Chan EH, Chalamalasetty RB, Nigg EA, Barr FA (2006) KIF14 and citron kinase act together to promote efficient cytokinesis. J Cell Biol 172:363–372
Gupton SL, Anderson KL, Kole TP, Fischer RS, Ponti A, Hitchcock-DeGregori SE, Danuser G, Fowler VM, Wirtz D, Hanein D, Waterman-Storer CM (2005) Cell migration without a lamellipodium: translation of actin dynamics into cell movement mediated by tropomyosin. J Cell Biol 168:619–631
Hall A (1998) Rho GTPases and the actin cytoskeleton. Science 279:509–514
Harlan JE, Yoon HS, Hajduk PJ, Fesik SW (1995) Structural characterization of the interaction between a pleckstrin hamalogy domain and phosphatidylinositol 4,5-bisphosphate. Biochemistry 34:9859–9864
Hartshorne DL, Ito M, Erdodi F (1998) Myosin light chain phosphatase: subunit composition, interactions and regulation. J Muscle Res Cell Motil 19:325–341
Hartwig JH, Bokoch GM, Carpenter CL, Janmey PA, Taylor LA, Toker A, Stossel TP (1995) Thrombin receptor ligation and activated Rac uncap actin filament barbed ends through phosphoinositide synthesis in permeabilized human platelets. Cell 82:643–653
Hirao M, Sato N, Kondo T, Yonemura S, Monden M, Sasaki T, Takai Y, Tsukita Sh, Sa T (1996) Regulation mechanism of ERM (ezrin/redixin/moesin) protein/plasma membrane association: possible involvement of phosphatidylinositol turnover and Rho-dependent signaling pathway. J Cell Biol 135:37–51
Huang TY, DerMardirossian C, Bokoch GM (2006) Cofilin phosphatases and regulation of actin dynamics. Curr Opin Cell Biol 18:26–31
Huttelmaier S, Mayboroda O, Harbeck B, Jarchau T, Jockusch BM, Rudiger M (1998) The interaction of the cell-contact proteins VASP and vinculin is regulated by phosphatidylinositol 4,5-biphosphate. Curr Biol 8:479–488
Ichetovkin I, Grant W, Condeelis J (2002) Cofilin produces newly polymerized actin filaments that are preferred for dendritic nucleation by the Arp2/3 complex. Curr Biol 12:79–84
Itoh RE, Kurokawa K, Ohba Y, Yoshizaki H, Mochizuki N, Matsuda M (2002) Activation of Rac and Cdc42 video imaged by fluorescent resonance energy transfer-based single-molecule probes in the membrane of living cells. Mol Cell Biol 22:6582–6591
Kaibuchi K, Kuroda S, Amano M (1999) Regulation of the cytoskeleton and cell adhesion by the Rho family GTPases in mammalian cells. Annu Rev Biochem 68:459–486
Katoh K, Kano Y, Amano M, Onishi H, Kaibuchi K, Fujiwara K (2001) Rho-kinase-mediated contraction of isolated stress fibers. J Cell Biol 153:569–583
Kawano Y, Fukata Y, Oshiro N, Amano M, Nakamura T, Ito M, Matsumura F, Inagaki M, Kaibuchi K (1999) Phosphorylation of myosin-binding subunit (MBS) of myosin phosphatase by Rho-kinase *in vivo*. J Biol Cell 147:1023–1038

Kimura K, Ito M, Amano M, Chihara K, Fukata Y, Nakafuku M, Yamamori B, Feng J, Nakano T, Okawa K, Iwamatsu A, Kaibuchi K (1996) Regulation of myosin phosphatase by Rho and Rho-associated kinase (Rho-kinase). Science 273:245–248

Kłopocka W, Pomorski P (1996) Cytoplasmic calcium transients in *Amoeba proteus* during induction of pinocytotic and non-pinocytotic rosettes. Acta Protozool 35:169–175

Kłopocka W, Rędowicz MJ (2003) Effect of Rho family GTP-binding on *Amoeba proteus*. Protoplasma 220:163–172

Kłopocka W, Rędowicz MJ (2004) Rho/Rho-dependent kinase affects locomotion and actin-myosin II activity of *Amoeba proteus*. Protoplasma 224:113–121

Kłopocka W, Moraczewska J, Rędowicz MJ (2005) Characterisation of the Rac/PAK pathway in *Amoeba proteus*. Protoplasma 224:77–84

Klosses WB, Shattil SJ, Pampori N, Schwartz MA (2001) Rac recruits high-affinity integrin alphavbeta3 to lamellipodia in endothelial cell migration. Nat Cell Biol 3:316–320

Korczyński J, Sobierajska K, Krzemiński P, Wasik A, Wypych D, Pomorski P, Kłopocka W (2011) Is the MLC phosphorylation essential for the recovery from ROCK inhibition in glioma C6 cells? Acta Biochim Pol 58:125–130

Korczyński J, Sobierajska K, Pomorski P, Kłopocka W (in preparation) The role of $\alpha_v\beta_5$ integrins in glioma C6 cells recovery from ROCK inhibition

Kosako H, Yoshida T, Matsumura F, Ishizaki T, Narumiya S, Inagaki M (2000) Rho-kinase/ROCK is involved in cytokinesis through the phosphorylation of myosin light chain and not ezrin/radixin/moezin proteins at the cleavage furrow. Oncogene 19:6059–6064

Kozma R, Sarner S, Ahmed S, Lim L (1997) Rho family GTPases and neuronal growth cone remodeling: relationship between increased complexity induced by Cdc42Hs, Rac1, and acetylocholine and collapse induced by RhoA and lysophosphatidic acid. Mol Cell Biol 17:1201–1211

Liao Z, Seye CI, Weisman GA, Erb L (2007) The $P2Y_2$ nucleotide receptor requires interaction with αv integrins to access and activate G_{12}. J Cell Sci 120:1654–1662

Liu J, Liao Z, Camden J, Griffin KD, Garrad RC, Santiago-Pérez LI, Gonzáles FA, Seye CI, Weisman GA, Erb L (2004) Src homology 3 binding sites in the $P2Y_2$ nucleotide receptor interact with Src and regulate activities of Src, proline-rich tyrosine kinase 2, and growth factor receptors. J Biol Chem 279:8212–8218

Luo L (2000) Rho GTPases in neuronal morphogenesis. Nat Rev Neurosci 1:173–180

Lustig KD, Weisman GA, Turner JT, Garrad R, Shiau AK, Erb L (1996) P2U purinoreceptors: cDNA cloning, signal transduction mechanisms and structure-function analysis. In: Chadwick DJ, Goode JA (eds), P2 purinoceptors: localization, function and transduction mechanisms. Wiley, New York, p 193–204.

Mackay DJ, Esch F, Furthmayr H, Hall A (1997) Rho- and rac-dependent assembly of focal adhesion complexes and actin filaments in permeabilized fibroblasts: an essential role for ezrin/radixin/moesin proteins. J Cell Biol 138:927–938

Madaule P, Eda M, Watanabe N, FujisawaK MT, Bito H, Ishizaki T, Narumiya S (1998) Role of citron kinase as a target of the small GTPase Rho in cytokinesis. Nature 394:491–494

Maddox AS, Burridge K (2003) RhoA is required for cortical retraction and rigidity during mitotic cell rounding. J Cell Biol 160:255–265

Maekawa M, Ishizaki T, Boku S, Watanabe N, Fujita A, Iwamatsu A, Obinata T, Ohashi K, Mizuno K, Narumiya S (1999) Signaling from Rho to the actin cytoskeleton through protein kinases ROCK and LIM-kinase. Science 285:895–898

Matsui T, Maeda M, Doi Y, Yonemura S, Amano M, Kaibuchi K, Sa T, Tsukita Sh (1998) Rho-kinase phosphorylates COOH-terminal threonines of ezrin/radixin/moesin (ERM) proteins and regulates their head-to-tail association. J Cell Biol 140:647–657

Matsumura F (2005) Regulation of myosin II during cytokinesis in higher eukaryotes. Trends Cell Biol 15:371–377

Matsumura F, Ono S, Yamakita Y, Totsukawa G, Yamashiro S (1998) Specific localization of serine 19 phosphorylated myosin II during cell locomotion and mitosis of cultured cells. J Cell Biol 140:119–129

Mermall V, Post PL, Mooseker MS (1998) Unconventional myosin in cell movement, membrane traffic, and signal transduction. Science 279:527–533

Miki H, Miura K, Takenawa T (1996) N-WASP, a novel actin-depolymerizing protein, regulates the cortical cytoskeletal rearrangement in a PIP2-dependent manner downstream of tyrosine kinases. EMBO J 15:5326–5335

Miki H, Jamaguchi H, Suetsugu S, Takenawa T (2000) IRSp53 is an essential intermediate between Rac and WAVE in the regulation of membrane ruffling. Nature 408:732–735

Mouneimne G, Soon L, DesMarais V, Sidani M, Song X, Yip SC, Ghosh M, Eddy R, Backer JM, Condeelis J (2004) Phospholipase C and cofilin are required for carcinoma cell directionality in response to EGF stimulation. J Cell Biol 166:697–708

Moussavi RS, Kelley CA, Adelstein RS (1993) Phosphorylation of vertebrate nonmuscle and smooth muscle myosin heavy chains and light chains. Mol Cell Biochem 127–128:219–227

Niwa R, Nagata-Ohashi K, Takeichi M, Mizuno K, Uemura T (2002) Control of actin reorganization by Slingshot, a family of phosphatase that dephosphorylate of ADF/cofilin. Cell 108:233–246

Nobes CD, Hall A (1995) Rho, Rac and Cdc42 regulate the assembly of multimolecular focal complexes associated with actin stress fibers, lamellipodia and filopodia. Cell 81:53–62
Nobes CD, Hall A (1999) Rho GTPases control polarity, protrusion and adhesion during cell movement. J Cell Biol 144:1235–1244
Ohta Y, Kousaka K, Nagata-Ohashi K, Ohashi K, Muramoto A, Shima Y, Niwa R, Uemura T, Mizuno K (2003) Differential activities, subcellular distribution and tissue expression patterns of three members of Slingshot family phosphatases that dephosphorylate cofilin. Genes Cells 8:811–824
Pak CW, Flynn KC, Bamburg JR (2008) Actin-binding proteins take the reins in growth cones. Nat Rev Neurosci 9:136–147
Peterson TS, Camden JM, Wang Y, Seye CI, Wood WG, Sun GY, Erb L, Petris MJ, Weisman GA (2010) P2Y2 nucleotide receptor-mediated responses in brain cells. Mol Neurobiol 41:356–366
Piekny A, Werner M, Glotzer M (2005) Cytokinesis: welcome to the Rho zone. Trends Cell Biol 15:651–658
Pletjushkina OJ, Rajfur Z, Pomorski P, Oliver TN, Vasiliev JM, Jacobson KA (2001) Induction of cortical oscillations in spreading cells by depolymerization of microtubules. Cell Motil Cytoskeleton 48:235–244
Pollard TD, Blanchoin L, Mulins RD (2000) Molecular mechanisms controlling actin filament dynamics in nonmuscle cells. Annu Rev Biophys Biomol Struct 29:545–576
Pomorski P, Watson JM, Haskill S, Jacobson KA (2004) How adhesion, migration, and cytoplasmic calcium transients influence interleukin-1beta mRNA stabilization in human monocytes. Cell Motil Cytoskeleton 57:143–157
Pomorski P, Krzemiński P, Wasik A, Wierzbicka K, Barańska J, Kłopocka W (2007) Actin dynamics in *Amoeba proteus* motility. Protoplasma 231:31–41
Ramachandran C, Patil RV, Combrink K, Sharif NA, Srinivas SP (2011) Rho-Rho kinase pathway in the actomyosin contraction and cell-matrix adhesion in immortalized human trabecular meshwork. Mol Vis 17:1877–1890
Ridley AJ (2001) Rho GTPases and cell migration. J Cell Sci 114:2713–2722
Ridley AJ, Hall A (1992) The small GTP-binding protein rho regulates the assembly of focal adhesions and actin stress fibers in response to growth factors. Cell 70:389–399
Ridley AJ, Comoglio PM, Hall A (1995) Regulation of scatter factor/hepatocyte growth factor responses by Ras, Rac, and Rho in MDCK cells. Mol Cell Biol 15:1110–1122
Ridley AJ, Schwartz MA, Burridge K, Firtel RA, Ginsberg MH, Borisy G, Parsons JT, Horwitz AR (2003) Cell migration: integrating signals from front to back. Science 302:1704–1709
Riento K, Ridley AJ (2003) Rocks: multifunctional kinases in cell behaviour. Nat Rev Mol Cell Biol 4:446–456
Royal I, Lamarche-Vane N, Lamorte L, Kaibuchi K, Park M (2000) Activation of cdc42, rac PAK, and rho-kinase in response to hepatocyte growth factor differentially regulates cell colony spreading and dissociation. Mol Biol Cell 11:1709–1725
Salhia B, Rutten F, Nakada M, Beaudry C, Berens M, Kwan A, Rutka JT (2005) Inhibition of Rho-kinase affects astrocytoma morphology, motility, and invasion through activation of Rac1. Cancer Res 65:8792–8800
Sanders LC, Matsumura F, Bokoch GM, de Lanerolle P (1999) Inhibition of myosin light chain kinase by p21-activated kinase. Science 283:2083–2085
Schmitz AA, Govek EE, Bottner B, Van Aelst L (2000) Rho GTPases: signaling, migration, and invasion. Exp Cell Res 261:1–12
Sechi AS, Wehland J (2000) The actin cytoskeleton and plasma membrane connection: PtdIns(4,5)P_2 influences cytoskeletal protein activity at the plasma membrane. J Cell Sci 113:3685–3695
Shestakova EA, Wyckoff J, Jones J, Singer RH, Condeelis J (1999) Correlation of beta-actin messenger RNA localization with metastatic potential in rat adenocarcinoma cell lines. Cancer Res 59:1202–1205
Sidani M, Wessels D, Mouneimne G, Ghosh M, Goswami S, Sarmiento C, Wang W, Kuhl S, El-Sibai M, Backer JM, Eddy R, Soll D, Condeelis J (2007) Cofilin determines the migration behavior and turning frequency of metastatic cancer cells. J Cell Biol 179:777–791
Sohn RH, Chen J, Koblan KS, Bray PF, Goldschmidt-Clermont PJ (1995) Localization of a binding side for phosphatidylinositol 4,5-biphosphate on human profiling. J Biol Chem 270:21114–21120
Somlyo AP, Somlyo AV (2003) Ca^{2+} sensitivity of smooth muscle and nonmuscle myosin II: modulated by G proteins, kinases, and myosin phosphatase. Physiol Rev 83:1325–1358
Song X, Chen X, Yamaguchi H, Mouneimne G, Condeelis JS, Eddy RJ (2006) Initiation of cofilin activity in response to EGF is uncoupled from cofilin phosphorylation and dephosphorylation in carcinoma cells. J Cell Sci 119:2871–2881
Stockem W, Kłopocka W (1988) Ameboid movement and related phenomena. Int Rev Cytol 112:137–183
Sun CX, Magalhaes MAO, Glogauer M (2007) Rac1 and Rac2 differentially regulate actin free barbed end formation downstream of the fMLP receptor. J Cell Biol 179:239–245
Suplat-Wypych D, Dygas A, Baranska J (2010) 2′, 3′-O-(4-benzoylbenzoyl)-ATP-mediated calcium signaling in rat glioma C6 cells: role of the P2Y(2) nucleotide receptor. Purinergic Signal 6:317–325
Sutton TA, Mang HE, Atkinson SJ (2001) Rho-kinase regulates myosin activation in MDCK cells during recovery after ATP depletion. Am J Physiol Renal Physiol 281:F810–F818

Takenawa T, Itoh T (2001) Phosphoinositides, key molecules for regulation of actin cytoskeletal organization and membrane traffic from the plasma membrane. Biochim Biophys Acta 1533:190–206

Targos B, Pomorski P, Krzemiński P, Barańska J, Rędowicz MJ, Kłopocka W (2006) Effect of Rho-associated kinase inhibition on Actin cytoskeleton structure and calcium response in glioma C6 cells. Acta Biochim Pol 53:825–831

Toker A (2002) Phosphoinositides and signal transduction. Cell Mol Life Sci 59:761–779

Tolias KF, Hartwig JH, Ishihara H, Shibasaki Y, Cantley LC, Carpenter CL (2000) Type Ialpha phosphatidylinositol -4-phasphate 5-kinase mediates Rac-dependent actin assembly. Curr Biol 10:153–156

Totsukawa G, Yamakita Y, Yamashiro S, Hartshorne DJ, Sasaki Y, Matsumura F (2000) Distinct roles of ROCK (Rho-kinase) and MLCK in spatial regulation of MLC phosphorylation for assembly of stress fibers and focal adhesions in 3T3 fibroblasts. J Cell Biol 150:797–806

Tsuji T, Ishizaki T, Okamoto M, Higashida C, Kimura K, Furuyashiki T, Arakawa Y, Birge RB, Nakamoto T, Hirai H, Narumiya S (2002) ROCK and mDia1 antagonize in Rho-dependent Rac activation in Swiss 3T3 fibroblasts. J Cell Biol 157:819–830

Van Aelst L, D'Souza-Schorey C (1997) Rho GTPases and signaling network. Genes Dev 11:2295–2322

van Leeuwen FN, Kain HE, Kammen RA, Michiels F, Kranenburg OW, Collard JG (1997) The guanine nucleotide exchange factor Tiam1 affects neuronal morphology: opposing rols for the small GTPases Rac and Rho. J Cell Biol 139:797–807

van Leeuwen FN, van Delft S, Kain HE, van der Kammen RA, Collard JG (1999) Rac regulates phosphorylation of the myosin-II heavy chain, actinomyosin disassembly and cell spreading. Nat Cell Biol 1:242–248

van Rhee AM, Jacobson KA, Garrad R, Weisman GA, Erb L (1998) P2 receptor modeling and identification of ligand binding sites. In: Turner JT, Weisman GA, Fedan JS (eds) P2 nucleotide receptors. Humana Press, Totowa, pp 135–166

van Rheenen J, Song X, van Roosmalen W, Cammer M, Chen X, DesMarais V, Yip S-C, Backer JM, Eddy RJ, Condeelis JS (2007) EGF-induced PIP2 hydrolysis releases and activates cofilin locally in carcinoma cells. J Cell Biol 179:1247–1259

van Rheenen J, Condeelis J, Glogauer M (2009) A common cofilin activity cycle in invasive tumor cells and inflammatory cells. J Cell Sci 122:305–311

Wang M, Kong Q, Gonzalez FA, Sun G, Erb L, Seye C, Weisman GA (2005) $P2Y_2$ nucleotide receptor interaction with α_v integrin mediates astrocyte migration. J Neurochem 95:630–640

Watanabe N, Kato T, Fujita A, Ishizaki T, Narumiya S (1999) Cooperation between mDia1 and ROCK in Rho-induced actin reorganization. Nat Cell Biol 1:136–143

Welch MD, Mullins RD (2000) Cellular control of actin nucleation. Annu Rev Cell Dev Biol 18:247–288

Wong WT, Faulkner-Jones B, Sanes JR, Wong RO (2000) Rapid dendritic remodeling in the developing retina: dependence on neurotransmission and reciprocal regulation by Rac and Rho. J Neurosci 20:5024–5036

Worthylake RA, Lemoine S, Watson JM, Burridge K (2001) RhoA is required for monocyte tail retraction during transendothelial migration. J Cell Biol 154:147–160

Yamashiro S, Totsukawa G, Yamakita Y, Sasaki Y, Madaule P, Ishizaki T, Narumiya S, Matsumura F (2003) Citron kinase, a Rho-dependent kinase, induces diphosphorylation of regulatory light chain of Myosin II. Mol Biol Cell 14:1745–1756

Yang N, Higuchi O, Ohashi K, Nagata K, Wada A, Kangawa K, Nishida E, Mizuno K (1998) Cofilin phosphorylation by LIM-kinase1 and its role in Rac-mediated actin reorganization. Nature 393:809–812

Chapter 7
Signaling Determinants of Glioma Cell Invasion

Aneta Kwiatkowska and Marc Symons

Abstract Tumor cell invasiveness is a critical challenge in the clinical management of glioma patients. In addition, there is accumulating evidence that current therapeutic modalities, including anti-angiogenic therapy and radiotherapy, can enhance glioma invasiveness. Glioma cell invasion is stimulated by both autocrine and paracrine factors that act on a large array of cell surface-bound receptors. Key signaling elements that mediate receptor-initiated signaling in the regulation of glioblastoma invasion are Rho family GTPases, including Rac, RhoA and Cdc42. These GTPases regulate cell morphology and actin dynamics and stimulate cell squeezing through the narrow extracellular spaces that are typical of the brain parenchyma. Transient attachment of cells to the extracellular matrix is also necessary for glioblastoma cell invasion. Interactions with extracellular matrix components are mediated by integrins that initiate diverse intracellular signalling pathways. Key signaling elements stimulated by integrins include PI3K, Akt, mTOR and MAP kinases. In order to detach from the tumor mass, glioma cells secrete proteolytic enzymes that cleave cell surface adhesion molecules, including CD44 and L1. Key proteases produced by glioma cells include uPA, ADAMs and MMPs. Increased understanding of the molecular mechanisms that control glioma cell invasion has led to the identification of molecular targets for therapeutic intervention in this devastating disease.

Keywords Glioblastoma • Invasion • Migration • Rho GTPase • PI3K • Akt • Protease

Abbreviations

ATX	Autotaxin
BEHAB	Brain-enriched hyaluronic acid binding protein
DG	Dentate gyrus
DOCK180	Dedicator of cytokinesis 180
ECM	Extracellular matrix

A. Kwiatkowska (✉) • M. Symons
Center for Oncology and Cell Biology, Laboratory for Brain Tumor Biology,
Feinstein Institute for Medical Research, 350 Community Dr, 11030 Manhasset, NY, USA
e-mail: akwiatkows@nshs.edu; msymons@nshs.edu

J. Barańska (ed.), *Glioma Signaling*, Advances in Experimental Medicine and Biology 986,
DOI 10.1007/978-94-007-4719-7_7, © Springer Science+Business Media Dordrecht 2013

ELMO1	Engulfment and cell motility-1
Gab1	Grb-2 associated binder-1
GAP	GTPase activating protein
GBM	*Glioblastoma multiforme*
GDI	Guanine nucleotide dissociation inhibitor
GEF	Guanine nucleotide exchange factor
HGG	High grade glioma
LGG	Low grade glioma
LPA	Lysophosphatidic acid
LPC	Lysophosphatidylcholine
MCP-1	Monocyte chemotactic protein-1
mDia	Mammalian homolog of Drosophila diaphanous
MMP	Matrix metalloproteinase
MT1-MMP	Membrane type metalloproteinase 1
NB	Non-neoplastic brain
NSC	Neural stem cell
PDGFR	Platelet-derived growth factor receptor
PI3K	Phosphatidylinositol 3-kinase
PDK1	PI3K-dependent kinase 1
PH	Pleckstrin homology domain
PTEN	Phosphatase and tensin homolog deleted on chromosome ten
$p130^{Cas}$	Crk-associated substrate
ROCK	Rho-associated coiled-coil forming kinase
RTK	Receptor tyrosine kinase
SPARC	Secreted protein acidic and rich in cystein
SVZ	Subventricular zone
TAM	Tumor associated macrophage
TGFβ-1	Transforming growth factor-β-1
TIMP	Tissue inhibitor of metalloproteinases
TN-C	Tenascin-C
TSP-1	Thrombospondin-1
VEGF	Vascular endothelial growth factor
VEGFR-1	Vascular endothelial growth factor receptor-1

7.1 Invasiveness of Glioma Cells

Malignant gliomas are characterized by a high proliferation rate, increased angiogenesis and diffusive growth. There is rarely a clear border between the tumor and the surrounding brain parenchyma. This complicates complete surgical resection, and as a consequence, usually within months after surgery, recurrent neoplasms are established in the proximity of the resection zone.

The pattern of glioma cell migration in the brain is not random. Tumor cells infiltrate the brain parenchyma as individual cells or isolated clusters, distributed mainly along blood vessels (perivascular zone), fiber tracts and subependyma (Farin et al. 2006; Giese et al. 2003; Scherer 1940). Although it has been shown that C6 rat glioma cells can intercalate between endothelial cells and astrocyte end feet, or in some cases displace astrocytes from entothelial cells, they rarely invade the blood vessel lumen (Farin et al. 2006). This is consistent with the well-established clinical observation that gliomas hardly metastasize to other organs or the spinal cord (Armstrong et al. 2011; Birbilis et al. 2010; Gotway et al. 2011; Schonsteiner et al. 2011).

In the resting adult brain, migrating neural stem cells (NSC) mainly originate from two niches, the subventricular zone (SVZ) and the dentate gyrus (DG) of the hippocampus. From there, these neural progenitors migrate towards the olfactory bulb (OB) or granular cell layer of the DG, respectively. There is evidence that the adult SVZ continues to generate glial progenitor cells. However, the vast majority of these progenitor cells reside outside the neurogenic niches and usually do not migrate (Cayre et al. 2009). Progenitor cell migration can be stimulated however by pathological conditions such as inflammation or stroke (Cayre et al. 2009; Zhang et al. 2005). Time-lapse microscopy analysis of neural and glial progenitor migration revealed that these cells are moving in a unique two-step process: continuous extension of long leading protrusions followed by saltatory movement of the cell body (Bellion et al. 2005; Cayre et al. 2009; Kakita and Goldman 1999).

Interestingly, the pattern of glioma cell migration strongly resembles the pattern of glial progenitor cell migration during normal brain development (Beadle et al. 2008; Farin et al. 2006; Kakita and Goldman 1999). Moreover, it was shown that the saltatory mode of migration reflects the requirement for the nucleus to squeeze through the small extracellular spaces that characterize the brain parenchyma (Beadle et al. 2008). This nuclear squeezing is dependent on myosin-based contractility, as it is inhibited by both, blebbistatin, an inhibitor of myosin II, and Y27632, a small molecule inhibitor of Rho-associated coiled-coil forming kinase (ROCK), a Rho effector protein that controls myosin II activation. Thus, studying the mechanisms that drive progenitor cell migration during brain development should facilitate our understanding of the signaling pathways that are involved in glioma cell dissemination.

There is growing evidence that antiangiogenic therapy prolongs progression-free survival (Norden et al. 2009). Unfortunately, treatment with bevacizumab (an antibody against vascular endothelial growth factor, VEGF) or cediranib (a VEGF receptor tyrosine kinase inhibitor) have resulted in little improvement in overall survival (de Groot et al. 2010). It has been reported that bevacizumab treatment results in a shift to more infiltrative tumor growth (de Groot et al. 2010; Lucio-Eterovic et al. 2009). This behavior is recapitulated in experiments using a model of primary human glioblastoma cells in the rat, in which treatment with bevacizumab was accompanied by a strong increases in the number of invading glioblastoma cells and in the distance that they travel from the tumor core (Keunen et al. 2011). Thus, combining bevacizumab treatment with an anti-invasion therapy may be beneficial.

7.2 Factors That Control Glioma Invasion

Both the development of glioma as well as their invasive behavior is strongly controlled by the local microenvironment. Factors secreted by tumor cells diffuse into the peritumoral stroma affecting the local tissue. In response, cells in brain parenchyma secrete ligands that stimulate enhanced glioma invasion and/or change the local microenvironment into a more permissive one for tumor progression (Hoelzinger et al. 2007).

7.2.1 Autocrine Factors

Cells residing in the brain are embedded in extracellular matrix (ECM) primarily composed of hyaluronan and proteoglycans. The latter include brevican (brain enriched hyaluronic acid binding protein), neurocan, as well as the glycoproteins SPARC (secreted protein acidic and rich in cystein, also known as osteonectin), tenascin-C (TN-C) and thrombospondin-1 (TSP-1). Collagens, laminins and fibronectins, which are widely found in other tissues, are present only in the proximity of blood vessels in the brain (Bellail et al. 2004). Importantly, overexpression of hyaluronan, vitronectin, osteopontin, tenascin-C and BEHAP correlates with tumor grade (Delpech et al. 1993; Higuchi et al. 1993; Jaworski et al. 1996; Mahesparan et al. 2003; Saitoh et al. 1995; Toy et al. 2009; Viapiano et al. 2003). ECM components play an important role in the regulation of signaling pathways that are responsible for tumor growth, proliferation, adhesion, migration and angiogenesis (Akiyama et al. 2001; Higuchi et al. 1993; Matusan-Ilijas et al. 2008; Zagzag et al. 1995, 1996). Therefore, it is likely that glioma progression is in part mediated by alterations in ECM composition. This is illustrated by the finding that experimental inhibition of osteopontin expression by U87 glioblastoma cells causes a significant reduction in the number of migrating cells *in vitro* and slower tumor growth *in vivo*, as knock down of osteopontin in U87 cells reduce the proliferation of cells within experimental glioma tumors (Lamour et al. 2010).

Glioma cells also secrete factors that, upon binding to their cognate receptor tyrosine kinases, contribute to enhanced tumor cell proliferation and motility. Important factors are epidermal growth factor (EGF), transforming growth factor α (TGFα), heparin-binding epidermal growth factor (HB-EGF), platelet derived growth factor (PDGF) and hepatocyte growth factor/scatter factor (HGF/SF) (Brockmann et al. 2003; Hoelzinger et al. 2007; Koochekpour et al. 1997; Ramnarain et al. 2006; Shih and Holland 2006). A detailed discussion about receptor tyrosine kinase signaling in glioma can be found in other chapters in this book (mainly in Chap. 8).

In addition to conventional autocrine signaling, recent data imply bioactive phospholipids in the regulation of glioblastoma dissemination. A good example is autotaxin (ATX), an enzyme with lysophospholipase D activity that converts lysophosphatidylcholine (LPC) into lysophosphatidic acid (LPA). Glioblastoma cells *in vivo* are exposed to various plasma components when the blood brain barrier (BBB) is disrupted (Seitz and Wechsler 1987; Wolff and Boker 1989). One such component is LPC, which is present in plasma at a high concentration (100–300 μM) (Kishimoto et al. 2002). Interestingly, in most analyzed glioblastoma tissues and glioma cell lines, autotaxin and LPA_1 receptor are highly expressed (Kishi et al. 2006). In addition, there is more ATX expressed in glioblastoma cells in the invading rim in comparison to those in the tumor core (Hoelzinger et al. 2005). Experimental overexpression of ATX enhances cell migration both *in vitro* and in *ex vivo* brain slices (Kishi et al. 2006; Hoelzinger et al. 2008). Conversely, inhibition of ATX expression leads to decreased invasiveness of cells in a three-dimensional collagen spheroid invasion assay in response to LPC (Hoelzinger et al. 2008). Taken together, these data strongly suggest a role for ATX in glioblastoma invasion.

7.2.2 Paracrine Factors

Microglia and macrophages can constitute up to 30 % of the total number of cells in glioblastomas, anaplastic astrocytomas and rodent gliomas (Badie et al. 2002; Badie and Schartner 2000; Charles et al. 2011; Roggendorf et al. 1996). For many years, tumor-associated macrophages (TAMs) were considered as a part of the immune response against the tumor or a nonspecific reaction evoked by local damage (Badie et al. 2002; Flugel et al. 1999; Watters et al. 2005). Although the role of microglia and TAMs in brain tumors is not fully understood, recent studies suggest that microglia/macrophages may be attracted by tumor-secreted factors such as monocyte chemotactic protein-1 (MCP-1) (Platten et al. 2003), in order to promote tumor growth and dissemination into the brain parenchyma (Badie and Schartner 2001; Charles et al. 2011; Markovic et al. 2009; Platten et al. 2003; Sliwa et al. 2007; Watters et al. 2005; Wesolowska et al. 2008; Zhai et al. 2011). It is thought that microglia, in response to glioma stimulation, produce diverse factors, including matrix metalloproteinases such as membrane type matrix metalloproteinase-1 (MT1-MMP) that contribute to ECM degradation and the processing of growth factors (Markovic et al. 2009). Moreover, microglia produce cytokines such as transforming growth factor β-1 (TGFβ-1) that promote tumor cell proliferation and migration (Watters et al. 2005; Wesolowska et al. 2008). In addition, TAMs have been shown to activate NF-κB transcription factor-dependent production of interleukin-8 (IL-8) by gliomas (Hong et al. 2009). This chemokine also stimulates tumor cell migration (Wakabayashi et al. 2004). Interestingly, T cells are rarely seen in gliomas (Morimura et al. 1990) which correlates well with compromised microglia-mediated antigen presentation in these tumors (Badie et al. 2002; Flugel et al. 1999). In conclusion, the current literature suggests that glioblastoma tumors re-educate microglia/macrophages from an inflammatory phenotype to an anti-inflammatory and pro-tumor phenotype (Gabrusiewicz et al. 2011).

The presence of a neoplasm in the brain affects the function of parenchymal cells. As mentioned above, glioma cells stereotactically inoculated into the rat or mouse brain accumulate around blood vessels, where they displace astrocytic end feet from the endothelial cells. Astrocytes that withdraw their processes from the vascular wall become reactive (Nagano et al. 1993; Zagzag et al. 2000). Reactive astrocytes have been shown to secrete urokinase-type plasminogen activator (uPA) *in vitro*. uPA is a serine protease that converts plasminogen produced by glioma cells into active plasmin. Plasmin, in turn, activates pro-metalloproteinase-2 (pro-MMP-2) that is secreted by astrocytes (Le et al. 2003), thereby increasing local proteolytic activity. Thus, brain tumor dissemination is complex and relies on interactions between several cell types in a way that is not readily recapitulated using *in vitro* assays.

Expression of chemokine receptor 4 (CXCR4) and its ligand chemokine ligand 12 (CXCL12, also known as SDF-1) is increased in human astrocytomas (Barbero et al. 2002; Bajetto et al. 2006) and significantly more CXCR4 is expressed in invasive tumor foci as compared to the non-invasive tumor core (Ehtesham et al. 2006). Moreover, SDF-1 produced by the endothelium has been shown to stimulate U87 glioblastoma invasion *in vitro*. Under these conditions, glioblastoma cells produce more MMP-9 and cathepsins (another class of proteases, see below Sect. 7.4) (Kenig et al. 2010), enzymes that promote glioma cell invasion by cleavage of ECM components as well as activation of pro-enzymes present in the extracellular space (Kobayashi et al. 1991; Mai et al. 2002). The observations that inhibiting CXCR4, either by CXCR4-neutralizing antibodies, CXCR4-directed siRNA technology or AMD3100, a CXCR4-specific inhibitor, impaired glioma cell invasion *in vitro*, strongly supports a role for SDF-1 in the invasive behavior of glioma (Ehtesham et al. 2006; Hong et al. 2006).

7.3 Signaling Mechanisms That Control Glioma Invasion

7.3.1 Integrins

Cell migration requires anchoring of the leading edge to the ECM and release of cell attachment at the rear (Ridley et al. 2003). Integrins are heterodimeric transmembrane receptors, composed of α and β subunits. There are 8 β and 18 α subunits, combinations of which determine substrate specificity (D'Abaco and Kaye 2007). Integrins participate in bidirectional signaling across the plasma membrane. Integrins can be stimulated to bind ligands by intracellular signaling (inside-out signaling) or become activated upon interaction with extracellular ligands (outside-in signaling). Conformational changes within integrins enable association with and activation of diverse cytoplasmic adaptor proteins (Fagerholm et al. 2004; Hynes 2002). Integrins lack catalytic activity, and recruitment of these adaptor proteins is thought to be regulated by phosphorylation of the integrin cytoplasmic tails (D'Abaco and Kaye 2007; Fagerholm et al. 2004).

Integrins have been proposed to play a key role in glioma biology including cell migration (D'Abaco and Kaye 2007). Many studies have shown overexpression of the β1 subunit in malignant gliomas in comparison to normal brain tissue and β1-blocking antibodies decrease glioma cell migration and invasion *in vitro* (Paulus et al. 1993; Rooprai et al. 1999; Tysnes et al. 1996). Experimental overexpression of α6 subunit in U87 glioma cells bearing a high level of β1 subunits increases glioma cell migration and invasion *in vitro*, enhances dissemination of glioma cells *in vivo* and the formation of infiltrative foci at the margin of tumors established in nude mice (Delamarre et al. 2009). In addition, immunohistochemistry studies have demonstrated that αvβ3 and αvβ5 are overexpressed in both glioma cells and tumor vasculature and that their expression is correlated with tumor grade (Bello et al. 2001a; Stupp and Ruegg 2007). Interestingly, stimulation of human glioblastoma cells with either TGF-β1 or TGF-β2 leads to an increase of αvβ3 at the cell surface and enhanced glioma cell migration (Platten et al. 2000). TGF-β-dependent stimulation of glioblastoma cell migration was shown to be abrogated by echistatin (a 49 amino acid peptide that binds integrins and blocks downstream signaling) or an αvβ3 neutralizing antibody (Platten et al. 2000).

αvβ3 and αvβ5 are the first integrins targeted to suppress tumor angiogenesis. Three classes of integrin inhibitors are under investigation: monoclonal antibodies targeting the extracellular domain of αvβ3 (e.g. vitaxin, phase II clinical trials completed) (Stupp and Ruegg 2007; Tucker 2006), synthetic peptides containing an RGD sequence recognized by both αvβ3 and αvβ5 (e.g. cilengitide, phase II clinical trials completed) (Gilbert et al. 2012; Reardon et al. 2008; Tucker 2006), and an RGD peptidomimetic antagonist of αvβ3 (e.g. S247) (Abdollahi et al. 2005). Clinical evidence shows modest antitumor activity of cilengitide (Gilbert et al. 2012; Reardon et al. 2008). However, as for other antiangiogenic drugs, the targeting of integrins may be most effective in combination with other therapeutic modalities, such as radiotherapy, especially as αvβ3 expression in endothelial cells is increased by radiation (Abdollahi et al. 2005; Gilbert et al. 2012; Stupp and Ruegg 2007).

7.3.2 Rho GTPases

Rho GTPases constitute a family of 22 members in humans and regulate a large number of the cellular functions, such as actin organization, cell migration, invasion (Burridge and Wennerberg 2004; Bustelo et al. 2007; Schmitz et al. 2000), cell proliferation and survival (Croft and Olson 2006; Gomez del Pulgar et al. 2005). They switch between the GTP-bound active state and the GDP-bound inactive state. The GTPase cycle is tightly controlled by three groups of regulators: guanine nucleotide exchange factors (GEFs), GTPase activating proteins (GAPs) and guanine nucleotide dissociation inhibitors (GDIs) (Rossman et al. 2005; Schmidt and Hall 2002). Rho GTPases are activated by GEFs which promote exchange of GDP to GTP. Subsequently, GTPases are inactivated by interaction with GAPs, that stimulate intrinsic GTP hydrolysis (Moon and Zheng 2003). RhoGDIs play a dual role in the regulation of Rho proteins (Garcia-Mata et al. 2011). On the one hand, they inhibit spontaneous release of GDP, thus clamping GTPases in the inactive state, on the other, by binding to the prenyl groups of GTPases, they prevent association with and facilitate extraction of GTPases from membranes. Interaction with additional proteins, called "GDI displacement factors" releases the GDI from the GTPase, thereby facilitating access to GEFs (Dransart et al. 2005).

RhoA, Rac1 and Cdc 42 are the best characterized of the Rho GTPases (Heasman and Ridley 2008; Raftopoulou and Hall 2004). Rac1 controls the formation of lamellipodia, which are flat, actin-rich membrane protrusions at the cell periphery. RhoA regulates actomyosin contractility, formation of focal adhesions and stress fibers, and retraction of the tail of the cell during cell migration. Cdc42 regulates the formation of filopodia (thin, finger-like membrane protrusions) and is a key control element in the regulation of cell polarization.

Rac proteins (Rac1, Rac2 and Rac3) are highly homologous, but differ in their tissue distribution (Heasman and Ridley 2008). Rac1 is ubiquitously expressed, Rac2 is specific for hematopoietic cells and Rac3 is abundantly expressed in neural tissues (Burridge and Wennerberg 2004). Although no significant change was observed in Rac1 mRNA expression levels across astrocytoma grades, in a large set of glioblastoma tumors, Rac1 mRNA was shown to be elevated in tumors of patients with shorter survival (Salhia et al. 2008). Interestingly, immunohistochemical analysis revealed a strong increase in Rac1 protein expression with tumor grade (Salhia et al. 2008). These findings are consistent with an earlier proteomic study showing that Rac1 protein levels are increased in high-grade (85 %) versus low-grade (20 %) gliomas and correlate with poor survival (Iwadate et al. 2005). Taken together, these data suggest that Rac1 is regulated at the translational and/or protein stability level. Notably, Rac1 has been shown to display marked plasma membrane localization in a fraction of glioblastoma tumor samples, but not in low grade astrocytomas (Salhia et al. 2008). Plasma membrane localization of Rac1 reflects a high activation state, suggesting that this GTPase may contribute to the malignant behavior of glioblastomas.

siRNA-mediated depletion of Rac1 or Rac3 significantly decreases glioblastoma cell invasion in a Matrigel invasion assay (Chan et al. 2005). Interestingly, in contrast to depletion of Rac1, Rac3 depletion only slightly inhibits glioblastoma migration, implying that Rac1 and Rac3 may be involved in different mechanism that contribute to cell invasion (Chan et al. 2005). Depletion of Rac1 also significantly inhibits glioblastoma cell invasion in *ex vivo* brain slices (Chuang et al. 2004), underlining the importance of this GTPase in glioblastoma invasiveness.

There are a number of Rac effectors (proteins that bind to active Rac and relay its functions) that control cell migration and invasion (Burridge and Wennerberg 2004), although their role in glioma invasion still largely remains to be characterized. The Rac effector synaptojanin 2 (Malecz et al. 2000), a phosphatidylinositol 5-phosphatase, has been shown to regulate glioblastoma cell migration and invasion *in vitro*. Synaptojanin 2 localizes to both invadopodia and lamellipodia and is thought to control the formation of these structures (Chuang et al. 2004).

In addition to overexpression of Rho GTPases, aberrant expression or genetic alterations of upstream regulators has been detected in a variety of human cancer types (Gomez del Pulgar et al. 2005). In particular, the Rho GEFs Ect2, Vav3, Trio and SWAP-70 display increased expression at the message level in brain tumors when compared to normal brain tissue and expression is correlated with poor patient survival (Salhia et al. 2008; Seol et al. 2009; Tu et al. 2010). Notably, depletion of any of these GEFs significantly inhibits glioblastoma cell migration and invasion (Salhia et al. 2008; Seol et al. 2009).

Dedicator of cytokinesis 180 (DOCK180) and engulfment and cell motility 1 (ELMO) form a bipartite GEF that activates Rac proteins (Cote and Vuori 2007; Lu and Ravichandran 2006). Depletion of either ELMO1 or DOCK180 strongly reduces Rac1 activation and glioblastoma cell invasion (Jarzynka et al. 2007). Interestingly, both DOCK180 and ELMO proteins display increased expression in the invading tumor rim compared to the tumor core of human glioma specimens.

An additional mechanism that could contribute to an increase in Rho GTPase activation levels in cancer is the downregulation of GAPs. One example in gliomas is the Rac GAP β2-chimaerin, which shows high levels in normal brain and low-grade astrocytomas in comparison to malignant gliomas (Yuan et al. 1995). Thus, loss of β2-chimaerin may contribute to the increase in Rac activation levels in glioblastoma.

There are two major effector proteins of RhoA, RhoB and RhoC GTPases: mammalian homolog of Drosophila diaphanous (mDia) and Rho-associated coiled-coil forming kinase (ROCK) (Narumiya et al. 2009; Wheeler and Ridley 2004). mDia belongs to the formin protein family and catalyzes actin nucleation and polymerization. mDia depletion in glioma cells interferes with microtubule stabilization, cell polarization, focal adhesion turnover and results in attenuated cell migration (Yamana et al. 2006). Analysis of the underlying mechanism revealed that mDia controls actin-dependent c-Src recruitment to focal adhesions, phosphorylation of the adaptor protein Crk-associated substrate ($p130^{Cas}$), Rac activation and subsequent focal adhesion disassembly. mDia also promotes the accumulation of Cdc42 and adenomatous polyposis coli (APC) at the front of the cell, which may provide a mechanism for the role of mDia in cell polarization.

ROCK phosphorylates and inactivates myosin phosphatase and directly phosphorylates myosin light chain. These two actions of ROCK stimulate actomyosin contractility. ROCK also phosphorylates and activates LIM kinase that in turn phosphorylates and inactivates the actin filament severing protein cofilin. As mentioned above, ROCK has been implicated in nuclear squeezing and glioblastoma cell invasion in *ex vivo* brain slices (Beadle et al. 2008). ROCK is also necessary for glioma cell migration along myelinated retinal axons *in vitro*, where ROCK inhibition was shown to result in shorter lamellipodia and non-polarized extension of filopodia (Oellers et al. 2009). In contrast, inhibition of ROCK has been shown to increase the migratory behavior of glioma cells under less physiological conditions, such as a two-dimensional cell migration in a radial migration assay and a transwell Matrigel invasion assay (Salhia et al. 2005), underlining the importance of physiologically relevant experimental conditions to assess the role of actomyosin contractility in cell invasion.

7.3.3 PI3K and Phospholipid Signaling

Phosphatidylinositol 3-kinases are lipid kinases that phosphorylate the 3 position of the inositol ring of phosphatidylinositols and phosphoinositides (Endersby and Baker 2008). Class IA PI3Ks are heterodimers composed of a regulatory subunit (five isoforms encoded by three genes p85α, p55α, p50α (*PIK3R1*), p85β (*PIKR2*) and p55γ (*PIKR3*)) and a catalytic subunit (p110α, p110β and p110δ). The three catalytic subunit isoforms are encoded by the *PIK3CA*, *PIK3CB* and *PIK3CD* genes (Furnari et al. 2007).

PI3Ks are regulated through the inhibitory effect of the regulatory subunit on the catalytic subunit (Yu et al. 1998). Upon direct binding of the regulatory subunit to phosphorylated receptor tyrosine kinases (RTKs), including c-Met, VEGFR and PDGFR (Escobedo et al. 1991; Igarashi et al. 1998; Ponzetto et al. 1993), p110 inhibition is released (Bader et al. 2005).

PI3K is also regulated by Src family kinases. An interesting example is that binding of CD95L (also known as FasL) to CD95 on glioblastoma cells recruits Yes (a member of Src kinase family) to the receptor. Yes, in turn, recruits the p85 subunit of PI3K leading to PI3K activation and enhanced glioblastoma cells migration *in vitro*. Interestingly, neutralization of CD95L in a murine intracranial model of GBM as well as treatment of C6 glioma cells with FasL Interfering Protein (FIP) reduces the number of invading cells (Kleber et al. 2008; Wisniewski et al. 2010). It is remarkable that CD95L, which promotes apoptosis in many cells, also stimulates their invasiveness. Notably, in general, glioblastoma tumors are resistant to CD95-induced apoptosis and even induce CD95L expression in the surrounding tissue. These findings suggest that targeting the CD95L/CD95 system may be beneficial in the treatment of glioblastoma tumors, as no CD95 is expressed in the healthy brain (Kleber et al. 2008).

Class I PI3Ks preferentially convert phosphatidylinositol-4,5-bisphosphate ($PI(4,5)P_2$) to phosphatidylinositol-3,4,5-trisphosphate ($PI(3,4,5)P_3$) (Martelli et al. 2010). The formation of $PI(3,4,5)P_3$ and $PI(3,4)P_2$ triggers the recruitment of proteins with pleckstrin homology (PH) domains to the plasma membrane, including PI3K-dependent kinase 1 (PDK1), Akt (Stambolic and Woodgett 2006) and GEFs (Fleming et al. 2000; Shinohara et al. 2002). A negative regulator of PI3K signaling is phosphatase and tensin homolog deleted on chromosome ten (PTEN). PTEN-mediated hydrolysis of $P(3,4,5)P_3$ (Stambolic et al. 1998) counteracts PI3K-dependent stimulation of many cellular functions including cell survival, proliferation and invasion.

Deregulation of PI3K function leads to tumorigenesis (Jaiswal et al. 2009). This observation is important in glioma biology as RTKs are often overexpressed or mutated in gliomas (Kapoor and O'Rourke 2003; Parsons et al. 2008; TCGA 2008). PTEN is also extensively deregulated in glioblastomas. PTEN mutations occur in 25 % of patients and loss of 10q, which includes PTEN, occurs in 70 % of patients (Endersby and Baker 2008). Epigenetic gene silencing by methylation of the PTEN promoter also has been reported (Baeza et al. 2003). Notably, GBM patients with inactivated PTEN have a shorter survival time (Ermoian et al. 2002).

Additionaly, PI3K signaling is deregulated in glioblastomas as a consequence of genetic aberrations within the *PI3K* gene. Mutations in *PIK3CA* have been described in primary human glioblastomas from both adult and pediatric patients with a frequency between 7 and 20 % (Gallia et al. 2006; Parsons et al. 2008; TCGA 2008). Several of these mutations have been shown to be kinase activating in other human malignancies. Interestingly, mutations in the *PIK3R1* gene have been found in 8–10 % of studied glioblastomas (Parsons et al. 2008; TCGA 2008). The clustering of these mutations around residues that serve as contact points with p110 strongly suggest that these mutations may relieve the inhibitory effect of p85α on p110α, thereby rendering PI3K constitutively active (TCGA 2008).

Importantly, a chemical inhibitor of PI3K (PX-866) has been shown to suppress glioblastoma cells invasiveness and VEGF secretion and diminished tumor growth, thereby prolonging animal survival (Koul et al. 2010). Thus, PI3K inhibitors may be beneficial for patients who overexpress the PI3K or carry an activating mutation in *PIK3CA* or *PIK3R1* genes.

7.3.4 Akt Kinase

The Akt family of kinases comprises Akt1 (PKBα), Akt2 (PKBβ) and Akt3 (PKBγ) that are encoded by three independent genes. All isoforms of Akt kinases are activated in a PI3K-dependent manner (Matheny and Adamo 2009). PI(3,4,5)P_3 and PI(3,4)P_2 recruit Akt from the cytoplasm to the plasma membrane via the N-terminal PH domain of the kinase. Membrane recruitment of Akt results in conformational changes that permit its subsequent phosphorylation and full activation (Franke et al. 1995; King et al. 1997; Matheny and Adamo 2009; Milburn et al. 2003). First, mammalian target of rapamycin complex 2 (mTORC2) phosphorylates Ser-473 in the activation loop of Akt (Huang and Manning 2009; Partovian et al. 2008; Sarbassov et al. 2005), that facilitates subsequent PDK1-dependent phosphorylation of Thr-308 (Wick et al. 2000). mTOR participates in two signaling complexes, mTORC1 and mTORC2.

Interestingly, Akt signaling is governed by negative feedback loop interactions. Akt activates mTORC1 by phosphorylation and inactivation of tuberus sclerosis 2 (TSC2). mTORC1 in turn stimulates the activity p70S6 kinase (p70S6K), that can inhibit Akt by a dual mechanism. First, p70S6K phosphorylates insulin receptor substrate 1 (IRS-1), leading to its proteasomal degradation. IRS-1 is an adaptor protein of the insulin receptor and the insulin-like growth factor-1 receptor (IGF-1R), both of which are major activators of the PI3K/Akt pathway. Thus, p70S6K-mediated inhibition of IRS-1 results in inhibition of Akt signaling (Easton et al. 2006; Martelli et al. 2010). A second negative feedback loop involves p70S6K-stimulated phosphorylation of Rictor, the core component of mTORC2, which negatively regulates mTORC2 dependent activation of Akt (Dibble et al. 2009). Thus, inhibition of mTOR can enhance Akt kinase activity (O'Reilly et al. 2006), which could explain the moderate antitumor activity of mTOR inhibitors (Chang et al. 2005; Galanis et al. 2005). Figure 7.1 shows model of Akt activation and mTORC1-dependent negative feedback mechanisms.

The kinase activity of Akt may be additionally increased via direct interaction with phosphatidylinositol 3-kinase enhancer A (PIKE-A), a protein that is amplified in variety of human cancer cells, including glioblastomas. PIKE-A preferentially binds to activated Akt, and siRNA-mediated depletion of PIKE-A diminishes Akt phosphorylation (Ahn et al. 2004).

Akt activation is terminated via dephosphorylation by protein phosphatase 2A (PP2A) and PH domain leucine-rich repeat protein phosphatase (PHLPPα) (Brazil and Hemmings 2001; Fayard et al. 2005; Gao et al. 2005; Meier et al. 1998; Millward et al. 1999). Interestingly, expression of PP2A regulatory subunit Aα has been shown to be reduced in a significant fraction of both glioblastomas and oligodendrogliomas, thereby deregulating this phosphatase (Colella et al. 2001).

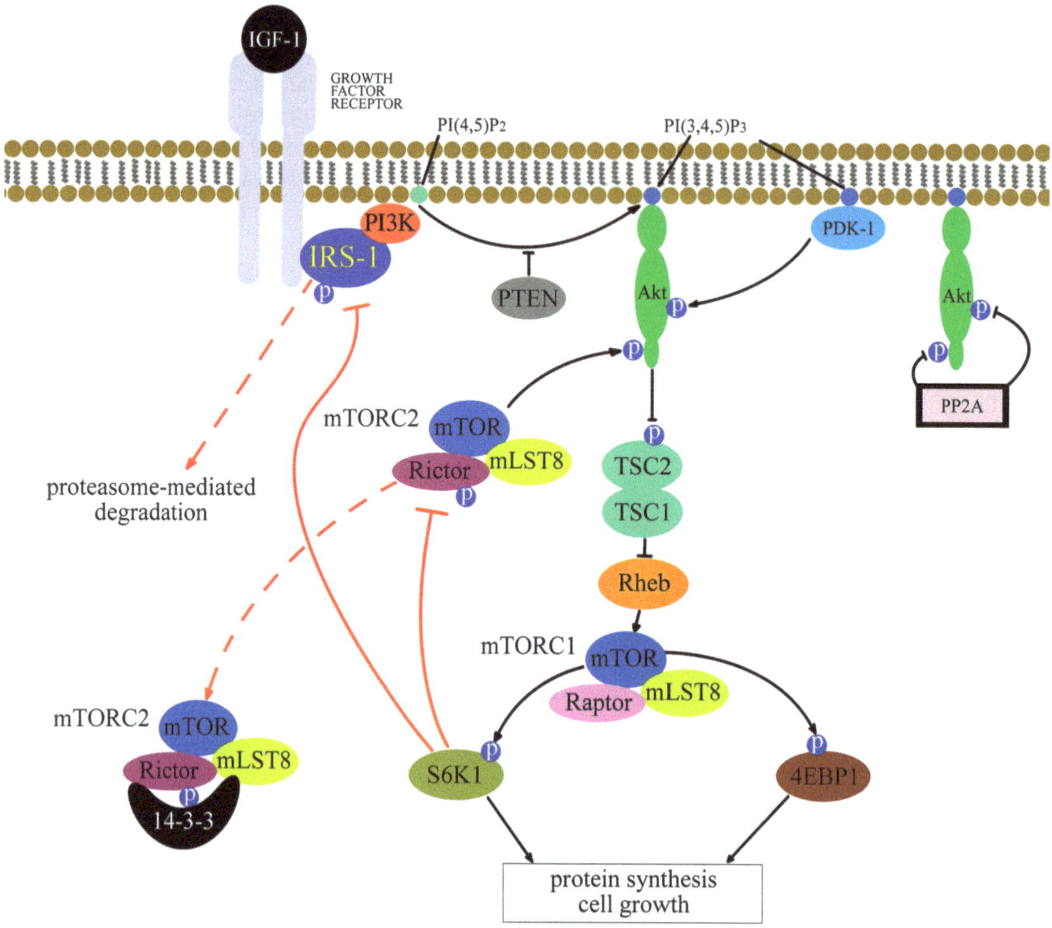

Fig. 7.1 Model of Akt kinase activation and mTORC1-dependent negative feedback mechanisms. Ligand binding induces IGF-1R-dependent association of IRS-1 with PI3K, thereby stimulating production of PI(3,4,5)P_3 by PI3K. PI(3,4,5)P_3 recruits Akt and PDK1. Subsequently, Akt is phosphorylated by mTORC2 and PDK1, leading to full activation. Akt phosphorylates and inhibits TSC2, leading to Rheb- dependent activation of mTOR. In turn, mTORC1 phosphorylates S6K1 kinase and 4EBP1, stimulating protein synthesis. Activation of mTORC1 stimulates two parallel negative-feedback loops (*red pathways*) that inhibit Akt. S6K-1 phosphorylates IRS-1 (*red line*) and directs IRS-1 to proteasomal degradation (*red dashed arrow*). A second negative loop was proposed where S6K1 phosphorylates the Rictor subunit of mTORC2, thereby promoting 14-3-3 binding to Rictor. S6K-1-dependent phosphorylation of Rictor negatively regulates the ability of mTORC2 to phosphorylate Akt. Akt activation is terminated via dephosphorylation by PP2A

Initially, Akt kinase was known primarily for its role in regulation of cell survival and cell cycle progression (Brunet et al. 1999; Datta et al. 1997; del Peso et al. 1997; Diehl et al. 1998; Kennedy et al. 1999; Li et al. 2002; Shin et al. 2002; Viglietto et al. 2002; Zhou et al. 2001). However, evidence has accumulated that Akt plays a key role in regulation of the invasive glioma phenotype (Molina et al. 2010; Pu et al. 2004). Indeed, invasive glioblastoma cells have a higher level of phosphorylated Akt in comparison with cells isolated from the tumor core in a model of human invasive GBM established in the mouse brain (Molina et al. 2010).

The molecular mechanisms that mediate Akt-stimulated cell migration are still being explored (Stambolic and Woodgett 2006). Phosphorylated Akt localizes to lamellipodia of moving cells, where it colocalizes with Rac and Cdc42. Constitutive activation of Rac1 or Cdc42 increases Akt phosphorylation in fibroblasts, and inhibition of Akt inhibits cell migration stimulated by Rac or Cdc42 (Higuchi et al. 2001), thus Akt may be activated via GTPase-dependent PI3K activation (Murga et al. 2002). Akt can stimulate cell migration in a number of ways (Stambolic and Woodgett 2006), including the role of p70S6K in actin reorganization (Qian et al. 2004) and direct phosphorylation of Girdin/Akt phosphorylation enhancer (APE) protein (Enomoto et al. 2005; Zhang et al. 2009). Girdin can crosslink actin filaments and anchor cortical actin to the plasma membrane. Phosphorylated Girdin relocalizes to the leading edge of moving cells and promotes short-branched actin filaments (Enomoto et al. 2005).

A comprehensive study has evaluated the role of all three Akt isoforms in gliomagenesis using a model system driven by common glioma abnormalities: loss of function of PTEN and p53 protein and expression of EGFRvIII receptor in primary murine astrocytes (PMA). The results showed that Akt3 regulates anchorage-independent growth of transformed astrocytes and human glioma cells. Additionally Akt3, but not Akt1 or Akt2, knockdown reduces the ability of PMA to invade matrigel (Endersby et al. 2011). Previous studies discussed the critical role of Akt2 in regulation of glioma invasion (Pu et al. 2004; Zhang et al. 2009, 2010). Thus, depending on the glioma model used, the specific functions of the various Akt isoforms may vary. It also remains unclear which Akt isoform plays a major role downstream of PI3K in gliomas and whether the respective isoforms interact with distinct binding partners depending on the cell setting.

The expression of Akt1 is similar in gliomas and normal control tissue (Mure et al. 2010). Akt3 mRNA and protein decrease with increasing grade of malignancy (Mure et al. 2010), contrary to Akt2 expression that increases with tumor grade (Mure et al. 2010; Wang et al. 2010; Zhang et al. 2010). Although Akt3 expression levels in malignant glioma are significantly reduced compared to normal tissue, its kinase activity is equal to that of Akt2, and approximately 2-fold higher than that of Akt1 (Mure et al. 2010), thereby compensating for its decrease in expression.

7.4 Proteases

ECM proteins as well as parenchymal cells are natural obstacles for migrating tumor cells. In order to invade brain tissue, glioma cells detach from the tumor mass and reorganize the ECM by complex proteolytic mechanisms and expression of ectopic ECM components (Nakada et al. 2007). Glioblastomas overexpress a number of proteases, including uPA, matrix metalloproteinases (MMPs), ADAMs and cathepsins (Fillmore et al. 2001; Fukuda et al. 2005; Nakada et al. 1999; Pagenstecher et al. 2001; Rempel et al. 1994; Sivaparvathi et al. 1996a, b, c; Yamamoto et al. 1994; Zhao et al. 2008).

The regulation of proteolytic activity is complex and often involves cross-talk between different classes of proteases. One example is uPA, a serine protease that is synthesized as an inactive propeptide – zymogen (known as pro-uPA), which binds to the uPA receptor (uPAR) on the plasma membrane and becomes cleaved by active plasmin. uPA catalyses the conversion of nonactive plasminogen into plasmin, thus establishing a positive feedback loop. Notably, binding of uPA to its receptor provides localized proteolytic activity (Pillay et al. 2007). uPa directly cleaves and activates pro-MMP-9 in glioblastomas (Zhao et al. 2008) and experimental depletion of both uPA and uPAR from glioblastoma cells suppresses invasion *in vitro,* intracerebral tumor formation in nude mice as well as growth of subcutaneously pre-established tumors (Gondi et al. 2003). In summary, the uPA-uPAR axis controls many signaling pathways that contribute to the malignant behavior of glioblastoma. Importantly, uPA expression and activity are elevated in anaplastic astrocytomas (AA) and GBMs in comparison to non-neoplastic brain (NB) or low grade glioma (LGG) (Landau et al. 1994; Yamamoto et al. 1994). Thus, targeting these proteases may be therapeutically beneficial.

Interdependent activation of proteases also occurs in the family of matrix metalloproteinases, which comprises 25 enzymes that bind Zn^{2+} ions in their active site. Based on substrate specificity, structure and subcellular localization matrix metalloproteinases are grouped into collagenases, gelatinases, stromelysins and membrane metalloproteinases. Like uPA, metalloproteinases are synthesized in cells as zymogens that have to be proteolytically processed to become fully active. The number of active enzymes in the cell is regulated on the level of gene expression, protein secretion and activation or inhibition by tissue inhibitors of metalloproteinases (TIMPs) (Ra and Parks 2007).

MT1-MMP is a key metalloproteinase that contains a transmembrane sequence and cytoplasmic domain (Ra and Parks 2007). The transmembrane/cytoplasmic domain is responsible for localizing the enzyme to invadopodia (Nakahara et al. 1997), actin-rich structures to which most proteolytic activity of invading cells is localized (Frittoli et al. 2011; Ridley 2011; Symons 2008). The MT1-MMP cytoplasmic domain is also necessary for dynamin-dependent removal of the enzyme from the cell surface via clathrin-mediated endocytosis (Jiang et al. 2001), an additional mechanism of MT-MMP-1 regulation.

Processing of pro-MMP-2 involves the formation of a MT1-MMP/TIMP-2 complex at the cell surface (Zucker et al. 1998). This complex binds pro-MMP-2 via interactions between C-terminal of TIMP-2 and the hemopexin C-domain of pro-MMP-2 (Overall et al. 2000). Subsequently, an additional molecule of MT1-MMP cleaves and activates MMP-2 (Fillmore et al. 2001; Ra and Parks 2007) (Fig. 7.2). This mechanism of activation requires a proper balance in the expression level of pro-MMP-2, TIMP-2 and MT1-MMP. In line with this mechanism is the finding that CD95L (FasL) stimulates MMP-2 activity in rat glioma cells via NF-κB-driven transcription of TIMP-2. Inhibition of TIMP-2 expression decrease overall MMP-2 activity and leads to accumulation of inactive pro-MMP-2 (Wisniewski et al. 2010). Alternative mechanisms of MMP-2 activation have also been suggested (Mazzieri et al. 1997; Monea et al. 2002; Morrison et al. 2001).

Interestingly, MMP-2 is subject to autocatalytic processing, resulting in the formation of the hemopexin fragment (PEX). As for MMP-2 itself, the level of PEX increases with tumor grade (Bello et al. 2001b). Paradoxically, PEX inhibits glioblastoma cell migration, invasion and proliferation and induces apoptosis. It also strongly inhibits angiogenesis, most likely, through its binding to $\alpha v\beta 3$. Intraperitoneal administration of PEX inhibits tumor growth by 99 % in both subcutaneous and intracranial human glioma xenografts mouse models, with no sign of toxicity (Bello et al. 2001b), suggesting that PEX is a promising therapeutic candidate. The anti-tumor and anti-angiogenic effects of PEX are reminiscent to those of endostatin (Cao 2001) and suggest that the effects of PEX on the malignant behavior of glioblastoma is overridden by powerful proinvasive and proangiogenic factors.

MT1-MMP, MMP-2 and MMP-9 are three major MMPs overexpressed in glioblastoma in comparison with non-neoplastic brain (Fillmore et al. 2001; Nakada et al. 1999; Pagenstecher et al. 2001; Zhang et al. 2010). Upregulation of expression of those MMPs in HGG can be explained, at least in part, by hyperactivation of the PI3K/Akt signaling pathway in those tumors. In line with this, inhibition of PI3K or Akt reduces MMP expression in glioblastoma cells (Kwiatkowska et al. 2011). In addition, downregulation of Akt-dependent MMPs production is associated with inhibition of glioma cell invasion *in vitro* and a reduction in the number of satellite tumors and tumor volume in an *in vivo* model of glioma (Pu et al. 2004; Zhang et al. 2009; 2010).

Proteases also regulate glioma cell invasion by promoting receptor shedding from the plasma membrane (Nagano et al. 2004; Okamoto et al. 1999a, b; Yang et al. 2011). Notably, ADAM-10, which belongs to the multidomain membrane-anchored protein family called adamalysins, catalyses proteolysis of the L1 receptor. Notably, increased surface expression of ADAM-10 on migrating glioblastoma cells correlates with loss of surface L1 (Yang et al. 2011). Experimental attenuation of L1 expression in glioblastoma cells reduces migration velocity *in vitro* and suppresses invasion of tumor cells into chick embryonic brain. Migration can be restored upon adding L1 ectodomain to migrating cells (Yang et al. 2011). In addition, ADAM-10 and ADAM-17, in response to different stimuli, cleave the CD44 hyaluronan receptor. Depletion of either adamalysin suppresses CD44 ectodomain shedding and strongly inhibits cancer cell migration on hyaluronan (Nagano et al. 2004).

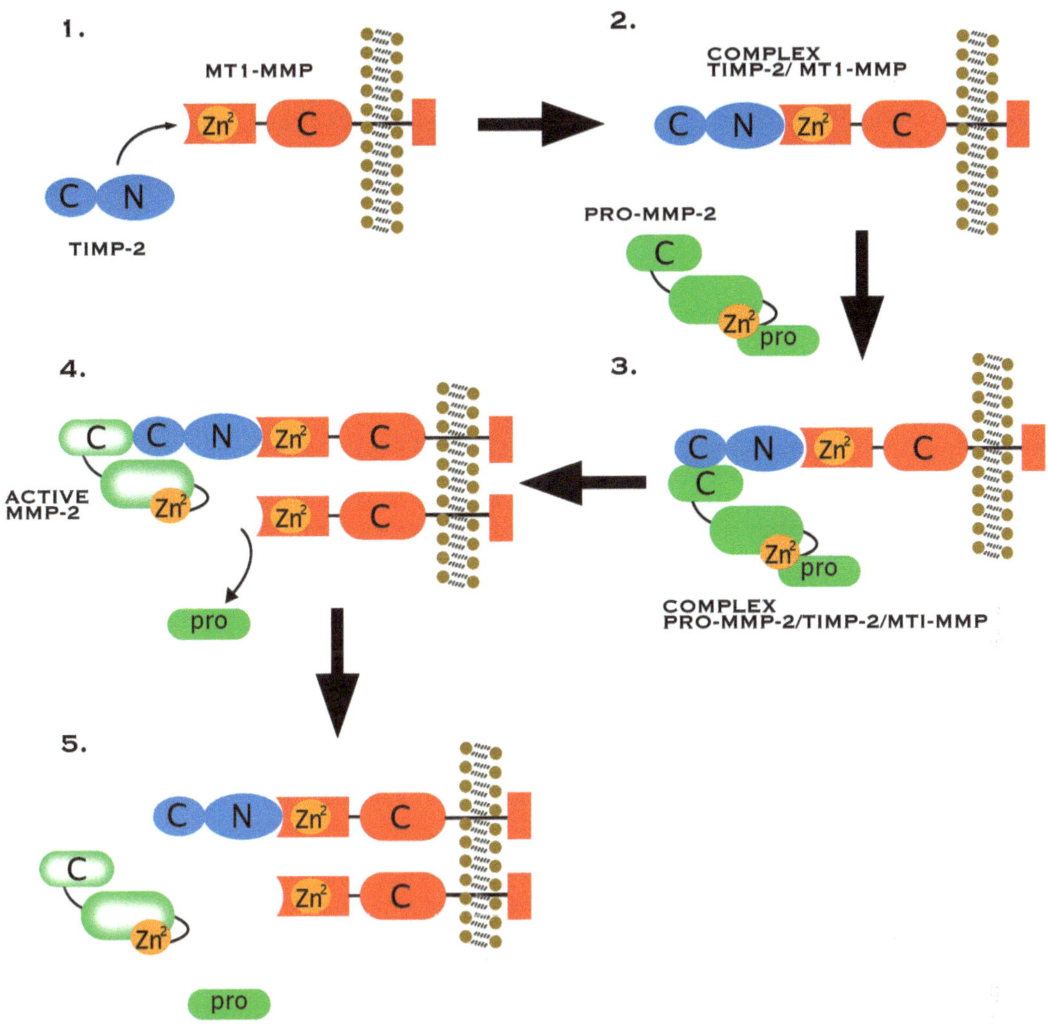

Fig. 7.2 Mechanism of MMP-2 activation. Processing of pro-MMP-2 requires formation of a MT1-MMP and TIMP-2 complex (*1–2*). This complex functions as a receptor for binding of pro-MMP-2 (*3*). Subsequently, an additional MT1-MMP molecule cleaves pro-MMP-2 (*4*). Active MMP-2 is released into the intercellular space (*5*). Modified from Brauer 2006

Moreover, overexpression of ADAM-17 in U87 glioma cells results in increased invasion into Matrigel and tumor growth, whereas inhibition of ADAM-17 reduces tumor growth. Interestingly, specific ablation of ADAM-17 decreases activation of EGFR/PI3K/Akt pathway, probably by decreasing TGFα shedding (Zheng et al. 2011). Thus, ADAM proteases stimulate cell invasion by multiple mechanisms, including ECM degradation, receptor shedding and growth factor activation.

7.5 Conclusions and Future Directions

Extensive *in vitro* and *in vivo* studies have revealed the complexity of signaling mechanism that drive dissemination of malignant glioma cells into surrounding brain tissue. The hope is that better understanding of critical signaling elements will help to identify molecular targets for therapeutic intervention. In addition, we anticipate that specific targeting of the invasive behavior of glioblastoma will have limited therapeutic benefit and that anti-invasion strategies will have to be combined with additional therapeutic modalities, such as chemo- or radio-therapy. The discovery of genetically distinct subclasses within HGG patients is an important step toward introducing personalized therapies (Huse et al. 2011). Thus, we anticipate that specific combination regimens will be introduced to treat respective HGG subgroups.

References

Abdollahi A, Griggs DW, Zieher H, Roth A, Lipson KE, Saffrich R, Grone HJ, Hallahan DE, Reisfeld RA, Debus J, Niethammer AG, Huber PE (2005) Inhibition of alpha(v)beta3 integrin survival signaling enhances antiangiogenic and antitumor effects of radiotherapy. Clin Cancer Res 11:6270–6279

Ahn JY, Rong R, Kroll TG, Van Meir EG, Snyder SH, Ye K (2004) PIKE (phosphatidylinositol 3-kinase enhancer)-A GTPase stimulates Akt activity and mediates cellular invasion. J Biol Chem 279:16441–16451

Akiyama Y, Jung S, Salhia B, Lee S, Hubbard S, Taylor M, Mainprize T, Akaishi K, van Furth W, Rutka JT (2001) Hyaluronate receptors mediating glioma cell migration and proliferation. J Neurooncol 53:115–127

Armstrong TS, Prabhu S, Aldape K, Hossan B, Kang S, Childress A, Tolentino L, Gilbert MR (2011) A case of soft tissue metastasis from glioblastoma and review of the literature. J Neurooncol 103:167–172

Bader AG, Kang S, Zhao L, Vogt PK (2005) Oncogenic PI3K deregulates transcription and translation. Nat Rev Cancer 5:921–929

Badie B, Schartner JM (2000) Flow cytometric characterization of tumor-associated macrophages in experimental gliomas. Neurosurgery 46:957–961, discussion 961–952

Badie B, Schartner J (2001) Role of microglia in glioma biology. Microsc Res Tech 54:106–113

Badie B, Bartley B, Schartner J (2002) Differential expression of MHC class II and B7 costimulatory molecules by microglia in rodent gliomas. J Neuroimmunol 133:39–45

Baeza N, Weller M, Yonekawa Y, Kleihues P, Ohgaki H (2003) PTEN methylation and expression in glioblastomas. Acta Neuropathol 106:479–485

Bajetto A, Barbieri F, Dorcaratto A, Barbero S, Daga A, Porcile C, Ravetti JL, Zona G, Spaziante R, Corte G, Schettini G, Florio T (2006) Expression of CXC chemokine receptors 1–5 and their ligands in human glioma tissues: role of CXCR4 and SDF1 in glioma cell proliferation and migration. Neurochem Int 49:423–432

Barbero S, Bajetto A, Bonavia R, Porcile C, Piccioli P, Pirani P, Ravetti JL, Zona G, Spaziante R, Florio T, Schettini G (2002) Expression of the chemokine receptor CXCR4 and its ligand stromal cell-derived factor 1 in human brain tumors and their involvement in glial proliferation in vitro. Ann N Y Acad Sci 973:60–69

Beadle C, Assanah MC, Monzo P, Vallee R, Rosenfeld SS, Canoll P (2008) The role of myosin II in glioma invasion of the brain. Mol Biol Cell 19:3357–3368

Bellail AC, Hunter SB, Brat DJ, Tan C, Van Meir EG (2004) Microregional extracellular matrix heterogeneity in brain modulates glioma cell invasion. Int J Biochem Cell Biol 36:1046–1069

Bellion A, Baudoin JP, Alvarez C, Bornens M, Metin C (2005) Nucleokinesis in tangentially migrating neurons comprises two alternating phases: forward migration of the Golgi/centrosome associated with centrosome splitting and myosin contraction at the rear. J Neurosci 25:5691–5699

Bello L, Francolini M, Marthyn P, Zhang J, Carroll RS, Nikas DC, Strasser JF, Villani R, Cheresh DA, Black PM (2001a) Alpha(v)beta3 and alpha(v)beta5 integrin expression in glioma periphery. Neurosurgery 49:380–389, discussion 390

Bello L, Lucini V, Carrabba G, Giussani C, Machluf M, Pluderi M, Nikas D, Zhang J, Tomei G, Villani RM, Carroll RS, Bikfalvi A, Black PM (2001b) Simultaneous inhibition of glioma angiogenesis, cell proliferation, and invasion by a naturally occurring fragment of human metalloproteinase-2. Cancer Res 61:8730–8736

Birbilis TA, Matis GK, Eleftheriadis SG, Theodoropoulou EN, Sivridis E (2010) Spinal metastasis of glioblastoma multiforme: an uncommon suspect? Spine (Phila Pa 1976) 35:E264–269

Brauer (2006) MMPs- Role in cardiovascular development and disease. Front Biosci 11:447–478

Brazil DP, Hemmings BA (2001) Ten years of protein kinase B signalling: a hard Akt to follow. Trends Biochem Sci 26:657–664

Brockmann MA, Ulbricht U, Gruner K, Fillbrandt R, Westphal M, Lamszus K (2003) Glioblastoma and cerebral microvascular endothelial cell migration in response to tumor-associated growth factors. Neurosurgery 52:1391–1399, discussion 1399

Brunet A, Bonni A, Zigmond MJ, Lin MZ, Juo P, Hu LS, Anderson MJ, Arden KC, Blenis J, Greenberg ME (1999) Akt promotes cell survival by phosphorylating and inhibiting a Forkhead transcription factor. Cell 96:857–868

Burridge K, Wennerberg K (2004) Rho and Rac take center stage. Cell 116:167–179
Bustelo XR, Sauzeau V, Berenjeno IM (2007) GTP-binding proteins of the Rho/Rac family: regulation, effectors and functions in vivo. Bioessays 29:356–370
Cao Y (2001) Endogenous angiogenesis inhibitors and their therapeutic implications. Int J Biochem Cell Biol 33:357–369
Cayre M, Canoll P, Goldman JE (2009) Cell migration in the normal and pathological postnatal mammalian brain. Prog Neurobiol 88:41–63
Chan AY, Coniglio SJ, Chuang YY, Michaelson D, Knaus UG, Philips MR, Symons M (2005) Roles of the Rac1 and Rac3 GTPases in human tumor cell invasion. Oncogene 24:7821–7829
Chang SM, Wen P, Cloughesy T, Greenberg H, Schiff D, Conrad C, Fink K, Robins HI, De Angelis L, Raizer J, Hess K, Aldape K, Lamborn KR, Kuhn J, Dancey J, Prados MD (2005) Phase II study of CCI-779 in patients with recurrent glioblastoma multiforme. Invest New Drugs 23:357–361
Charles NA, Holland EC, Gilbertson R, Glass R, Kettenmann H (2011) The brain tumor microenvironment. Glia 59:1169–1180
Chuang YY, Tran NL, Rusk N, Nakada M, Berens ME, Symons M (2004) Role of synaptojanin 2 in glioma cell migration and invasion. Cancer Res 64:8271–8275
Colella S, Ohgaki H, Ruediger R, Yang F, Nakamura M, Fujisawa H, Kleihues P, Walter G (2001) Reduced expression of the Aalpha subunit of protein phosphatase 2A in human gliomas in the absence of mutations in the Aalpha and Abeta subunit genes. Int J Cancer 93:798–804
Cote JF, Vuori K (2007) GEF what? Dock180 and related proteins help Rac to polarize cells in new ways. Trends Cell Biol 17:383–393
Croft DR, Olson MF (2006) The Rho GTPase effector ROCK regulates cyclin A, cyclin D1, and p27Kip1 levels by distinct mechanisms. Mol Cell Biol 26:4612–4627
D'Abaco GM, Kaye AH (2007) Integrins: molecular determinants of glioma invasion. J Clin Neurosci 14:1041–1048
Datta SR, Dudek H, Tao X, Masters S, Fu H, Gotoh Y, Greenberg ME (1997) Akt phosphorylation of BAD couples survival signals to the cell-intrinsic death machinery. Cell 91:231–241
de Groot JF, Fuller G, Kumar AJ, Piao Y, Eterovic K, Ji Y, Conrad CA (2010) Tumor invasion after treatment of glioblastoma with bevacizumab: radiographic and pathologic correlation in humans and mice. Neuro Oncol 12:233–242
del Peso L, Gonzalez-Garcia M, Page C, Herrera R, Nunez G (1997) Interleukin-3-induced phosphorylation of BAD through the protein kinase Akt. Science 278:687–689
Delamarre E, Taboubi S, Mathieu S, Berenguer C, Rigot V, Lissitzky JC, Figarella-Branger D, Ouafik L, Luis J (2009) Expression of integrin alpha6beta1 enhances tumorigenesis in glioma cells. Am J Pathol 175:844–855
Delpech B, Maingonnat C, Girard N, Chauzy C, Maunoury R, Olivier A, Tayot J, Creissard P (1993) Hyaluronan and hyaluronectin in the extracellular matrix of human brain tumour stroma. Eur J Cancer 29A:1012–1017
Dibble CC, Asara JM, Manning BD (2009) Characterization of Rictor phosphorylation sites reveals direct regulation of mTOR complex 2 by S6K1. Mol Cell Biol 29:5657–5670
Diehl JA, Cheng M, Roussel MF, Sherr CJ (1998) Glycogen synthase kinase-3beta regulates cyclin D1 proteolysis and subcellular localization. Genes Dev 12:3499–3511
Dransart E, Olofsson B, Cherfils J (2005) RhoGDIs revisited: novel roles in Rho regulation. Traffic 6:957–966
Easton JB, Kurmasheva RT, Houghton PJ (2006) IRS-1: auditing the effectiveness of mTOR inhibitors. Cancer Cell 9:153–155
Ehtesham M, Winston JA, Kabos P, Thompson RC (2006) CXCR4 expression mediates glioma cell invasiveness. Oncogene 25:2801–2806
Endersby R, Baker SJ (2008) PTEN signaling in brain: neuropathology and tumorigenesis. Oncogene 27:5416–5430
Endersby R, Zhu X, Hay N, Ellison DW, Baker SJ (2011) Nonredundant functions for Akt isoforms in astrocyte growth and gliomagenesis in an orthotopic transplantation model. Cancer Res 71:4106–4116
Enomoto A, Murakami H, Asai N, Morone N, Watanabe T, Kawai K, Murakumo Y, Usukura J, Kaibuchi K, Takahashi M (2005) Akt/PKB regulates actin organization and cell motility via Girdin/APE. Dev Cell 9:389–402
Ermoian RP, Furniss CS, Lamborn KR, Basila D, Berger MS, Gottschalk AR, Nicholas MK, Stokoe D, Haas-Kogan DA (2002) Dysregulation of PTEN and protein kinase B is associated with glioma histology and patient survival. Clin Cancer Res 8:1100–1106
Escobedo JA, Navankasattusas S, Kavanaugh WM, Milfay D, Fried VA, Williams LT (1991) cDNA cloning of a novel 85 kd protein that has SH2 domains and regulates binding of PI3-kinase to the PDGF beta-receptor. Cell 65:75–82
Fagerholm SC, Hilden TJ, Gahmberg CG (2004) P marks the spot: site-specific integrin phosphorylation regulates molecular interactions. Trends Biochem Sci 29:504–512
Farin A, Suzuki SO, Weiker M, Goldman JE, Bruce JN, Canoll P (2006) Transplanted glioma cells migrate and proliferate on host brain vasculature: a dynamic analysis. Glia 53:799–808
Fayard E, Tintignac LA, Baudry A, Hemmings BA (2005) Protein kinase B/Akt at a glance. J Cell Sci 118:5675–5678

Fillmore HL, VanMeter TE, Broaddus WC (2001) Membrane-type matrix metalloproteinases (MT-MMPs): expression and function during glioma invasion. J Neurooncol 53:187–202
Fleming IN, Gray A, Downes CP (2000) Regulation of the Rac1-specific exchange factor Tiam1 involves both phosphoinositide 3-kinase-dependent and -independent components. Biochem J 351:173–182
Flugel A, Labeur MS, Grasbon-Frodl EM, Kreutzberg GW, Graeber MB (1999) Microglia only weakly present glioma antigen to cytotoxic T cells. Int J Dev Neurosci 17:547–556
Franke TF, Yang SI, Chan TO, Datta K, Kazlauskas A, Morrison DK, Kaplan DR, Tsichlis PN (1995) The protein kinase encoded by the Akt proto-oncogene is a target of the PDGF-activated phosphatidylinositol 3-kinase. Cell 81:727–736
Frittoli E, Palamidessi A, Disanza A, Scita G (2011) Secretory and endo/exocytic trafficking in invadopodia formation: the MT1-MMP paradigm. Eur J Cell Biol 90:108–114
Fukuda ME, Iwadate Y, Machida T, Hiwasa T, Nimura Y, Nagai Y, Takiguchi M, Tanzawa H, Yamaura A, Seki N (2005) Cathepsin D is a potential serum marker for poor prognosis in glioma patients. Cancer Res 65:5190–5194
Furnari FB, Fenton T, Bachoo RM, Mukasa A, Stommel JM, Stegh A, Hahn WC, Ligon KL, Louis DN, Brennan C, Chin L, DePinho RA, Cavenee WK (2007) Malignant astrocytic glioma: genetics, biology, and paths to treatment. Genes Dev 21:2683–2710
Gabrusiewicz K, Ellert-Miklaszewska A, Lipko M, Sielska M, Frankowska M, Kaminska B (2011) Characteristics of the alternative phenotype of microglia/macrophages and its modulation in experimental gliomas. PLoS One 6:e23902
Galanis E, Buckner JC, Maurer MJ, Kreisberg JI, Ballman K, Boni J, Peralba JM, Jenkins RB, Dakhil SR, Morton RF, Jaeckle KA, Scheithauer BW, Dancey J, Hidalgo M, Walsh DJ (2005) Phase II trial of temsirolimus (CCI-779) in recurrent glioblastoma multiforme: a North Central Cancer Treatment Group Study. J Clin Oncol 23:5294–5304
Gallia GL, Rand V, Siu IM, Eberhart CG, James CD, Marie SK, Oba-Shinjo SM, Carlotti CG, Caballero OL, Simpson AJ, Brock MV, Massion PP, Carson BS Sr, Riggins GJ (2006) PIK3CA gene mutations in pediatric and adult glioblastoma multiforme. Mol Cancer Res 4:709–714
Gao T, Furnari F, Newton AC (2005) PHLPP: a phosphatase that directly dephosphorylates Akt, promotes apoptosis, and suppresses tumor growth. Mol Cell 18:13–24
Garcia-Mata R, Boulter E, Burridge K (2011) The ‘invisible hand’: regulation of RHO GTPases by RHOGDIs. Nat Rev Mol Cell Biol 12:493–504
Giese A, Bjerkvig R, Berens ME, Westphal M (2003) Cost of migration: invasion of malignant gliomas and implications for treatment. J Clin Oncol 21:1624–1636
Gilbert MR, Kuhn J, Lamborn KR, Lieberman F, Wen PY, Mehta M, Cloughesy T, Lassman AB, Deangelis LM, Chang S, Prados M (2012) Cilengitide in patients with recurrent glioblastoma: the results of NABTC 03–02, a phase II trial with measures of treatment delivery. J Neurooncol 106:147–153
Gomez del Pulgar T, Benitah SA, Valeron PF, Espina C, Lacal JC (2005) Rho GTPase expression in tumourigenesis: evidence for a significant link. Bioessays 27:602–613
Gondi CS, Lakka SS, Yanamandra N, Siddique K, Dinh DH, Olivero WC, Gujrati M, Rao JS (2003) Expression of antisense uPAR and antisense uPA from a bicistronic adenoviral construct inhibits glioma cell invasion, tumor growth, and angiogenesis. Oncogene 22:5967–5975
Gotway MB, Conomos PJ, Bremner RM (2011) Pleural metastatic disease from glioblastoma multiforme. J Thorac Imaging 26:W54–58
Heasman SJ, Ridley AJ (2008) Mammalian Rho GTPases: new insights into their functions from in vivo studies. Nat Rev Mol Cell Biol 9:690–701
Higuchi M, Ohnishi T, Arita N, Hiraga S, Hayakawa T (1993) Expression of tenascin in human gliomas: its relation to histological malignancy, tumor dedifferentiation and angiogenesis. Acta Neuropathol 85:481–487
Higuchi M, Masuyama N, Fukui Y, Suzuki A, Gotoh Y (2001) Akt mediates Rac/Cdc42-regulated cell motility in growth factor-stimulated cells and in invasive PTEN knockout cells. Curr Biol 11:1958–1962
Hoelzinger DB, Mariani L, Weis J, Woyke T, Berens TJ, McDonough WS, Sloan A, Coons SW, Berens ME (2005) Gene expression profile of glioblastoma multiforme invasive phenotype points to new therapeutic targets. Neoplasia 7:7–16
Hoelzinger DB, Demuth T, Berens ME (2007) Autocrine factors that sustain glioma invasion and paracrine biology in the brain microenvironment. J Natl Cancer Inst 99:1583–1593
Hoelzinger DB, Nakada M, Demuth T, Rosensteel T, Reavie LB, Berens ME (2008) Autotaxin: a secreted autocrine/ paracrine factor that promotes glioma invasion. J Neurooncol 86:297–309
Hong X, Jiang F, Kalkanis SN, Zhang ZG, Zhang XP, DeCarvalho AC, Katakowski M, Bobbitt K, Mikkelsen T, Chopp M (2006) SDF-1 and CXCR4 are up-regulated by VEGF and contribute to glioma cell invasion. Cancer Lett 236:39–45
Hong TM, Teng LJ, Shun CT, Peng MC, Tsai JC (2009) Induced interleukin-8 expression in gliomas by tumor-associated macrophages. J Neurooncol 93:289–301
Huang J, Manning BD (2009) A complex interplay between Akt, TSC2 and the two mTOR complexes. Biochem Soc Trans 37:217–222

Huse JT, Phillips HS, Brennan CW (2011) Molecular subclassification of diffuse gliomas: seeing order in the chaos. Glia 59:1190–1199

Hynes RO (2002) Integrins: bidirectional, allosteric signaling machines. Cell 110:673–687

Igarashi K, Isohara T, Kato T, Shigeta K, Yamano T, Uno I (1998) Tyrosine 1213 of Flt-1 is a major binding site of Nck and SHP-2. Biochem Biophys Res Commun 246:95–99

Iwadate Y, Sakaida T, Saegusa T, Hiwasa T, Takiguchi M, Fujimoto S, Yamaura A (2005) Proteome-based identification of molecular markers predicting chemosensitivity to each category of anticancer agents in human gliomas. Int J Oncol 26:993–998

Jaiswal BS, Janakiraman V, Kljavin NM, Chaudhuri S, Stern HM, Wang W, Kan Z, Dbouk HA, Peters BA, Waring P, Dela Vega T, Kenski DM, Bowman KK, Lorenzo M, Li H, Wu J, Modrusan Z, Stinson J, Eby M, Yue P, Kaminker JS, de Sauvage FJ, Backer JM, Seshagiri S (2009) Somatic mutations in p85alpha promote tumorigenesis through class IA PI3K activation. Cancer Cell 16:463–474

Jarzynka MJ, Hu B, Hui KM, Bar-Joseph I, Gu W, Hirose T, Haney LB, Ravichandran KS, Nishikawa R, Cheng SY (2007) ELMO1 and Dock180, a bipartite Rac1 guanine nucleotide exchange factor, promote human glioma cell invasion. Cancer Res 67:7203–7211

Jaworski DM, Kelly GM, Piepmeier JM, Hockfield S (1996) BEHAB (brain enriched hyaluronan binding) is expressed in surgical samples of glioma and in intracranial grafts of invasive glioma cell lines. Cancer Res 56:2293–2298

Jiang A, Lehti K, Wang X, Weiss SJ, Keski-Oja J, Pei D (2001) Regulation of membrane-type matrix metalloproteinase 1 activity by dynamin-mediated endocytosis. Proc Natl Acad Sci U S A 98:13693–13698

Kakita A, Goldman JE (1999) Patterns and dynamics of SVZ cell migration in the postnatal forebrain: monitoring living progenitors in slice preparations. Neuron 23:461–472

Kapoor GS, O'Rourke DM (2003) Receptor tyrosine kinase signaling in gliomagenesis: pathobiology and therapeutic approaches. Cancer Biol Ther 2:330–342

Kenig S, Alonso MB, Mueller MM, Lah TT (2010) Glioblastoma and endothelial cells cross-talk, mediated by SDF-1, enhances tumour invasion and endothelial proliferation by increasing expression of cathepsins B, S, and MMP-9. Cancer Lett 289:53–61

Kennedy SG, Kandel ES, Cross TK, Hay N (1999) Akt/Protein kinase B inhibits cell death by preventing the release of cytochrome c from mitochondria. Mol Cell Biol 19:5800–5810

Keunen O, Johansson M, Oudin A, Sanzey M, Rahim SA, Fack F, Thorsen F, Taxt T, Bartos M, Jirik R, Miletic H, Wang J, Stieber D, Stuhr L, Moen I, Rygh CB, Bjerkvig R, Niclou SP (2011) Anti-VEGF treatment reduces blood supply and increases tumor cell invasion in glioblastoma. Proc Natl Acad Sci U S A 108:3749–3754

King WG, Mattaliano MD, Chan TO, Tsichlis PN, Brugge JS (1997) Phosphatidylinositol 3-kinase is required for integrin-stimulated AKT and Raf-1/mitogen-activated protein kinase pathway activation. Mol Cell Biol 17:4406–4418

Kishi Y, Okudaira S, Tanaka M, Hama K, Shida D, Kitayama J, Yamori T, Aoki J, Fujimaki T, Arai H (2006) Autotaxin is overexpressed in glioblastoma multiforme and contributes to cell motility of glioblastoma by converting lysophosphatidylcholine to lysophosphatidic acid. J Biol Chem 281:17492–17500

Kishimoto T, Soda Y, Matsuyama Y, Mizuno K (2002) An enzymatic assay for lysophosphatidylcholine concentration in human serum and plasma. Clin Biochem 35:411–416

Kleber S, Sancho-Martinez I, Wiestler B, Beisel A, Gieffers C, Hill O, Thiemann M, Mueller W, Sykora J, Kuhn A, Schreglmann N, Letellier E, Zuliani C, Klussmann S, Teodorczyk M, Grone HJ, Ganten TM, Sultmann H, Tuttenberg J, von Deimling A, Regnier-Vigouroux A, Herold-Mende C, Martin-Villalba A (2008) Yes and PI3K bind CD95 to signal invasion of glioblastoma. Cancer Cell 13:235–248

Kobayashi H, Schmitt M, Goretzki L, Chucholowski N, Calvete J, Kramer M, Gunzler WA, Janicke F, Graeff H (1991) Cathepsin B efficiently activates the soluble and the tumor cell receptor-bound form of the proenzyme urokinase-type plasminogen activator (Pro-uPA). J Biol Chem 266:5147–5152

Koochekpour S, Jeffers M, Rulong S, Taylor G, Klineberg E, Hudson EA, Resau JH, Vande Woude GF (1997) Met and hepatocyte growth factor/scatter factor expression in human gliomas. Cancer Res 57:5391–5398

Koul D, Shen R, Kim YW, Kondo Y, Lu Y, Bankson J, Ronen SM, Kirkpatrick DL, Powis G, Yung WK (2010) Cellular and in vivo activity of a novel PI3K inhibitor, PX-866, against human glioblastoma. Neuro Oncol 12:559–569

Kwiatkowska A, Kijewska M, Lipko M, Hibner U, Kaminska B (2011) Downregulation of Akt and FAK phosphorylation reduces invasion of glioblastoma cells by impairment of MT1-MMP shuttling to lamellipodia and downregulates MMPs expression. Biochim Biophys Acta 1813:655–667

Lamour V, Le Mercier M, Lefranc F, Hagedorn M, Javerzat S, Bikfalvi A, Kiss R, Castronovo V, Bellahcene A (2010) Selective osteopontin knockdown exerts anti-tumoral activity in a human glioblastoma model. Int J Cancer 126:1797–1805

Landau BJ, Kwaan HC, Verrusio EN, Brem SS (1994) Elevated levels of urokinase-type plasminogen activator and plasminogen activator inhibitor type-1 in malignant human brain tumors. Cancer Res 54:1105–1108

Le DM, Besson A, Fogg DK, Choi KS, Waisman DM, Goodyer CG, Rewcastle B, Yong VW (2003) Exploitation of astrocytes by glioma cells to facilitate invasiveness: a mechanism involving matrix metalloproteinase-2 and the urokinase-type plasminogen activator-plasmin cascade. J Neurosci 23:4034–4043

Li Y, Dowbenko D, Lasky LA (2002) AKT/PKB phosphorylation of p21Cip/WAF1 enhances protein stability of p21Cip/WAF1 and promotes cell survival. J Biol Chem 277:11352–11361
Lu M, Ravichandran KS (2006) Dock180-ELMO cooperation in Rac activation. Methods Enzymol 406:388–402
Lucio-Eterovic AK, Piao Y, de Groot JF (2009) Mediators of glioblastoma resistance and invasion during antivascular endothelial growth factor therapy. Clin Cancer Res 15:4589–4599
Mahesparan R, Read TA, Lund-Johansen M, Skaftnesmo KO, Bjerkvig R, Engebraaten O (2003) Expression of extracellular matrix components in a highly infiltrative in vivo glioma model. Acta Neuropathol 105:49–57
Mai J, Sameni M, Mikkelsen T, Sloane BF (2002) Degradation of extracellular matrix protein tenascin-C by cathepsin B: an interaction involved in the progression of gliomas. Biol Chem 383:1407–1413
Malecz N, McCabe PC, Spaargaren C, Qiu R, Chuang Y, Symons M (2000) Synaptojanin 2, a novel Rac1 effector that regulates clathrin-mediated endocytosis. Curr Biol 10:1383–1386
Markovic DS, Vinnakota K, Chirasani S, Synowitz M, Raguet H, Stock K, Sliwa M, Lehmann S, Kalin R, van Rooijen N, Holmbeck K, Heppner FL, Kiwit J, Matyash V, Lehnardt S, Kaminska B, Glass R, Kettenmann H (2009) Gliomas induce and exploit microglial MT1-MMP expression for tumor expansion. Proc Natl Acad Sci U S A 106:12530–12535
Martelli AM, Evangelisti C, Chiarini F, McCubrey JA (2010) The phosphatidylinositol 3-kinase/Akt/mTOR signaling network as a therapeutic target in acute myelogenous leukemia patients. Oncotarget 1:89–103
Matheny RW Jr, Adamo ML (2009) Current perspectives on Akt Akt-ivation and Akt-ions. Exp Biol Med (Maywood) 234:1264–1270
Matusan-Ilijas K, Behrem S, Jonjic N, Zarkovic K, Lucin K (2008) Osteopontin expression correlates with angiogenesis and survival in malignant astrocytoma. Pathol Oncol Res 14:293–298
Mazzieri R, Masiero L, Zanetta L, Monea S, Onisto M, Garbisa S, Mignatti P (1997) Control of type IV collagenase activity by components of the urokinase-plasmin system: a regulatory mechanism with cell-bound reactants. EMBO J 16:2319–2332
Meier R, Thelen M, Hemmings BA (1998) Inactivation and dephosphorylation of protein kinase Balpha (PKBalpha) promoted by hyperosmotic stress. EMBO J 17:7294–7303
Milburn CC, Deak M, Kelly SM, Price NC, Alessi DR, Van Aalten DM (2003) Binding of phosphatidylinositol 3,4,5-trisphosphate to the pleckstrin homology domain of protein kinase B induces a conformational change. Biochem J 375:531–538
Millward TA, Zolnierowicz S, Hemmings BA (1999) Regulation of protein kinase cascades by protein phosphatase 2A. Trends Biochem Sci 24:186–191
Molina JR, Hayashi Y, Stephens C, Georgescu MM (2010) Invasive glioblastoma cells acquire stemness and increased Akt activation. Neoplasia 12:453–463
Monea S, Lehti K, Keski-Oja J, Mignatti P (2002) Plasmin activates pro-matrix metalloproteinase-2 with a membrane-type 1 matrix metalloproteinase-dependent mechanism. J Cell Physiol 192:160–170
Moon SY, Zheng Y (2003) Rho GTPase-activating proteins in cell regulation. Trends Cell Biol 13:13–22
Morimura T, Neuchrist C, Kitz K, Budka H, Scheiner O, Kraft D, Lassmann H (1990) Monocyte subpopulations in human gliomas: expression of Fc and complement receptors and correlation with tumor proliferation. Acta Neuropathol 80:287–294
Morrison CJ, Butler GS, Bigg HF, Roberts CR, Soloway PD, Overall CM (2001) Cellular activation of MMP-2 (gelatinase A) by MT2-MMP occurs via a TIMP-2-independent pathway. J Biol Chem 276:47402–47410
Mure H, Matsuzaki K, Kitazato KT, Mizobuchi Y, Kuwayama K, Kageji T, Nagahiro S (2010) Akt2 and Akt3 play a pivotal role in malignant gliomas. Neuro Oncol 12:221–232
Murga C, Zohar M, Teramoto H, Gutkind JS (2002) Rac1 and RhoG promote cell survival by the activation of PI3K and Akt, independently of their ability to stimulate JNK and NF-kappaB. Oncogene 21:207–216
Nagano N, Sasaki H, Aoyagi M, Hirakawa K (1993) Invasion of experimental rat brain tumor: early morphological changes following microinjection of C6 glioma cells. Acta Neuropathol 86:117–125
Nagano O, Murakami D, Hartmann D, De Strooper B, Saftig P, Iwatsubo T, Nakajima M, Shinohara M, Saya H (2004) Cell-matrix interaction via CD44 is independently regulated by different metalloproteinases activated in response to extracellular Ca(2+) influx and PKC activation. J Cell Biol 165:893–902
Nakada M, Nakamura H, Ikeda E, Fujimoto N, Yamashita J, Sato H, Seiki M, Okada Y (1999) Expression and tissue localization of membrane-type 1, 2, and 3 matrix metalloproteinases in human astrocytic tumors. Am J Pathol 154:417–428
Nakada M, Nakada S, Demuth T, Tran NL, Hoelzinger DB, Berens ME (2007) Molecular targets of glioma invasion. Cell Mol Life Sci 64:458–478
Nakahara H, Howard L, Thompson EW, Sato H, Seiki M, Yeh Y, Chen WT (1997) Transmembrane/cytoplasmic domain-mediated membrane type 1-matrix metalloprotease docking to invadopodia is required for cell invasion. Proc Natl Acad Sci U S A 94:7959–7964
Narumiya S, Tanji M, Ishizaki T (2009) Rho signaling, ROCK and mDia1, in transformation, metastasis and invasion. Cancer Metastasis Rev 28:65–76

Norden AD, Drappatz J, Wen PY (2009) Antiangiogenic therapies for high-grade glioma. Nat Rev Neurol 5:610–620
O'Reilly KE, Rojo F, She QB, Solit D, Mills GB, Smith D, Lane H, Hofmann F, Hicklin DJ, Ludwig DL, Baselga J, Rosen N (2006) mTOR inhibition induces upstream receptor tyrosine kinase signaling and activates Akt. Cancer Res 66:1500–1508
Oellers P, Schroer U, Senner V, Paulus W, Thanos S (2009) ROCKs are expressed in brain tumors and are required for glioma-cell migration on myelinated axons. Glia 57:499–509
Okamoto I, Kawano Y, Matsumoto M, Suga M, Kaibuchi K, Ando M, Saya H (1999a) Regulated CD44 cleavage under the control of protein kinase C, calcium influx, and the Rho family of small G proteins. J Biol Chem 274:25525–25534
Okamoto I, Kawano Y, Tsuiki H, Sasaki J, Nakao M, Matsumoto M, Suga M, Ando M, Nakajima M, Saya H (1999b) CD44 cleavage induced by a membrane-associated metalloprotease plays a critical role in tumor cell migration. Oncogene 18:1435–1446
Overall CM, Tam E, McQuibban GA, Morrison C, Wallon UM, Bigg HF, King AE, Roberts CR (2000) Domain interactions in the gelatinase A.TIMP-2.MT1-MMP activation complex. The ectodomain of the 44-kDa form of membrane type-1 matrix metalloproteinase does not modulate gelatinase A activation. J Biol Chem 275:39497–39506
Pagenstecher A, Wussler EM, Opdenakker G, Volk B, Campbell IL (2001) Distinct expression patterns and levels of enzymatic activity of matrix metalloproteinases and their inhibitors in primary brain tumors. J Neuropathol Exp Neurol 60:598–612
Parsons DW, Jones S, Zhang X, Lin JC, Leary RJ, Angenendt P, Mankoo P, Carter H, Siu IM, Gallia GL, Olivi A, McLendon R, Rasheed BA, Keir S, Nikolskaya T, Nikolsky Y, Busam DA, Tekleab H, Diaz LA Jr, Hartigan J, Smith DR, Strausberg RL, Marie SK, Shinjo SM, Yan H, Riggins GJ, Bigner DD, Karchin R, Papadopoulos N, Parmigiani G, Vogelstein B, Velculescu VE, Kinzler KW (2008) An integrated genomic analysis of human glioblastoma multiforme. Science 321:1807–1812
Partovian C, Ju R, Zhuang ZW, Martin KA, Simons M (2008) Syndecan-4 regulates subcellular localization of mTOR Complex2 and Akt activation in a PKCalpha-dependent manner in endothelial cells. Mol Cell 32:140–149
Paulus W, Baur I, Schuppan D, Roggendorf W (1993) Characterization of integrin receptors in normal and neoplastic human brain. Am J Pathol 143:154–163
Pillay V, Dass CR, Choong PF (2007) The urokinase plasminogen activator receptor as a gene therapy target for cancer. Trends Biotechnol 25:33–39
Platten M, Wick W, Wild-Bode C, Aulwurm S, Dichgans J, Weller M (2000) Transforming growth factors beta(1) (TGF-beta(1)) and TGF-beta(2) promote glioma cell migration via Up-regulation of alpha(V)beta(3) integrin expression. Biochem Biophys Res Commun 268:607–611
Platten M, Kretz A, Naumann U, Aulwurm S, Egashira K, Isenmann S, Weller M (2003) Monocyte chemoattractant protein-1 increases microglial infiltration and aggressiveness of gliomas. Ann Neurol 54:388–392
Ponzetto C, Bardelli A, Maina F, Longati P, Panayotou G, Dhand R, Waterfield MD, Comoglio PM (1993) A novel recognition motif for phosphatidylinositol 3-kinase binding mediates its association with the hepatocyte growth factor/scatter factor receptor. Mol Cell Biol 13:4600–4608
Pu P, Kang C, Li J, Jiang H (2004) Antisense and dominant-negative AKT2 cDNA inhibits glioma cell invasion. Tumour Biol 25:172–178
Qian Y, Corum L, Meng Q, Blenis J, Zheng JZ, Shi X, Flynn DC, Jiang BH (2004) PI3K induced actin filament remodeling through Akt and p70S6K1: implication of essential role in cell migration. Am J Physiol Cell Physiol 286:C153–163
Ra HJ, Parks WC (2007) Control of matrix metalloproteinase catalytic activity. Matrix Biol 26:587–596
Raftopoulou M, Hall A (2004) Cell migration: Rho GTPases lead the way. Dev Biol 265:23–32
Ramnarain DB, Park S, Lee DY, Hatanpaa KJ, Scoggin SO, Otu H, Libermann TA, Raisanen JM, Ashfaq R, Wong ET, Wu J, Elliott R, Habib AA (2006) Differential gene expression analysis reveals generation of an autocrine loop by a mutant epidermal growth factor receptor in glioma cells. Cancer Res 66:867–874
Reardon DA, Fink KL, Mikkelsen T, Cloughesy TF, O'Neill A, Plotkin S, Glantz M, Ravin P, Raizer JJ, Rich KM, Schiff D, Shapiro WR, Burdette-Radoux S, Dropcho EJ, Wittemer SM, Nippgen J, Picard M, Nabors LB (2008) Randomized phase II study of cilengitide, an integrin-targeting arginine-glycine-aspartic acid peptide, in recurrent glioblastoma multiforme. J Clin Oncol 26:5610–5617
Rempel SA, Rosenblum ML, Mikkelsen T, Yan PS, Ellis KD, Golembieski WA, Sameni M, Rozhin J, Ziegler G, Sloane BF (1994) Cathepsin B expression and localization in glioma progression and invasion. Cancer Res 54:6027–6031
Ridley AJ (2011) Life at the leading edge. Cell 145:1012–1022
Ridley AJ, Schwartz MA, Burridge K, Firtel RA, Ginsberg MH, Borisy G, Parsons JT, Horwitz AR (2003) Cell migration: integrating signals from front to back. Science 302:1704–1709
Roggendorf W, Strupp S, Paulus W (1996) Distribution and characterization of microglia/macrophages in human brain tumors. Acta Neuropathol 92:288–293
Rooprai HK, Vanmeter T, Panou C, Schnull S, Trillo-Pazos G, Davies D, Pilkington GJ (1999) The role of integrin receptors in aspects of glioma invasion in vitro. Int J Dev Neurosci 17:613–623

Rossman KL, Der CJ, Sondek J (2005) GEF means go: turning on RHO GTPases with guanine nucleotide-exchange factors. Nat Rev Mol Cell Biol 6:167–180
Saitoh Y, Kuratsu J, Takeshima H, Yamamoto S, Ushio Y (1995) Expression of osteopontin in human glioma. Its correlation with the malignancy. Lab Invest 72:55–63
Salhia B, Rutten F, Nakada M, Beaudry C, Berens M, Kwan A, Rutka JT (2005) Inhibition of Rho-kinase affects astrocytoma morphology, motility, and invasion through activation of Rac1. Cancer Res 65:8792–8800
Salhia B, Tran NL, Chan A, Wolf A, Nakada M, Rutka F, Ennis M, McDonough WS, Berens ME, Symons M, Rutka JT (2008) The guanine nucleotide exchange factors trio, Ect2, and Vav3 mediate the invasive behavior of glioblastoma. Am J Pathol 173:1828–1838
Sarbassov DD, Guertin DA, Ali SM, Sabatini DM (2005) Phosphorylation and regulation of Akt/PKB by the rictor-mTOR complex. Science 307:1098–1101
Scherer HJ (1940) A critical review: the pathology of cerebral gliomas. J Neurol Psychiat 3:147–177
Schmidt A, Hall A (2002) Guanine nucleotide exchange factors for Rho GTPases: turning on the switch. Genes Dev 16:1587–1609
Schmitz AA, Govek EE, Bottner B, Van Aelst L (2000) Rho GTPases: signaling, migration, and invasion. Exp Cell Res 261:1–12
Schonsteiner SS, Bommer M, Haenle MM, Klaus B, Scheuerle A, Schmid M, Mayer-Steinacker R (2011) Rare phenomenon: liver metastases from glioblastoma multiforme. J Clin Oncol 29:E668–671
Seitz RJ, Wechsler W (1987) Immunohistochemical demonstration of serum proteins in human cerebral gliomas. Acta Neuropathol 73:145–152
Seol HJ, Smith CA, Salhia B, Rutka JT (2009) The guanine nucleotide exchange factor SWAP-70 modulates the migration and invasiveness of human malignant glioma cells. Transl Oncol 2:300–309
Shih AH, Holland EC (2006) Platelet-derived growth factor (PDGF) and glial tumorigenesis. Cancer Lett 232:139–147
Shin I, Yakes FM, Rojo F, Shin NY, Bakin AV, Baselga J, Arteaga CL (2002) PKB/Akt mediates cell-cycle progression by phosphorylation of p27(Kip1) at threonine 157 and modulation of its cellular localization. Nat Med 8:1145–1152
Shinohara M, Terada Y, Iwamatsu A, Shinohara A, Mochizuki N, Higuchi M, Gotoh Y, Ihara S, Nagata S, Itoh H, Fukui Y, Jessberger R (2002) SWAP-70 is a guanine-nucleotide-exchange factor that mediates signalling of membrane ruffling. Nature 416:759–763
Sivaparvathi M, Sawaya R, Chintala SK, Go Y, Gokaslan ZL, Rao JS (1996a) Expression of cathepsin D during the progression of human gliomas. Neurosci Lett 208:171–174
Sivaparvathi M, Sawaya R, Gokaslan ZL, Chintala SK, Rao JS (1996b) Expression and the role of cathepsin H in human glioma progression and invasion. Cancer Lett 104:121–126
Sivaparvathi M, Yamamoto M, Nicolson GL, Gokaslan ZL, Fuller GN, Liotta LA, Sawaya R, Rao JS (1996c) Expression and immunohistochemical localization of cathepsin L during the progression of human gliomas. Clin Exp Metastasis 14:27–34
Sliwa M, Markovic D, Gabrusiewicz K, Synowitz M, Glass R, Zawadzka M, Wesolowska A, Kettenmann H, Kaminska B (2007) The invasion promoting effect of microglia on glioblastoma cells is inhibited by cyclosporin A. Brain 130:476–489
Stambolic V, Woodgett JR (2006) Functional distinctions of protein kinase B/Akt isoforms defined by their influence on cell migration. Trends Cell Biol 16:461–466
Stambolic V, Suzuki A, de la Pompa JL, Brothers GM, Mirtsos C, Sasaki T, Ruland J, Penninger JM, Siderovski DP, Mak TW (1998) Negative regulation of PKB/Akt-dependent cell survival by the tumor suppressor PTEN. Cell 95:29–39
Stupp R, Ruegg C (2007) Integrin inhibitors reaching the clinic. J Clin Oncol 25:1637–1638
Symons M (2008) Cell biology: watching the first steps of podosome formation. Curr Biol 18:R925–927
TCGA (2008) Comprehensive genomic characterization defines human glioblastoma genes and core pathways. Nature 455:1061–1068
Toy H, Yavas O, Eren O, Genc M, Yavas C (2009) Correlation between osteopontin protein expression and histological grade of astrocytomas. Pathol Oncol Res 15:203–207
Tu Y, Lu J, Fu J, Cao Y, Fu G, Kang R, Tian X, Wang B (2010) Over-expression of neuroepithelial-transforming protein 1 confers poor prognosis of patients with gliomas. Jpn J Clin Oncol 40:388–394
Tucker GC (2006) Integrins: molecular targets in cancer therapy. Curr Oncol Rep 8:96–103
Tysnes BB, Larsen LF, Ness GO, Mahesparan R, Edvardsen K, Garcia-Cabrera I, Bjerkvig R (1996) Stimulation of glioma-cell migration by laminin and inhibition by anti-alpha3 and anti-beta1 integrin antibodies. Int J Cancer 67:777–784
Viapiano MS, Matthews RT, Hockfield S (2003) A novel membrane-associated glycovariant of BEHAB/brevican is up-regulated during rat brain development and in a rat model of invasive glioma. J Biol Chem 278:33239–33247

Viglietto G, Motti ML, Bruni P, Melillo RM, D'Alessio A, Califano D, Vinci F, Chiappetta G, Tsichlis P, Bellacosa A, Fusco A, Santoro M (2002) Cytoplasmic relocalization and inhibition of the cyclin-dependent kinase inhibitor p27(Kip1) by PKB/Akt-mediated phosphorylation in breast cancer. Nat Med 8:1136–1144

Wakabayashi K, Kambe F, Cao X, Murakami R, Mitsuyama H, Nagaya T, Saito K, Yoshida J, Seo H (2004) Inhibitory effects of cyclosporin A on calcium mobilization-dependent interleukin-8 expression and invasive potential of human glioblastoma U251MG cells. Oncogene 23:6924–6932

Wang G, Kang C, Pu P (2010) Increased expression of Akt2 and activity of PI3K and cell proliferation with the ascending of tumor grade of human gliomas. Clin Neurol Neurosurg 112:324–327

Watters JJ, Schartner JM, Badie B (2005) Microglia function in brain tumors. J Neurosci Res 81:447–455

Wesolowska A, Kwiatkowska A, Slomnicki L, Dembinski M, Master A, Sliwa M, Franciszkiewicz K, Chouaib S, Kaminska B (2008) Microglia-derived TGF-beta as an important regulator of glioblastoma invasion–an inhibition of TGF-beta-dependent effects by shRNA against human TGF-beta type II receptor. Oncogene 27:918–930

Wheeler AP, Ridley AJ (2004) Why three Rho proteins? RhoA, RhoB, RhoC, and cell motility. Exp Cell Res 301:43–49

Wick MJ, Dong LQ, Riojas RA, Ramos FJ, Liu F (2000) Mechanism of phosphorylation of protein kinase B/Akt by a constitutively active 3-phosphoinositide-dependent protein kinase-1. J Biol Chem 275:40400–40406

Wisniewski P, Ellert-Miklaszewska A, Kwiatkowska A, Kaminska B (2010) Non-apoptotic Fas signaling regulates invasiveness of glioma cells and modulates MMP-2 activity via NFkappaB-TIMP-2 pathway. Cell Signal 22:212–220

Wolff M, Boker DK (1989) Immunohistochemical demonstration of immunoglobulins and albumin in human brain tumors. Clin Neuropathol 8:72–78

Yamamoto M, Sawaya R, Mohanam S, Bindal AK, Bruner JM, Oka K, Rao VH, Tomonaga M, Nicolson GL, Rao JS (1994) Expression and localization of urokinase-type plasminogen activator in human astrocytomas in vivo. Cancer Res 54:3656–3661

Yamana N, Arakawa Y, Nishino T, Kurokawa K, Tanji M, Itoh RE, Monypenny J, Ishizaki T, Bito H, Nozaki K, Hashimoto N, Matsuda M, Narumiya S (2006) The Rho-mDia1 pathway regulates cell polarity and focal adhesion turnover in migrating cells through mobilizing Apc and c-Src. Mol Cell Biol 26:6844–6858

Yang M, Li Y, Chilukuri K, Brady OA, Boulos MI, Kappes JC, Galileo DS (2011) L1 stimulation of human glioma cell motility correlates with FAK activation. J Neurooncol 105:27–44

Yu J, Zhang Y, McIlroy J, Rordorf-Nikolic T, Orr GA, Backer JM (1998) Regulation of the p85/p110 phosphatidylinositol 3′-kinase: stabilization and inhibition of the p110alpha catalytic subunit by the p85 regulatory subunit. Mol Cell Biol 18:1379–1387

Yuan S, Miller DW, Barnett GH, Hahn JF, Williams BR (1995) Identification and characterization of human beta 2-chimaerin: association with malignant transformation in astrocytoma. Cancer Res 55:3456–3461

Zagzag D, Friedlander DR, Miller DC, Dosik J, Cangiarella J, Kostianovsky M, Cohen H, Grumet M, Greco MA (1995) Tenascin expression in astrocytomas correlates with angiogenesis. Cancer Res 55:907–914

Zagzag D, Friedlander DR, Dosik J, Chikramane S, Chan W, Greco MA, Allen JC, Dorovini-Zis K, Grumet M (1996) Tenascin-C expression by angiogenic vessels in human astrocytomas and by human brain endothelial cells in vitro. Cancer Res 56:182–189

Zagzag D, Amirnovin R, Greco MA, Yee H, Holash J, Wiegand SJ, Zabski S, Yancopoulos GD, Grumet M (2000) Vascular apoptosis and involution in gliomas precede neovascularization: a novel concept for glioma growth and angiogenesis. Lab Invest 80:837–849

Zhai H, Heppner FL, Tsirka SE (2011) Microglia/macrophages promote glioma progression. Glia 59:472–485

Zhang PB, Liu Y, Li J, Kang QY, Tian YF, Chen XL, Zhao JJ, Shi QD, Song TS, Qian YH (2005) Ependymal/subventricular zone cells migrate to the peri-infarct region and differentiate into neurons and astrocytes after focal cerebral ischemia in adult rats. J First Mil Med Univ 25:1201–1206

Zhang B, Gu F, She C, Guo H, Li W, Niu R, Fu L, Zhang N, Ma Y (2009) Reduction of Akt2 inhibits migration and invasion of glioma cells. Int J Cancer 125:585–595

Zhang J, Han L, Zhang A, Wang Y, Yue X, You Y, Pu P, Kang C (2010) AKT2 expression is associated with glioma malignant progression and required for cell survival and invasion. Oncol Rep 24:65–72

Zhao Y, Lyons CE Jr, Xiao A, Templeton DJ, Sang QA, Brew K, Hussaini IM (2008) Urokinase directly activates matrix metalloproteinases-9: a potential role in glioblastoma invasion. Biochem Biophys Res Commun 369:1215–1220

Zheng X, Jiang F, Katakowski M, Lu Y, Chopp M (2011) ADAM17 promotes glioma cell malignant phenotype. Mol Carcinog 51:150–164

Zhou BP, Liao Y, Xia W, Spohn B, Lee MH, Hung MC (2001) Cytoplasmic localization of p21Cip1/WAF1 by Akt-induced phosphorylation in HER-2/neu-overexpressing cells. Nat Cell Biol 3:245–252

Zucker S, Drews M, Conner C, Foda HD, DeClerck YA, Langley KE, Bahou WF, Docherty AJ, Cao J (1998) Tissue inhibitor of metalloproteinase-2 (TIMP-2) binds to the catalytic domain of the cell surface receptor, membrane type 1-matrix metalloproteinase 1 (MT1-MMP). J Biol Chem 273:1216–1222

Chapter 8
Receptor Tyrosine Kinases: Principles and Functions in Glioma Invasion

Mitsutoshi Nakada, Daisuke Kita, Lei Teng, Ilya V. Pyko, Takuya Watanabe, Yutaka Hayashi, and Jun-ichiro Hamada

Abstract Protein tyrosine kinases are enzymes that are capable of adding a phosphate group to specific tyrosines on target proteins. A receptor tyrosine kinase (RTK) is a tyrosine kinase located at the cellular membrane and is activated by binding of a ligand via its extracellular domain. Protein phosphorylation by kinases is an important mechanism for communicating signals within a cell and regulating cellular activity; furthermore, this mechanism functions as an "on" or "off" switch in many cellular functions. Ninety unique tyrosine kinase genes, including 58 RTKs, were identified in the human genome; the products of these genes regulate cellular proliferation, survival, differentiation, function, and motility. Tyrosine kinases play a critical role in the development and progression of many types of cancer, in addition to their roles as key regulators of normal cellular processes. Recent studies have revealed that RTKs such as epidermal growth factor receptor (EGFR), platelet-derived growth factor receptor (PDGFR), c-Met, Tie, Axl, discoidin domain receptor 1 (DDR1), and erythropoietin-producing human hepatocellular carcinoma (Eph) play a major role in glioma invasion. Herein, we summarize recent advances in understanding the role of RTKs in glioma pathobiology, especially the invasive phenotype, and present the perspective that RTKs are a potential target of glioma therapy.

Keywords Glioma • Glioblastoma • Invasion • Tyrosine kinase receptor • EGFR • PDGFR • c-Met • Tie • Axl • DDR1 • Eph • TrkA • Cross-talk • Tyrosine kinase inhibitor • Clinical trial

M. Nakada (✉) • D. Kita • I.V. Pyko • T. Watanabe • Y. Hayashi • J. Hamada
Department of Neurosurgery, Division of Neuroscience, Graduate School of Medical Science, Kanazawa University, 13-1 Takara-machi, Kanazawa, Ishikawa, 920-8640, Japan
e-mail: mnakada@med.kanazawa-u.ac.jp; kitad@ns.m.kanazawa-u.ac.jp; ilia11892@gmail.com; wat6713@gmail.com; yuh@ns.m.kanazawa-u.ac.jp; jhamada@ns.m.kanazawa-u.ac.jp

L. Teng
Department of Neurosurgery, Division of Neuroscience, Graduate School of Medical Science, Kanazawa University, 13-1 Takara-machi, Kanazawa, Ishikawa, 920-8641, Japan

Department of Neurosurgery, The First Clinical College of Harbin Medical University, 23 Youzheng Street, Nangang district, Harbin 150001, Heilongjiang Province, People's Republic of China
e-mail: tengleina@hotmail.com

J. Barańska (ed.), *Glioma Signaling*, Advances in Experimental Medicine and Biology 986, DOI 10.1007/978-94-007-4719-7_8, © Springer Science+Business Media Dordrecht 2013

Abbreviations

Ang	Angiopoietin
BBB	Blood brain barrier
CTGF	Connective tissue growth factor
DDR1	Discoidin domain receptor 1
EC	Endothelial cell
ECM	Extracellular matrix
EGFR	Epidermal growth factor receptor
Eph	Erythropoietin-producing human hepatocellular carcinoma
ERK	Extracellular signal-regulated kinase
FAK	Focal adhesion kinase
FGFR	Fibroblast growth factor receptor
Gas6	Growth arrest–specific gene 6
GBM	Glioblastoma multiforme
GPI	Glycosylphosphatidyl-inositol
HB-EGF	Heparin-binding EGF-like growth factor
HGF	Hepatocyte growth factor
HIF	Hypoxia inducible factor
IDH1	Isocitrate dehydrogenase-1
JAK	Janus kinase
MAPK	Mitogen-activated protein kinase
MEK	MAPK kinase
MMP	Matrix metalloproteinase
mAb	Monoclonal antibody
MT1-MMP	Membrane-type1-MMP
NGF	Nerve growth factor
NF-kB	Nuclear factor-kappa B
OS	Overall survival
PDGFR	Platelet derived growth factor receptor
PFS	Progression-free survival
PI3K	Phosphatidylinositol 3-kinase
PKC	Protein kinase C
PLC	Phospholipase C
PTEN	Phosphatase and tensin homolog deleted from chromosome 10
PTK	Protein tyrosine kinase
RTK	Receptor tyrosine kinase
STAT	Signal transducer and activator of transcription
TAMR	A member of the Tyro3, Axl, and Mer family of receptor tyrosine kinase
TCGA	The Cancer Genome Atlas
TGF-α	Transforming growth factor alpha
TKI	Tyrosine kinase inhibitor
TMZ	Temozolomide
TrkA	Neurotrophic tyrosine kinase receptor type 1
uPA	Urokinase-type plasminogen activator
VEGF	Vascular endothelial growth factor

8.1 Introduction

Glioblastoma (GBM) is an extremely aggressive, highly vascularized, infiltrative tumor with a median survival duration of 12–15 months after initial diagnosis (Nagane 2011). GBMs invade the surrounding brain tissue, making complete surgical excision highly improbable. Despite current therapeutic strategies, these tumors almost universally recur because of the invading GBM cells left after excision and are associated with a poor survival rate (Nakada et al. 2007).

Currently, numerous studies are attempting to decipher the molecular mechanism of invasion and to better understand the molecular mechanisms responsible for invasion processes (Chuang et al. 2004; Salhia et al. 2005; Nakada et al. 2007; Onishi et al. 2011). Among the many molecules that were reported as invasion promoters, members of the protein tyrosine kinase (PTK) family play a major role in modulating invasion (Nakada et al. 2007). The human protein kinase genome contains 518 protein kinase genes, including PTK genes that encode transmembrane receptor tyrosine kinases (RTKs) and soluble cytoplasmic tyrosine kinases that are also known as non-RTKs (Manning et al. 2002). More than 58 mammalian RTKs and 37 non-RTKs have been identified. RTKs contain an intracellular catalytic PTK domain and regulatory sequences, a transmembrane domain, and an extracellular ligand-binding domain. RTKs modulate a wide range of cellular events, including proliferation, migration, metabolism, differentiation, and apoptosis, under physiological as well as pathological conditions (Schlessinger 2000). The phosphorylation of tyrosine residues in RTKs is essential for maintaining cellular homeostasis and modulating gene expression in various intercellular and intracellular signaling pathways. Because the complex signaling network triggered by RTKs eventually leads to either activation or repression of various gene subsets, RTKs regulate intercellular communication and control cell proliferation, mitogenesis, survival, differentiation, motility, and metabolism (Schlessinger 2000). The individual cellular consequences of RTK activation are complex and depend on the cell type and the activated signal transduction pathway.

Several lines of experimental evidence have revealed that aberrant RTK activation frequently occurs during glioma initiation and progression and that these tumorigenic cascades may cooperate through multiple signaling cross-talks in the malignant transformation of cells, treatment resistance, and disease relapse (Table 8.1). Because GBMs actively synthesize a substantial variety of RTKs that contribute to invasion, a systematic approach to inhibiting RTKs is being undertaken as a treatment adjunct.

Table 8.1 RTK associated with glioma invasion

RTK	Ligand	Expression in GBM	Amplification in GBM	Downstream signaling
EGFR	EGF	High	40 %	PI3K/Akt, Ras/MAPK, FAK, MMP-2
PDGFR	PDGF	High	11 %	PI3K/Akt, Ras/MAPK, JAK/STAT
ERBB2	Neuregulin-1	High	8 %	PI3K/Akt, Ras/MAPK, PLC-γ, JAK/STAT, MT-MMP, FAK
c-Met	HGF	High	20 %	PI3K/Akt, Ras/MAPK
Tie	Ang	High	Not reported	MMP-2
Axl	Gas6	High	Not reported	PI3K/Akt
DDR1	Collagen	High	Not reported	MMP-2
Eph	Ephrin	High or low	Not reported	R-Ras, Rac1
Trk	NGF, CTGF	High or low	Not reported	NF-kB, ZEB-1

Designations:
Ang, angiopoietin; CTGF, connective tissue growth factor; DDR1, discoidin domain receptor 1; EGFR, epidermal growth factor receptor; Eph, erythropoietin-producing human hepatocellular carcinoma; FAK, focal adhesion kinase; Gas6, growth arrest–specific gene 6; HGF, hepatocyte growth factor; MAPK, Mitogen-activated protein kinase; PDGFR, platelet derived growth factor receptor; PI3K, phosphatidylinositol-3-kinase

In this chapter, the most recent advancements in the structural and functional characterization of invasion signal transduction elements of the RTK signaling network and the molecular mechanisms involved in glioma invasion are described. We provide an overview of this field, highlighting areas with the strongest research evidence for the translational potential of the use of tyrosine kinase inhibitors (TKIs).

8.2 EGFR/EGF

Epidermal growth factor receptor (EGFR) belongs to a large family of cell surface receptors with intrinsic protein tyrosine kinase activity. The EGFR family comprises four members designated EGFR (also known as ErbB1/HER1), ErbB2 (HER2/Neu), ErbB3 (HER3), and ErbB4 (HER4) (Jorissen et al. 2003). Six ligands are known to activate EGFR: EGF, transforming growth factor alpha (TGF-α), amphiregulin, betacellulin, heparin-binding EGF-like growth factor (HB-EGF), and epiregulin (Bogdan and Klambt 2001; Normanno et al. 2006). These ligands are secreted by glioma cells as well as by tumor microenvironmental cells such as microglias and reactive astrocytes (Hoelzinger et al. 2007). Upon binding of extracellular ligands, EGFR undergoes dimerization, resulting in trans-autophosphorylation of its cytoplasmic domain. EGFR can pair with another EGFR to form an active homodimer or with another member of the EGFR family to create a heterodimer (Yarden and Sliwkowski 2001). For example, EGFR can easily form heterodimers with ErbB2, which has a reduced internalization capacity compared with that of EGFR and thus prolongs EGFR signaling (Jones et al. 2006) (Fig. 8.1).

EGFR and its ligands are often over-expressed in human carcinomas. In GBM, EGFR gene amplification is the most frequent RTK alteration (approximately 40 %) (Libermann et al. 1985). EGFR overexpression and/or gene alteration is frequently observed in primary (*de novo*) GBM, which develops rapidly, has a short clinical history, and does not show evidence of less malignant precursor lesions (Ekstrand et al. 1992; Ohgaki et al. 2004; Wong et al. 1987). Genomic analysis by The Cancer Genome Atlas (TCGA) network revealed that *EGFR* aberration is related to the classical subtype of GBM (Network 2008). Amplification of the *EGFR* gene is also associated with structural alterations and the most common of these is called EGFR variant III (EGFRvIII). EGFRvIII is a mutant with an in-frame deletion of exons 2–7 from the extracellular region and can transmit constitutive growth signaling in a ligand-independent manner (Ekstrand et al. 1992; Yamazaki et al. 1988). EGFRvIII expression in glioma cells stimulates expression of TGF-α and HB-EGF, suggesting that EGFRvIII plays a role in generating an autocrine loop with wild-type EGFR expression (Ramnarain et al. 2006). Genetic alterations that affect EGFR signaling result in the activation of several downstream pathways such as the phosphatidylinositol 3-kinase (PI3K)/Akt and Ras/Raf/MEK (MAPK kinase)/MAPK (mitogen-activated protein kinase) pathways, which mediate cell proliferation, survival, and mobility (Kita et al. 2007; Ohgaki and Kleihues 2009). The PI3K/Akt pathway is negatively regulated by the tumor suppressor phosphatase and tensin homolog deleted from chromosome 10 (PTEN), which is mutated in 20–40 % of GBMs and is a hallmark of this disease (Ohgaki et al. 2004). PTEN dephosphorylates focal adhesion kinase (FAK), which is a key molecule for cell interaction with the extracellular matrix (ECM) (Gu et al. 1999; Kita et al. 2001; Tamura et al. 1998). Dephosphorylated FAK interferes with EGFR vIII-mediated glioma cell invasion, indicating that the EGFR-PTEN-FAK interaction plays an important role in glioma invasion (Cai et al. 2005) (Fig. 8.1).

Several reports have showed that GBM patients with EGFR overexpression or mutation have shorter survival, suggesting that EGFR alterations are associated with highly aggressive GBMs (Barker et al. 2001; Feldkamp et al. 1999; Shinojima et al. 2003). One report indicated that coexpression of EGFRvIII and PTEN was a positive indicator of responsiveness to EGFR inhibitors (e.g., gefitinib and erlotinib) in patients with GBM (Mellinghoff et al. 2005).

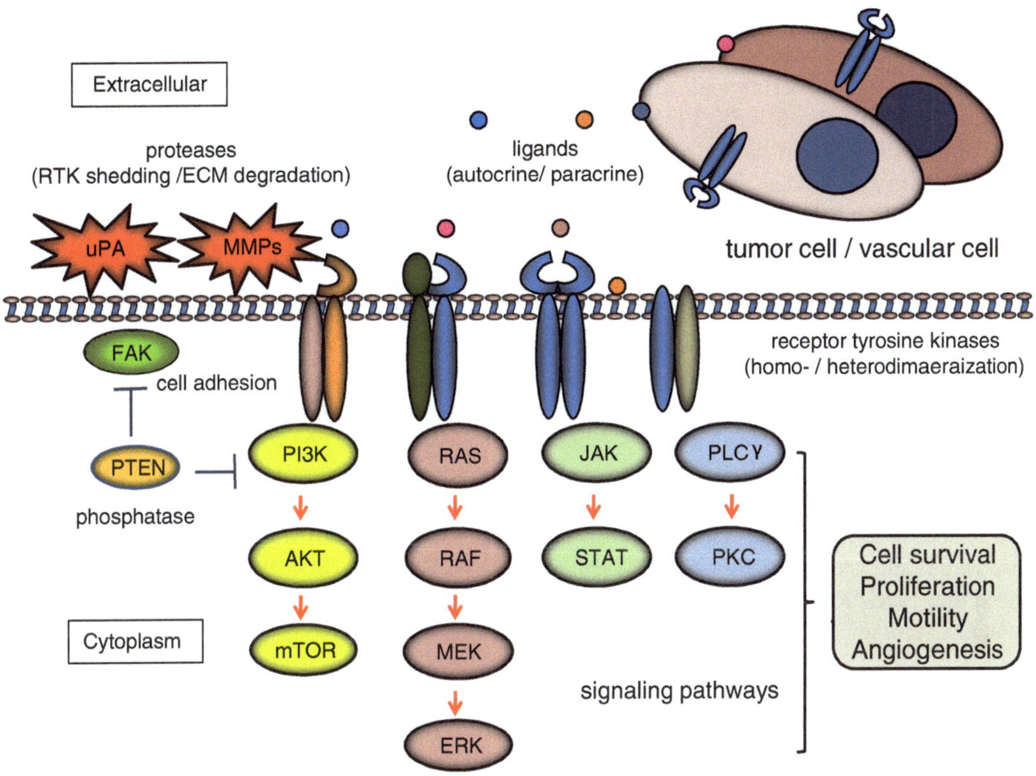

Fig. 8.1 Cell signaling pathways induced by receptors tyrosine kinase (RTK). Homo-/heterodimerization of RTKs are caused by their ligands in autocrine or paracrine fashion. The dimerized receptors can initiate signal transduction cascades involved in cell survival, proliferation, motility and angiogenesis and so on, e.g.: phosphatidylinositol 3-kinase (PI3K)/AKT/mTOR; RAS/RAF/mitogen-activated protein kinase (MEK)/extracellular signal-regulated kinase (ERK); Janus kinase (JAK)/signal transducer and activator of transcription (STAT); and phospholipase Cγ (PLC)γ/protein kinase C (PKC). Examples of cross-talk between RTKs' signaling and proteins associated with cell invasion (e.g. urokinase-type plasminogen activator (uPA), matrix metalloproteinase (MMP)s, focal adhesion kinase (FAK) and phosphatase and tensin homolog deleted from chromosome 10 (PTEN)) are demonstrated. Red arrows and blue bars indicate activation and suppression, respectively (ECM, extracellular matrix)

8.3 PDGFR/PDGF

Following EGFR/EGF signaling, aberrant platelet-derived growth factor receptor (PDGFR)/PDGF signaling is one of the hallmarks of GBM biology. Overexpression of PDGFR subtypes α and β and PDGF ligands A–D has been observed in glial tumors of all grades and is possibly associated with malignant progression (Fleming et al. 1992; Guha et al. 1995; Hermanson et al. 1992; Lokker et al. 2002; Nister et al. 1988; Ozawa et al. 2010). An experimental study revealed that glioma-like tumors can be induced after overproduction of PDGFB in the mouse brain (Uhrbom et al. 1998). Histochemical studies revealed that PDGFRα and PDGFA are expressed in glioma cells, whereas PDGFRβ and PDGFB have been found in the surrounding endothelial cells (ECs) (Hermanson et al. 1992; Plate et al. 1992). The expression of this receptor in blood vessels suggests that paracrine activation is also possible with respect to tumor cell migration and colony formation (Hoelzinger et al. 2007; Shih and Holland 2006). Although the expression of PDGFC and D ligands has also been demonstrated in gliomas, the clinical and biological significance of their expression has not been determined (Lokker et al. 2002). Ligand-receptor co-expressions in tumor cells allow for both autocrine and paracrine forms of activation. Some reports suggest that PDGFR/PDGF signaling thorough this autocrine/paracrine pair results in increased GBM cell motility *in vivo* (Cattaneo et al. 2006; Natarajan et al. 2006). Furthermore, these autocrine and/or paracrine loops can stimulate downstream signal transduction pathways including Ras/MAPK, PI3K/Akt and Janus kinase (JAK)/signal transducer and activator of transcription (STAT) and have pivotal roles in proliferation, differentiation, survival, and invasion (Blume-Jensen and Hunter 2001; Khoshyomn et al. 1999; Valius and Kazlauskas 1993) (Fig. 8.1).

The association between PDGFR expression and the prognosis of glioma patients is controversial. PDGFRα expression in low-grade gliomas was reportedly associated with both poor (Varela et al. 2004) and favorable (Ribom et al. 2002) prognosis. Furthermore, PDGFA expression, but not PDGFRα expression, was useful in predicting tumor grade in oligodendrogliomas (Majumdar et al. 2009). PDGFRα expression was not associated with survival in a series of high-grade pediatric gliomas and GBMs (Liang et al. 2008; Martinho et al. 2009). However, the absence of PDGFA expression was significantly associated with poor prognosis in patients with glioma (Martinho et al. 2009). PDGFR and ligand overexpression tend to be associated with loss of the TP53 tumor suppressor, which is characteristic of secondary GBMs that develop from less malignant precursors (Ohgaki and Kleihues 2009). According to the TCGA consortium, *PDGFRA* amplification (11 %) and isocitrate dehydrogenase-1 (*IDH1*) mutation are hallmarks of a proneural subtype of GBM, suggesting an association between this subtype and secondary GBM (Network 2008; Verhaak et al. 2010). Another study demonstrated that *PDGFRA* amplification was observed in 21 % of gliomas, although *PDGFRA*-activating mutations were not found (Martinho et al. 2009).

8.4 ERBB2

ErbB2 (HER2/Neu) belongs to the EGFR receptor family that contains the other three members: EGFR, ErbB3, and ErbB4 (see EGFR/EGF section). Although the intracellular tyrosine kinase domain of the EGFR family is highly conserved, none of the EGFR family ligands bind ErbB2 because of its extracellular region structure. Thus, the role of ErbB2 depends on the patterns of dimerization within the family (Normanno et al. 2006). For example, among all possible ErbB2-containing heterodimeric receptor complexes, the most potent signaling module in terms of cell proliferation and *in vitro* transformation is the ErbB2/ErbB3 heterodimer, even though ErbB3 lacks kinase activity (Citri et al. 2003). These homo- or heterodimers have various potencies for the induction of signaling pathways such as the Ras/Raf/MEK/MAPK pathway for proliferation, the PI3K/Akt pathway for survival (Ben-Levy et al. 1994; Prigent and Gullick 1994), the phospholipase C (PLC)-γ pathway for cell migration and invasion (Khoshyomn et al. 1999), and the STAT pathway for cell cycle regulation (Gao et al. 2010) (Fig. 8.1).

Forced ErbB2 expression can transform cells into the invasive phenotype in association with expression of membrane-type1 (MT1) matrix metalloproteinase (MMP), which is a key enzyme in glioma invasion (Miyamori et al. 2000; Nakada et al. 2001; Sato et al. 1994). In addition, post-transcriptional shedding of cell surface ErbB2 has been reported to be processed by proteases that can degrade the ECM (e.g., MMP2, MMP9, MMP13, and uPA) (Gondi et al. 2009; Spencer et al. 2000; Yong et al. 2010), indicating a strong association between ErbB2 expression and glioma invasion.

Although ErbB2 protein expression is mainly seen in high-grade gliomas (Andersson et al. 2004; Engelhard et al. 1995), evidence for the prognostic value of ErbB2 expression levels in GBMs is presently sparse (Gulati et al. 2010; Haapasalo et al. 1996; Hiesiger et al. 1993; Mineo et al. 2007; Schwechheimer et al. 1994). On the other hand, both ErbB2 and ErbB4 expression levels have been shown to predict prognosis in childhood medulloblastoma and ependymoma (Gilbertson et al. 1997; 2002). Although few reports have described mutation of the *ErbB2* gene (Stephens et al. 2004), a recent study of the TCGA consortium revealed that mutation was observed in 7 of 91 (8 %) GBMs (Network 2008; Verhaak et al. 2010).

Furthermore, ErbB2 overexpression has been well described in human breast cancers (20–30 %), which correlates with more aggressive tumors and a poorer prognosis (Hortobagyi 2005).

8.5 c-Met/HGF

Hepatocyte growth factor (HGF) was originally identified as a polypeptide growth factor for hepatocytes and is believed to play an important role in liver regeneration (Yamada et al. 1994). HGF functions as a mitogen for a variety of cell types and as a morphogen and motogen for some epithelial cells that express its receptor (Moriyama et al. 1995). The receptor protein for HGF, c-Met, is encoded by the *c-met* proto-oncogene, which has tyrosine kinase activity and was originally described as an activated oncogene in a human osteosarcoma cell line (Cooper et al. 1984). Recently, attention has been focused on the role of the HGF/c-Met system because of its multiple biological activities including motility, proliferation, survival, and morphogenesis.

Under normal conditions, HGF-induced c-Met activation is tightly regulated by paracrine ligand delivery, ligand activation at the target cell surface, and ligand activated receptor internalization and degradation (Cecchi et al. 2010). Despite its functions under normal conditions, HGF/c-Met signaling contributes to oncogenesis and tumor progression in several cancers and promotes aggressive cellular invasiveness that is strongly linked to tumor metastasis (Rosario and Birchmeier 2003; Zhang and Vande Woude 2003).

The HGF/c-Met pathway has been implicated in a wide variety of human malignancies. Overexpression of HGF and/or c-Met is frequently observed and amplification of the *c-met* gene has been reported in several tumor types (Burgess et al. 2006). Several studies have described the expression of HGF and c-met as well as HGF activator mRNAs in glioma cell lines and tissues, particularly GBM (Koochekpour et al. 1997; Moriyama et al. 1995). On the other hand, the expression of HGF and c-Met was low or barely detectable in low-grade astrocytoma and c-Met immunoreactivity was correlated with the histological grade of the tumor.

A few studies suggested that HGF and c-Met are expressed in human gliomas and that expression levels correlated with tumor grade (Moriyama et al. 1998b). Likewise, HGF expression was significantly higher in high-grade tumors than in low-grade tumors based on HGF content detection in samples of different clinical grades (Abounader and Laterra 2005; Lamszus et al. 1999). c-Met receptor expression has also been detected in malignant brain tumors including all gliomas, medulloblastomas, ependymomas, and schwannomas (Koochekpour et al. 1997; Moriyama et al. 1998b). The results of an overexpression study provided sufficient evidence implicating the HGF/c-Met pathway in brain tumorigenesis and malignant progression; this study demonstrated that HGF/c-Met plays important and critical roles in brain tumor formation and growth.

The HGF/c-Met system in vascular ECs works as signal transduction molecules/pathways in gliomas that mediate neovascularization. Numerous *in vivo* and *in vitro* studies have indicated that HGF and c-Met are expressed and functional in neuromicrovascular and brain tumor vascular cells. The HGF/c-Met system is highly activated in cultured neural microvascular ECs. In gliomas, HGF stimulates the proliferation of neuromicrovascular ECs by paracrine and autocrine mechanisms. Other studies demonstrated that HGF-dependent interactions between glioma cells, and between glioma cells and the endothelium, can contribute to the heterogeneous proliferative and angiogenic phenotypes of malignant gliomas *in vivo* (Abounader and Laterra 2005). Moriyama et al. explored the effect of HGF on vascular endothelial growth factor (VEGF) expression in c-Met-positive human glioma cell lines and their results suggest that HGF can act as an indirect angiogenic factor through autocrine induction of VEGF expression and secretion in malignant gliomas in addition to its direct angiogenic activities (Moriyama et al. 1998a). Taken together, these *in vitro* and *in vivo* findings suggest the multifunctional and multilevel involvement of the HGF/c-Met pathway in brain tumor angiogenesis as well as brain tumor growth.

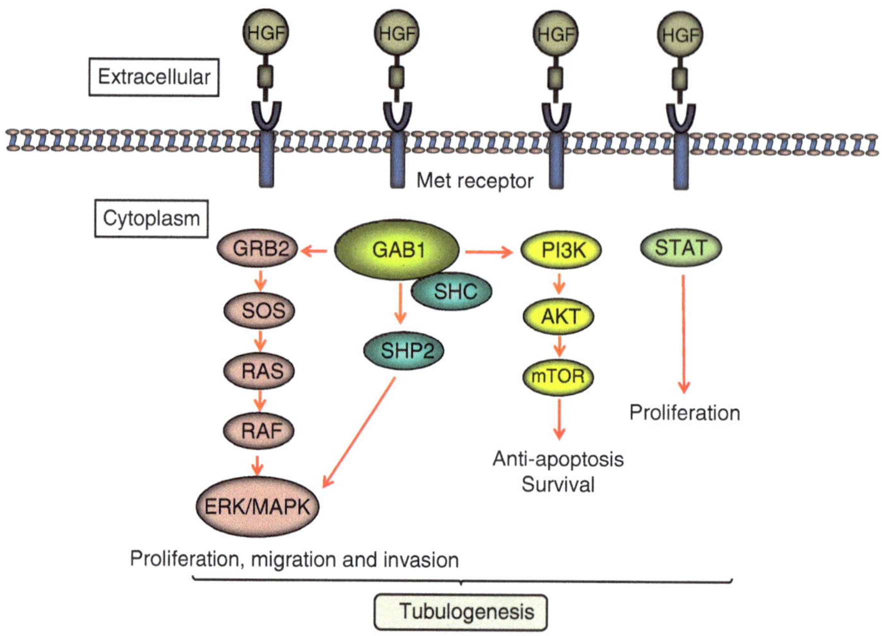

Fig. 8.2 Major signaling pathways by the receptor tyrosine kinase c-Met. Growth factor receptor-bound protein 2 (GRB2), Grb2-associated adaptor protein 1 (GAB1), hepatocyte growth factor (HGF), phosphatidylinositol 3-kinase (PI3K), son of sevenless (SOS), rat sarcoma oncogene homolog (RAS), extracellular receptor kinase (ERK), mitogen-activated protein kinase (MAPK), signal transducer and activator of transcription (STAT), SRC, SRC homology protein tyrosine phosphatase 2 (SHP2), SRC homology domain c-terminal adaptor homolog (SHC), mammalian target of rapamycin (mTOR)

Activation of the HGF/c-Met axis promotes proliferation and survival through a variety of downstream effectors including Gab1, Grb2, and PI3K (Abounader and Laterra 2005; Ponzetto et al. 1994). Following activation of the HGF/c-Met pathway, the RAS-MAPK signaling pathway plays an essential role in morphogenesis. The activation of c-Met prevents apoptosis through activation of PI3K and subsequent Akt activation. Cross-talk between the PI3K/Akt and the RAS-MAPK pathways has been implicated in promoting cell survival, migration, and invasion (Zeng et al. 2002) (Fig. 8.2).

8.6 Tie/Ang

The Angiopoietin/Tie system acts as a vascular-specific ligand/receptor system and plays an essential role in tumor-associated angiogenesis. The angiopoietin family includes four ligands, namely, angiopoietin-1 (Ang-1), angiopoietin-2 (Ang-2), angiopoietin-3 (Ang-3), and angiopoietin-4 (Ang-4), and two corresponding RTKs (Tie1 and Tie2). Ang-1 acts as an agonist, whereas Ang-2 acts as an antagonist, of the Tie2 receptor (Maisonpierre et al. 1997). Tie2 activation promotes vessel assembly and maturation by mediating EC survival signals and regulating the recruitment of mural cells. The Tie/Ang-1 system may function during vessel maturation and stabilization and Ang-1 is a prominent regulator of vascular development (Lee et al. 2009; Machein et al. 2004). In contrast, Ang-2 is produced by ECs and acts as an autocrine antagonist of Ang-1-mediated Tie2 activation. Ang-2 is an angiogenic factor that antagonizes Ang-1 activity by competitively inhibiting the binding of Ang-1 to its cognate endothelial receptor, Tie2, causing vasculature destabilization. Ang-2 also acts in concert with VEGF to regulate vessel growth (Hu et al. 2006) (Fig. 8.3).

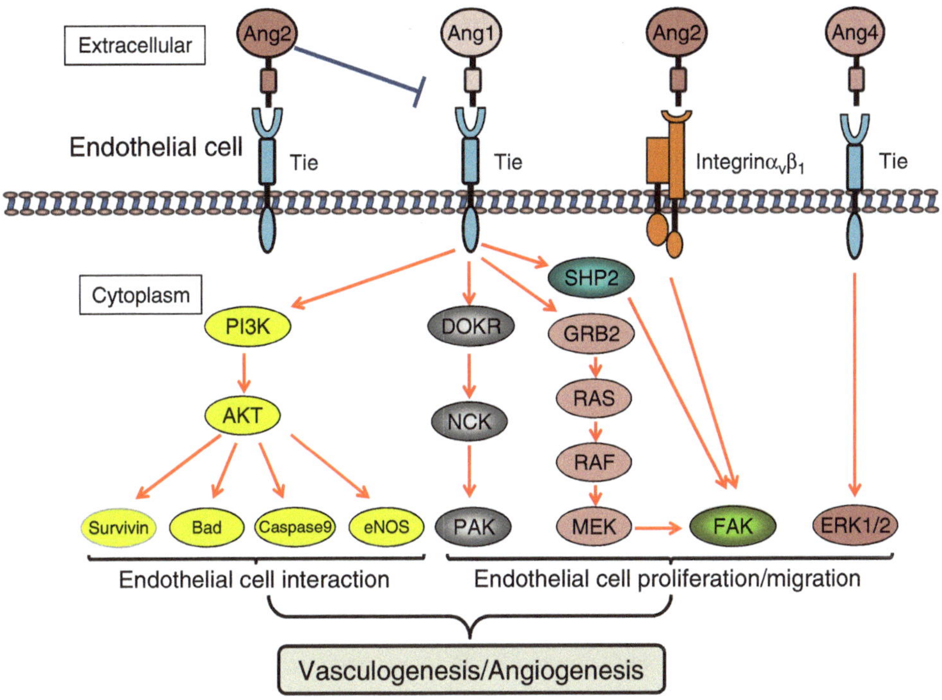

Fig. 8.3 The major Ang/Tie signaling pathways in glioma angiogenesis and migration. Angiopoietin (Ang), growth factor receptor-bound protein 2 (GRB2), phosphatidylinositol 3-kinase (PI3K), rat sarcoma oncogene homolog (RAS), SH2 domain-containing phosphatase (SHP2), focal adhesion kinase (FAK), endothelial nitric oxide synthase (eNOS), docking protein R (Dok-R), extracellular receptor kinase (ERK), p21-activated kinase (PAK)

The role of Ang-1 in tumor-associated angiogenesis remains controversial because some reports imply that Ang-1 induction impairs angiogenesis and inhibits tumor growth whereas other authors suggest that Ang-1 overexpression promotes tumor growth in some animal models (Hawighorst et al. 2002; Hayes et al. 2000; Machein et al. 2004). However, Machein et al. reported that Ang-1 can promote tumor angiogenesis in a rat glioma model (Machein et al. 2004).

Ang-2 immunoreactivity is higher in malignant gliomas than in low-grade gliomas, which documents a strong correlation between the expression of Ang-2 and increasing glioma tumor grade (Koga et al. 2001). Increasing data suggest that Ang-2 expression is negatively correlated with vessel maturation in malignant gliomas and that VEGF expression is positively correlated with vessel maturation in low-grade gliomas.

Conflicting results have been reported in the literature regarding the role of the Tie2/Ang system in tumor angiogenesis. The tyrosine kinase receptor Tie2 was initially reported as a specific vascular receptor present in both normal and tumoral ECs, including ECs in astrocytomas, and its levels correlate positively with increasing malignancy (Liu et al. 2010; Stratmann et al. 1998). Tie2 is also expressed in glioma cells and brain tumor stem cells present in malignant gliomas and its expression and activation increases with increasing astrocytoma malignancy grade (Zadeh et al. 2004b).

Studies on *in vivo* human glioma biopsies have showed that Ang-2, MMP-2, MT1-MMP, and laminin 5γ2 are co-overexpressed in the invasive areas but not in the central regions of the glioma tissues. *In vitro* data also demonstrated that Ang-2 promoted the expression and activation of MMP-2, MT1-MMP, and laminin 5γ2, which may be essential for malignant glioma invasiveness (Brinckerhoff and Matrisian 2002; Lohi 2001). In addition, inhibition of Tie2 activation significantly decreased GBM xenograft growth by disrupting tumor vascularity (Zadeh et al. 2004b).

Accumulating evidence suggests that Tie2 activation regulates angiogenesis in a highly context- and tissue-dependent manner and closely collaborates with VEGF and possibly with other angiogenesis regulators (Gale et al. 2002; Zadeh et al. 2004b). Activation of the Tie/Ang functional axis promotes glioma proliferation and migration through various downstream effectors, including FAK, VEGF-A, PI3K, $\alpha_{\gamma}\beta_{1,}$ and extracellular signal-regulated kinase 1/2 (ERK1/2) (Brinckerhoff and Matrisian 2002; Lee et al. 2009; Zadeh et al. 2004b). Following activation of the Tie/Ang pathway, the FAK signaling pathway plays an essential role in malignant cell migration. The activation of Tie-2 induces VEGF-A to bind its EC-specific receptors, VEGFR1 and VEGFR2, which induces internalization of vascular endothelial cadherin (Zadeh et al. 2004a).

Furthermore, Tie receptor-independent signaling and non-vascular Ang effects can produce other intracellular signaling outcomes. ECs can adhere to immobilized Ang via the $\alpha_{\gamma}\beta_{1}$ integrin and FAK signaling pathway to induce glioma cell invasion by stimulating MMP-2 expression (Hu et al. 2006). The activation of the Tie-2/Ang-4 functional axis promotes the *in vivo* growth of human GBM cells by promoting tumor angiogenesis and directly activating ERK1/2 in GBM cells (Brinckerhoff and Matrisian 2002).

Although the preceding studies focused on elucidating the role of Ang-2 in cancer biology, its exact role in glioma angiogenesis remains elusive. A recent report demonstrated that Ang-2 significantly enhances vascular growth and induces aberrant pathological changes in malignant astrocytomas; this report also found that Ang-2 is not consistently elevated throughout all growth stages of malignant astrocytomas (Zadeh et al. 2010).

Although Ang-1 and Ang-2 play known roles in tumor angiogenesis, it is unknown how Ang-4 affects GBM angiogenesis progression and the mechanism underlying its effects. Brunckhorst et al. found a novel mechanism of Ang-4 on glioma progression that directly activates the ERK1/2 kinase pathway (Brunckhorst et al. 2010). Taken together, these data indicate that the Tie-2/Ang-4 functional axis can be considered an attractive therapeutic target for GBM.

8.7 Axl/Gas6

The receptor tyrosine kinase Axl, a member of the Tyro3, Axl, and Mer family of receptor tyrosine kinases (TAMRs), is characterized by an extracellular domain consisting of two immunoglobulin-like domains in juxtaposition to two fibronectin type III domains (Janssen et al. 1991). TAMRs, in particular Axl, have transforming properties: overexpression of a truncated version of Axl in premalignant cells is sufficient to induce tumors in mice (Zhang et al. 1996). Growth arrest–specific gene 6 (Gas6), which is the natural ligand of Axl, was discovered because its expression is upregulated in fibroblasts under growth-arrest conditions (Manfioletti et al. 1993). Axl/Gas6 signaling has been shown to regulate survival, proliferation, and migration in a variety of cells *in vitro* including tumor-derived cell lines of epithelial, mesenchymal, and hematopoietic origin (Hafizi and Dahlback 2006).

Axl and Gas6 are overexpressed in human gliomas of malignancy grades WHO II to IV. In contrast, Axl staining was not detectable in non-neoplastic brain tissue and Gas6 was strongly expressed in neurons (Hutterer et al. 2008). The receptor tyrosine kinase Axl is a mediator of glioma growth and invasion. Axl is predominantly expressed in pseudopalisading glioma cells, which are characterized by an accumulation of tumor cells around necrotic areas. Furthermore, an accumulation of Axl-positive tumor cells was observed adjacent to microvascular neoformations, which is a characteristic feature of invading glioma tumor cells spreading along perivascular regions. Inhibition of Axl signaling by overexpression of a dominant-negative receptor mutant suppressed experimental gliomagenesis and resulted in long-term survival of mice after intracerebral glioma cell implantation when compared with Axl wild-type transfected tumor cells. Inhibition of Axl signaling interfered with cell proliferation (30 % inhibition versus Axl wild-type), glioma cell migration (90 % inhibition versus Axl wild-type), and invasion (79 % inhibition versus Axl wild-type) However, an analysis of tumor vessel density and diameter failed to reveal an attenuated tumor vasculature as an explanation for the reduced tumorigenesis in Axl dominant-negative cells (Vajkoczy et al. 2006). GBM patients with high Axl expression and Axl/Gas6 coexpression showed a significantly shorter time to tumor progression and poorer overall survival, indicating that Axl and Gas6 expression predict poor prognosis in GBM patients (Hutterer et al. 2008).

8.8 DDR1/Collagen

Discoidin domain receptor 1 (DDR1) tyrosine kinases constitute a family of non-integrin collagen receptors that contain a discoidin homology region in the ectodomain (Alves et al. 1995). DDR1, which is mainly expressed in epithelial cells, is primarily activated by collagens I to IV and VIII and facilitates cell adhesion (Vogel et al. 1997). DDR1 has five isoforms generated by alternative splicing: DDR1a, b, c, d and e. DDRs have been implicated in the expression of pro-inflammatory mediators and matrix-degrading enzymes and play an important role in migration, proliferation, ECM remodeling, and wound repair (Vogel et al. 2006).

In GBM, collagen IV is present in virtually all tumor vessels, in some giant glioma cells, and in tumor cells around vascular proliferations. Collagens VII and VIII are absent from normal brain but may be expressed in glioma tissues supporting DDR1 signaling (Senner et al. 2008). Furthermore, DDR1 is overexpressed in glioma. Although overexpression of either DDR1a or DDR1b in cell-based glioma models caused increased cell attachment, glioma cells overexpressing DDR1a exhibit enhanced invasion and migration concomitant with increased levels of MMP-2. Inhibition of MMP activity suppressed DDR1a-stimulated cell-invasion and inhibition of DDR1 reduced DDR1a-mediated invasion and enhanced adhesion of DDR1a and DDR1b overexpressing cells. DDR1a plays a critical role in inducing tumor cell adhesion and invasion, and this invasive phenotype is caused by activation of MMP-2 (Ram et al. 2006; Yamanaka et al. 2006) (Fig. 8.4). DDR1 expression is more closely correlated with survival than histological grade in gliomas, suggesting that DDR1 expression might be a better predictive factor of patient survival than WHO grading (Yamanaka et al. 2006).

8.9 Eph/Ephrin

Erythropoietin-producing human hepatocellular carcinoma (Eph) receptors comprise the largest family of RTKs in mammals. Eph receptors have been divided into an EphA subclass (nine members) and an EphB subclass (six members) on the basis of extracellular domain sequence homology and ligand affinity (Gale et al. 1996). Ephrin ligands are also transmembrane proteins and have been divided into two subclasses: glycosylphosphatidyl-inositol (GPI)-linked ephrin-As (five members), which are anchored to the cell membrane, and transmembrane ephrin-Bs (three members). Ephrin-As preferentially bind to EphA receptors, while ephrin-Bs preferentially bind to EphB receptors, although promiscuity has been observed.

Eph/ephrins form an essential cell-cell communication system capable of bi-directional intracellular signaling between adjacent cells, where receptor signaling is designated "forward" and ephrin signaling is "reverse" (Heroult et al. 2006; Kullander and Klein 2002). Eph/ephrin members are plentiful and their relationships are complex. Generally, Eph/ephrin interactions are repulsive because cells containing a given Eph are repelled by cells containing the corresponding Ephrin. Through this mechanism, the Eph/ephrin system plays a role in numerous biological processes including cell adhesion and migration during development, especially in the central nervous system (Wimmer-Kleikamp and Lackmann 2005). Recently, a role for the Eph/ephrin system has also emerged in cancer, especially in the area of invasive behavior (Campbell and Robbins 2008).

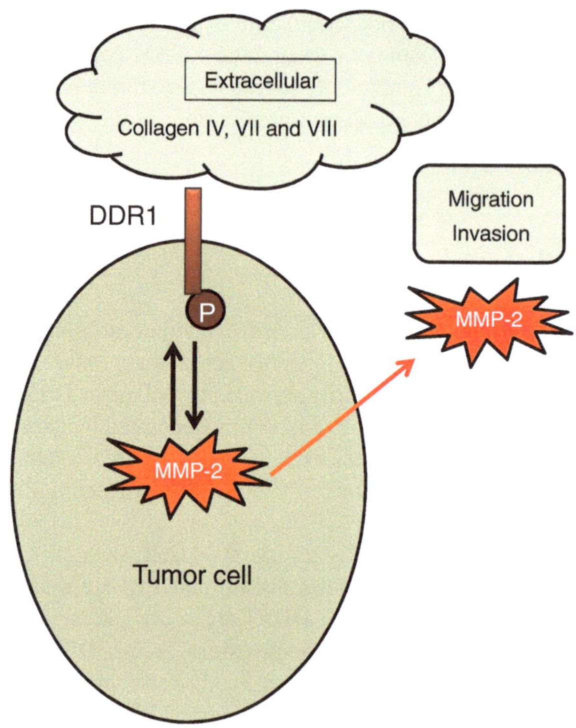

Fig. 8.4 DDR1 signaling in glioma. In GBM, DDR1 is overexpressed and activated by collagen IV which is present in virtually all tumor vessels, as well as in some giant glioma cells and in tumor cells around vascular proliferations; also collagens VII and VIII may be expressed in glioma tissues supporting DDR1 signaling. Glioma cells overexpressing DDR1a display enhanced migration and invasion associated with the increase levels of matrix metalloproteinase 2 (MMP-2). (DDR1, discoidin domain receptor 1)

Genes for Eph/ephrins are overexpressed or differentially expressed in numerous human cancers (Surawska et al. 2004). Mounting evidence documents a strong correlation between the expression and phosphorylation levels of many Eph/ephrin family members and increasing glioma tumor grade, suggesting that elevated Eph/ephrin expression levels may be diagnostic for GBM and reduced patient survival rates (Nakada et al. 2011). Based on the microarray data obtained from two distinct GBM cell phenotypes (invading cells and tumor core cells) collected from GBM specimens, pathway enrichment analysis indicated that EphB/ephrin-B is the most tightly linked system to the invading cell phenotype (Nakada et al. 2010). According to these results, it is likely that the Eph/ephrin system contributes to glioma invasion. In spite of the overexpression of Eph/ephrin members in glioma, no evidence of gene amplification or mutation has been reported.

Numerous Ephs and ephrins have been noted to influence or correlate with the malignant behavior of cancer cells. In glioma, EphA/ephrin-A was associated with proliferation (Wykosky et al. 2005) whereas EphB/ephrin-B was involved in invasion and neovascularization (Erber et al. 2006). EphB2, ephrin-B2, and ephrin-B3 mRNA levels were shown to significantly increase with histological grade in glioma and contribute invasive properties in GBM (Nakada et al. 2004, 2006, 2010). Additionally, a high ephrin-B2 level was shown to confer poor survival (Nakada et al. 2010). The co-expression of EphB2 and ephrin-B in GBM cells suggests the existence of an EphB/ephrin-B interaction through cell–cell contact in GBM. Data showing that glioma invasion is inhibited by blocking Eph/ephrin suggest that the Eph/ephrin system is a potential therapeutic target for invasive glioma.

Different Eph/ephrin molecules are conceivably linked to different intracellular signaling pathways in a cell-type-specific manner, which allows this system to perform a variety of functions. The key

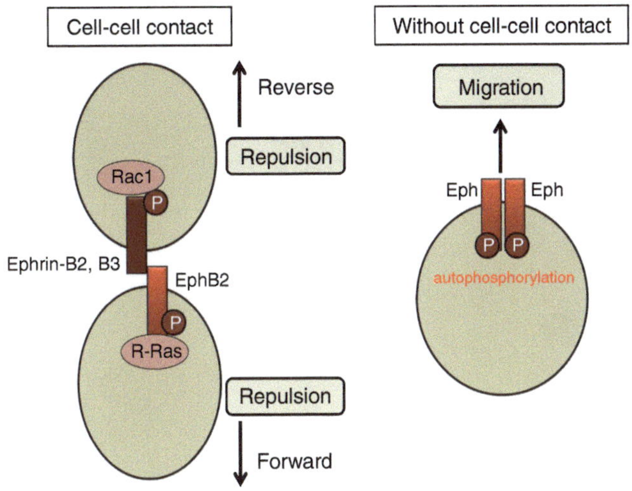

Fig. 8.5 Putative model of Eph/ephrin function in glioma. Signaling induced by cell-cell contact via EphB/ephrin-B induces repulsion. Ligand independent signaling via autophosphorylation of Eph without ephrin stimulation can also promote the invasion without cell-cell communication

signaling molecules in the invasion-signaling pathway induced by EphB/ephrin B in GBM appear to be small GTPases such as Rac and Ras. Ephrin-B2 and ephrin-B3 can activate EphB2 through cell–cell contact, inducing invasion via EphB2 forward signaling. A previous study showed that EphB2 plays a functional role in promoting GBM cell invasion by eliciting signaling through R-Ras and affecting integrin activity (Nakada et al. 2005). In contrast, reverse signaling of the ephrin-B2 and ephrin-B3 ligands was demonstrated to be an important factor in the regulation of glioma cell invasion through the Rac1 GTPase (Fig. 8.5).

A greater depth of investigation has occurred with respect to involvement of the Eph/ephrin subsystem in invasion. Fundamental effect of Eph/ephrin necessitates direct cell-cell contact. However, it was recently revealed that ligand-independent signaling via autophosphorylation of Eph without ephrin stimulation can promote invasion without cell-cell communication (Miao et al. 2009) (Fig. 8.5). EphB2 was overexpressed in invading GBM cells compared with stationary cells in the tumor core, suggesting that EphB2 is autophosphorylated in invading glioma cells (Nakada et al. 2004, 2006). This indicates that invading glioma cells overexpressing Eph/ephrin far from the tumor can migrate by themselves without interaction of other cells.

8.10 TrkA

Neurotrophic tyrosine kinase receptor type 1 (TrkA) is the high affinity receptor for nerve growth factor (NGF), neurotrophin-3, and neurotrophin-4/5. Phosphorylated TrkA may play an important role in mitotic spindle assembly because it is colocalized with α-tubulin at the mitotic spindle from prophase to anaphase, whereas in interphase, phosphorylated TrkA is localized on the membrane and processes of glioma cells (Zhang et al. 2005).

TrkA is strongly expressed in the subpopulation of highly infiltrating glioma cells *in vivo* but not in the glioma cells that remain within the bulk of the tumor. Thus, TrkA expression is dependent on both the cell type and the location within the tumor (Edwards et al. 2011). TrkA activation typically leads to the activation of survival- and growth-mediating pathways through the cytoplasmic proteins SHC, PI3K, and PLC-γ1 (Escalante et al. 2000; Meakin et al. 1999). GBM cell growth can be enhanced

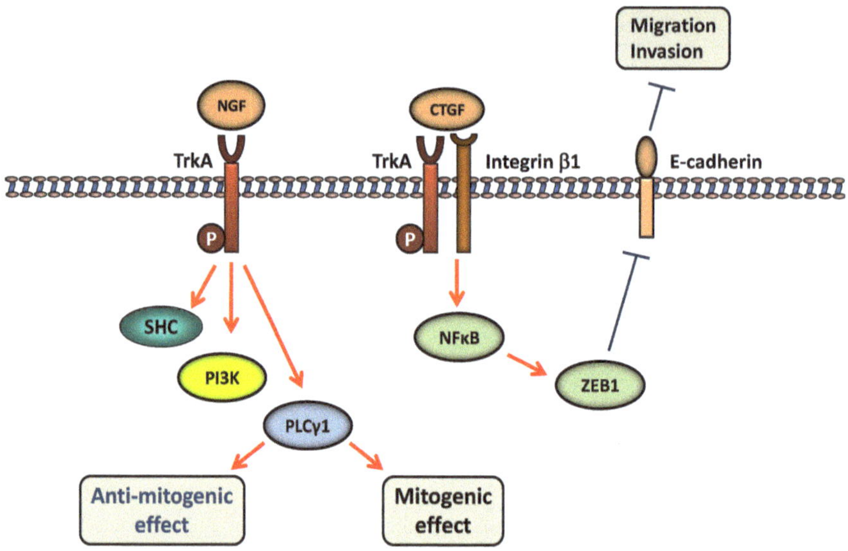

Fig. 8.6 TrkA signaling pathways in glioma. GBM cell growth can be enhanced by TrkA receptor (neurotrophic tyrosine kinase receptor type 1) phosphorylation activated by NGF (nerve growth factor). TrkA activation leads to the activation of survival and growth mediating pathways through cytoplasmic proteins SHC, PI3K and PLC-γ1. Switching between anti-mitogenic and mitogenic TrkA signaling is controlled by PLC-γ1 activity. Glioma initiating/stem cells require integrin β1–TrkA complex for connective tissue growth factor (CTGF) signaling; binding of the complex with CTGF causes NF-kB-mediated activation of the ZEB-1 promoter with subsequent induction of the ZEB-1 transcriptional repressor resulting in decreased expression of E-cadherin, accordingly enhancing glioma cell invasion and migration

by NGF acting via TrkA receptor phosphorylation (Singer et al. 1999). Switching between anti-mitogenic and mitogenic TrkA signaling is controlled by PLC-γ1 activity (Ye et al. 2000, 2002) (Fig. 8.6). Consequently, it was shown that TrkA expression in GBMs attenuates tumor progression *in vivo* (Rabin et al. 1998) by inducing differentiation of tumor cells from undifferentiated glioma to oligodendrocytes (Pflug et al. 2001).

Glioma initiating/stem cells require the cell surface protein receptors integrin β1 and TrkA, which constitute the integrin β1–TrkA complex, for connective tissue growth factor (CTGF) signaling. The integrin β1–TrkA complex can bind to CTGF and cause nuclear factor-kappa B (NF-kB)-mediated activation of the ZEB-1 promoter with subsequent induction of the ZEB-1 transcriptional repressor, resulting in decreased expression of E-cadherin and subsequent enhancement of glioma cell invasion and migration (Fig. 8.6). TrkA knockdown resulted in decreased CTGF-induced cell migration and the absence of tumor cell invasion into normal mice cortex. However, TrkA knockdown did not affect CTGF-induced proliferation or the clonogenicity of the glioma stem cells *in vitro*. Thus, CTGF, TrkA, and NF-kB may be potential therapeutic targets to alleviate tumor cell infiltration (Edwards et al. 2011).

8.11 Cross-Talk

As described above, many RTKs lead to the signaling of PI3K/Akt and Rac1 in invading glioma cells, suggesting that these molecules confer critical downstream signaling for invasion. A previous study showed coexpression of multiple activated RTKs in individual dissociated cells from a primary GBM (Stommel et al. 2007). Accordingly, multiple RTKs may be simultaneously or sequentially used by GBM cells to maintain invasion-signaling pathways via molecules such as PI3K/Akt and Rac1.

Given that individual cells express multiple RTKs, it is reasonable to speculate that these RTKs are interacting with each other. For example, the c-Met receptor is strongly phosphorylated as a function of EGFRvIII receptor levels, suggesting the presence of cross-talk between c-Met and EGFRvIII signaling, although the intermediary molecule has not yet been elucidated (Huang et al. 2007). The Axl RTK follows a similar phosphorylation response as a function of EGFRvIII levels (Huang et al. 2007). It was previously reported that EGFR and EphA2 are expressed in GBM and co-localize to the cell surface. EphA2 phosphorylation is dependent on EGFR activity and EphA2 downregulation inhibits EGFR phosphorylation, downstream signaling, and EGF-induced cell viability (Ramnarain et al. 2006). Previous studies reported that EphA4, whose expression is correlated with increasing glioma grade, forms a heteroreceptor complex with fibroblast growth factor receptor 1 (FGFR1) in glioma cells and that the EphA4-FGFR1 complex potentiated FGFR-mediated downstream signaling such as Akt/MAPK, Rac1, and Cdc42 pathways, resulting in the promotion of invasion (Fukai et al. 2008).

In such a multiple-input system with cross-talk among RTKs, a single-agent of anti-RTK inhibition might be incapable of sufficiently suppressing invasion signaling, resulting in insensitivity to invasion inhibition by any single agent and a lack of clinical efficacy. It is anticipated that combinations of drugs against different activated RTKs or single drugs with inhibitory activities against multiple activated RTKs will have more favorable outcomes.

8.12 Targeting Receptor Type Tyrosine Kinases

As mentioned in the previous paragraphs, TKIs that target multiple signaling pathways and critical growth factors essential for tumor progression have become a major focus of interest in various clinical studies. Although all studies remain experimental, several of these specific molecular agents that are most expectant or have been widely evaluated are reviewed below (Table 8.2).

8.12.1 Targeting EGFR

8.12.1.1 Gefitinib

Gefitinib is a low molecular weight, selective inhibitor of EGFR tyrosine kinase and was the first drug of this type. Although an early study suggested that gefitinib was active in patients with malignant gliomas (Franceschi et al. 2007; Mellinghoff et al. 2005), gefitinib therapy showed limited efficacy in multicenter phase II studies in patients with malignant glioma (Franceschi et al. 2007; Hegi et al. 2011; Rich et al. 2004), either for recurrent disease or as part of the initial treatment regimen. Because EGFR kinase domain mutations, which are associated with the greatest sensitivity to EGFR inhibitors (Pedersen et al. 2005), are uncommon in malignant gliomas (Lassman et al. 2005), it remains to be determined whether other molecular alterations involving EGFR signaling pathways occur selectively in treatment responders.

8.12.1.2 Erlotinib

Erlotinib is an orally active, reversible EGFR TKI. Because erlotinib is metabolized by the cytochrome P450 isoenzymes 3A4 (70 %) and CYP 1A2 (30 %), patients taking enzyme-inducing anti-epileptic drugs are not eligible for this treatment (Raizer et al. 2010). Similar to gefitinib, clinical studies examining the therapeutic efficacy of erlotinib have so far failed to demonstrate a major therapeutic break-through

Table 8.2 Summary of phase II studies in various tyrosine kinase inhibitors and related agents

Primary target	RTKI	Other target	Objects	Median OS (week)	OS (%)	Median PFS (week)	PFS (%)	References
EGFR	Gefitinib containing regimen							
			Newly diagnosed GB after RTx		54.2 (12 m)		16.2 (12 m)	(Uhm et al. 2011)
			Newly diagnosed pediatric brainstem glioma after RTx		56.4 (12 m)	32.1	20.9 (12 m)	(Pollack et al. 2011)
			Recurrent GB after RTx	24.6–25.1	50.0 (6 m)	8.4–17.0	13.0–14.3 (6 m)	(Franceschi et al. 2007; Hegi et al. 2011; Rich et al. 2004)
	Erlotinib containing regimen							
			Newly diagnosed GB after RTx	37.3	–	12.2	30.0 (6 m)	(Peereboom et al. 2010)
			Recurrent GB after RTx	26.0	11.4 (6 m)	8.7–31.2	3.0–11.4 (6 m)	(Brown et al. 2008; Lassman et al. 2005; Mellinghoff et al. 2005; Raizer et al. 2010; Van Den Bent et al. 2009)
			Recurrent AG after RTx	30.4	–	8.7	27.0 (6 m)	(Raizer et al. 2010)
	(125) I-Mab425							
			GB after RTx	63.1	62.5 (12 m), 25.5 (24 m)	–	–	(Li et al. 2010)
	Nimotuzumab							
			Recurrent pediatric brainstem glioma	82.0	–	–	–	(Massimino et al. 2011; Saurez et al. 2009)
			Recurrent malignant glioma after RTx	82.5	82.0 (6 m), 64.0 (12 m)	–	64.0 (6 m)	(Saurez et al. 2009)

	Cetuximab		Recurrent malignant glioma after RTx	22.0	37.9 (6 m)	8.2	7.3 (6 m)	(Neyns et al. 2009)
PDGFR	Imatinib containing regimen	BCR/Abl	Recurrent GB after RTx	48.9	–	14.4	3.0–27.0 (6 m)	(Raymond et al. 2008; Reardon et al. 2005; Wen et al. 2006)
			Recurrent AA after RTx	–	–	–	9.0–10.0 (6 m)	(Raymond et al. 2008; Wen et al. 2006)
	Sunitinib	VEGFR-2, Flt3, c-Kit	Recurrent malignant glioma after RTx	16.5	–	6.9	–	(Neyns et al. 2011)
Multi-kinases	Cediranib	All VEGFR subtypes, PDGFR-β, c-Kit	Recurrent GB after RTx	32.4		16.7	0.3 (6 m)	(Batchelor et al. 2010b)
	Sorafenib containing regimen	VEGFR-3, B-Raf, PDGFR-β, c-Kit, Ras, p38α	Newly diagnosed GB after RTx	52.1	–	26.0	16.0 (12 m)	(Hainsworth et al. 2010)
			Recurrent GB after RTx	41.5	–	6.4	–	(Reardon et al. 2011)
	Cabozantinib (XL184)	c-Met, RET, c-Kit, Flt3, Tie-2	Recurrent GB after RTx	–	–	16.0	–	(Wen et al. 2010)

Designations:
RTKI, receptor tyrosine kinase inhibitor; GB, glioblastoma; AA, anaplastic astrocytoma; AG, anaplastic glioma; RTx, radiation therapy; OS, overall survival; PFS, progression free survival; 6 m, six months; 12 m, twelve months; 24 m, twenty four months; EGFR, epidermal growth factor receptor; PDGFR, platelet derived growth factor receptor; VEGFR, vascular endothelial growth factor

in the setting of GBM (Haas-Kogan et al. 2005; Mellinghoff et al. 2005; Raizer et al. 2010; Van Den Bent et al. 2009), including newly diagnosed GBM treated with temozolomide (TMZ) and radiotherapy (Peereboom et al. 2010). Although Raizer and colleagues reported that the development of a rash during the initial erlotinib administration cycle correlates with survival in patients with non-progressive GBM after radiotherapy, the significance of this finding remains unclear (Raizer et al. 2010).

8.12.1.3 ^{125}I-mAb 425

^{125}I-mAb 425, an ^{125}I-labeled anti–EGFR 425 murine monoclonal antibody (mAb), is an IgG2a isotype developed from mice immunized with A-431 epidermoid carcinoma cells. mAb 425 binds to the tumor and produces anticarcinogenic effects mediated by direct cell growth inhibition, complement-dependent cytotoxicity, and activation of the humoral response. Initial clinical studies have shown survival benefits of adjuvant ^{125}I-425 mAb in GBM patients (Emrich et al. 2002; Quang and Brady 2004). In a phase II study of 192 patients with GBM treated with anti–EGFR ^{125}I-mAb 425 radioimmunotherapy, the reported survival was 15.7 months and treatment was safe and well tolerated (Li et al. 2010).

8.12.1.4 Nimotuzumab

Nimotuzumab (h-R3), a humanized monoclonal antibody directed against the EGFR, consequently inhibits tyrosine kinase activation. Recently, several reports of pediatric diffuse intrinsic pontine glioma treated with nimotuzumab showed a relatively favorable prognosis (Lam et al. 2009; Mateos et al. 2011; Saurez et al. 2009). Thus, nimotuzumab is currently being evaluated in conjunction with radiotherapy in a phase III trial in children with diffuse pontine glioma and adult GBM.

8.12.1.5 Cetuximab

Cetuximab, IMC-C225, is a chimeric (mouse/human) monoclonal IgG1 antibody that binds to EGFR with high specificity and affinity. Encouragingly, cetuximab enhanced the cytotoxicity produced by radiation therapy *in vivo* in EGFR-amplified GBM (Belda-Iniesta et al. 2006; Eller et al. 2005). In 2006, however, a phase II trial of cetuximab in 55 recurrent high-grade glioma patients revealed no significant correlation between response, survival, and EGFR amplification (Neyns et al. 2009).

8.12.1.6 Lapatinib

Lapatinib is a dual TKI that interrupts the HER2/neu (ErbB2) growth receptor pathway. It is used in therapy for HER2-positive breast cancer. However, a phase I/II trial of lapatinib in patients with relapsed GBM failed to show significant activity of this agent independent of the presence of EGFRvIII mutation and PTEN immunohistochemical status (Thiessen et al. 2010).

8.12.2 Targeting PDGFR

8.12.2.1 Imatinib

Imatinib mesylate (Gleevec; formerly known as STI571) is a potent inhibitor of the PDGFRα, PDGFRβ, Bcr-Abl, c-Fms, and c-Kit tyrosine kinases. Its antitumor activities in chronic myelogenous leukemia and gastrointestinal stromal tumors result from the inhibitions of Bcr-Abl (Druker et al.

2001) and c-Kit (Demetri et al. 2002), respectively. An initial study showed that imatinib mesylate plus hydroxyurea was well tolerated and associated with durable antitumor activity in some patients with recurrent GBM (Reardon et al. 2005). However, more extensive phase II and III studies have shown only minimal evidence of single agent activity (Dresemann et al. 2010; Raymond et al. 2008; Wen et al. 2006).

8.12.2.2 Sunitinib

Sunitinib malate is an oral small-molecule inhibitor of VEGFRs, PDGFRs, c-Kit, Flt3, and RET kinases (Chow and Eckhardt 2007). Sunitinib has substantial clinical activity against hypoxia-inducible factor (HIF)/VEGF-dependent, PDGFR-dependent, and KIT-dependent cancers. Recently, a phase II study of sunitinib failed to demonstrate a relevant clinical benefit from single-agent sunitinib (37.5 mg/day) in patients with alkylator-refractory recurrent glioma. Furthermore, substantial treatment-related toxicity was observed in several patients (Neyns et al. 2011). Although sunitinib affected the glioma vasculature in a small subgroup of patients, no objective tumor responses were observed regarding apparent reduced cerebral blood-flow and blood-volume within the lesion compared with the normal brain or reduced gadolinium enhancement of the tumor. A possible explanation for the failure of sunitinib to have meaningful clinical activity is the lower potency of sunitinib to selectively inhibit VEGF/VEGFR signaling within the tumor vasculature compared to the VEGF-targeted mAb bevacizumab (Friedman et al. 2009) or the more potent VEGFR-specific small-molecule TKI cediranib (Batchelor et al. 2010a, b).

8.12.2.3 Dasatinib

Dasatinib is an aminotriazole analog with high specificity for several kinases including Bcr-Abl, Src, c-Kit, PDGFRβ, and EphA2, which was approved for use in patients with chronic myelogenous leukemia after imatinib treatment and Philadelphia chromosome-positive acute lymphoblastic leukemia. Dasatinib appears to be a more potent inhibitor of Bcr-Abl than imatinib (Tokarski et al. 2006). Recently, investigators have shown that Src is frequently phosphorylated in GBM cell lines such as T98G and U87 compared with normal tissue and is activated in human GBM tumors (Du et al. 2009). These findings support the possible role of dasatinib for malignant gliomas, and future clinical trials will further assess the clinical value of SRC inhibition with dasatinib incorporating TMZ and other cytotoxic agents.

8.12.3 Multi-Kinase Inhibitors

8.12.3.1 Cediranib

Cediranib (AZD2171) is a potent oral inhibitor of all VEGF receptor tyrosine kinases and PDGF receptors (Batchelor et al. 2007). Cediranib caused a structural and functional normalization of tumor vasculature in all 16 patients with recurrent GBM and resulted in a significant reduction of tumor-associated vasogenic edema as measured by MRI techniques. This effect was paralleled by a potent steroid-sparing effect in most patients (Batchelor et al. 2007). In a subsequent phase II study, partial responses were seen in 57 % of patients based upon tumor measurements (Batchelor et al. 2010b), and the median progression-free survival (PFS) and the median overall survival (OS) were 32.4 weeks and 16.7 weeks, respectively (Table 8.2), with manageable toxicity. Based upon these results, a phase III study was conducted in 325 patients with recurrent GBM (Batchelor 2010a). Although cediranib treatment regimens resulted in a statistically significant decrease in steroid use and a reduced contrast enhancing area on neuroimaging, there was no statistically significant improvement in PFS, the primary endpoint of the trial.

8.12.3.2 Sorafenib

Sorafenib is a multi-target oral TKI with inhibitory effects on the VEGF receptor, PDGF receptor, and the Ras/Raf signaling pathway (Wilhelm et al. 2004). Despite potentially complementary direct and indirect mechanisms of anti-tumor activity, Reardon et al. demonstrated that sorafenib combined with daily TMZ has minimal activity as a salvage regimen for recurrent GBM patients in their single-arm phase II study (Reardon et al. 2011). Meanwhile, Hainsworth et al. also failed to demonstrate the efficacy of treatment when compared with the results expected with standard therapy in newly diagnosed GBM (Hainsworth et al. 2010). A plausible explanation for this low effectiveness is the low to moderate ability of sorafenib to penetrate into the brain. However, the blood brain barrier (BBB) is compromised in patients with GBM and it is unlikely that sorafenib would have lower penetration than bevacizumab, a large protein. In general, the addition of small molecule TKIs to chemotherapy has not consistently improved treatment results, even when both components of therapy have individual efficacy.

8.12.3.3 Vandetanib

Vandetanib (ZD6474) is a potent oral TKI for various RTKs, in particular VEGFR2 and EGFR. A phase I study of 35 children with newly diagnosed diffuse intrinsic pontine glioma treated with vandetanib reported 1- and 2-year OS outcomes of 37.5 % ± 10.5 % and 21.4 % ± 11 %, respectively. Three patients remained alive with PFS for more than 2 years. The recommended phase II dose of vandetanib in children is 145 mg/m^2 per day (Broniscer et al. 2010). Several phase II studies, including those in recurrent and newly diagnosed malignant glioma, are currently underway.

8.12.3.4 Cabozantinib (XL-184)

Recently, interim results have been reported from a phase II study of Cabozantinib (XL184) treatment in previously treated progressive GBM. Cabozantinib is an oral inhibitor of multiple RTKs that includes VEGFR2 as the main target followed by c-Met, RET, c-kit, Flt3, Tie-2, and Axl (Wen 2010; Yakes et al. 2011). In this study, the median PFS in antiangiogenic-naive cohorts was 16 weeks (Table 8.2). In total, 61 % of patients on corticosteroids at baseline had a reduction in corticosteroid dose of at least 50 %. The investigators concluded that XL184 demonstrates encouraging clinical activity in patients with progressive GBM. Cabozantinib, currently in phase III clinical trials, is a promising agent for inhibiting tumor angiogenesis and metastasis in glioma, especially in cases of dysregulated Met and VEGFR signaling.

8.13 Prospective

Future studies are required to more precisely establish the molecular mechanisms and specific downstream signaling elements that contribute to the cooperative or synergistic interactions of RTK signaling pathways in invading glioma cells. Further studies on the manipulation of the RTK systems involved in invasion will aid in engineering GBM therapies and in elucidating the complexity and additional functional implications of RTK systems. Moreover, it is of great therapeutic interest to define invasion-associated RTKs that could be targeted for blocking invasion. These works should help to develop novel potential pharmacological agents to modulate invasion processes and thereby counteract the activation of invasion-signaling pathways and promotion of invasion. The data obtained from GBM patients treated with TKIs should confirm the therapeutic benefit of TKIs and the safety of selectively targeting RTKs, alone or in combination with the current conventional therapies. The effort to combat GBM, along with emerging data regarding the underlying molecular invasion circuitry and the development of specific TKIs, will result in the development of an array of new treatment approaches.

References

Abounader R, Laterra J (2005) Scatter factor/hepatocyte growth factor in brain tumor growth and angiogenesis. Neuro Oncol 7:436–451

Alves F, Vogel W, Mossie K, Millauer B, Hofler H, Ullrich A (1995) Distinct structural characteristics of discoidin I subfamily receptor tyrosine kinases and complementary expression in human cancer. Oncogene 10:609–618

Andersson U, Guo D, Malmer B, Bergenheim AT, Brannstrom T, Hedman H, Henriksson R (2004) Epidermal growth factor receptor family (EGFR, ErbB2-4) in gliomas and meningiomas. Acta Neuropathol 108:135–142

Barker FG 2nd, Simmons ML, Chang SM, Prados MD, Larson DA, Sneed PK, Wara WM, Berger MS, Chen P, Israel MA, Aldape KD (2001) EGFR overexpression and radiation response in glioblastoma multiforme. Int J Radiat Oncol Biol Phys 51:410–418

Batchelor TT, Sorensen AG, Di Tomaso E, Zhang WT, Duda DG, Cohen KS, Kozak KR, Cahill DP, Chen PJ, Zhu M, Ancukiewicz M, Mrugala MM, Plotkin S, Drappatz J, Louis DN, Ivy P, Scadden DT, Benner T, Loeffler JS, Wen PY, Jain RK (2007) AZD2171, a pan-VEGF receptor tyrosine kinase inhibitor, normalizes tumor vasculature and alleviates edema in glioblastoma patients. Cancer Cell 11:83–95

Batchelor T, Mulholland J, Neyns B et al (2010a) A phase III randomized study comparing the efficacy of cediranib as monotherapy, and in combination with lomustine, with lomustine alone in recurrent glioblastoma patients. Ann Oncol 21:4

Batchelor TT, Duda DG, Di Tomaso E, Ancukiewicz M, Plotkin SR, Gerstner E, Eichler AF, Drappatz J, Hochberg FH, Benner T, Louis DN, Cohen KS, Chea H, Exarhopoulos A, Loeffler JS, Moses MA, Ivy P, Sorensen AG, Wen PY, Jain RK (2010b) Phase II study of cediranib, an oral pan-vascular endothelial growth factor receptor tyrosine kinase inhibitor, in patients with recurrent glioblastoma. J Clin Oncol 28:2817–2823

Belda-Iniesta C, Carpeno Jde C, Saenz EC, Gutierrez M, Perona R, Baron MG (2006) Long term responses with cetuximab therapy in glioblastoma multiforme. Cancer Biol Ther 5:912–914

Ben-Levy R, Paterson HF, Marshall CJ, Yarden Y (1994) A single autophosphorylation site confers oncogenicity to the Neu/ErbB-2 receptor and enables coupling to the MAP kinase pathway. EMBO J 13:3302–3311

Blume-Jensen P, Hunter T (2001) Oncogenic kinase signalling. Nature 411:355–365

Bogdan S, Klambt C (2001) Epidermal growth factor receptor signaling. Curr Biol 11:R292–295

Brinckerhoff CE, Matrisian LM (2002) Matrix metalloproteinases: a tail of a frog that became a prince. Nat Rev Mol Cell Biol 3:207–214

Broniscer A, Baker JN, Tagen M, Onar-Thomas A, Gilbertson RJ, Davidoff AM, Pai Panandiker AS, Leung W, Chin TK, Stewart CF, Kocak M, Rowland C, Merchant TE, Kaste SC, Gajjar A (2010) Phase I study of vandetanib during and after radiotherapy in children with diffuse intrinsic pontine glioma. J Clin Oncol 28:4762–4768

Brunckhorst MK, Wang H, Lu R, Yu Q (2010) Angiopoietin-4 promotes glioblastoma progression by enhancing tumor cell viability and angiogenesis. Cancer Res 70:7283–7293

Burgess T, Coxon A, Meyer S, Sun J, Rex K, Tsuruda T, Chen Q, Ho SY, Li L, Kaufman S, Mcdorman K, Cattley RC, Elliott G, Zhang K, Feng X, Jia XC, Green L, Radinsky R, Kendall R (2006) Fully human monoclonal antibodies to hepatocyte growth factor with therapeutic potential against hepatocyte growth factor/c-Met-dependent human tumors. Cancer Res 66:1721–1729

Cai XM, Tao BB, Wang LY, Liang YL, Jin JW, Yang Y, Hu YL, Zha XL (2005) Protein phosphatase activity of PTEN inhibited the invasion of glioma cells with epidermal growth factor receptor mutation type III expression. Int J Cancer 117:905–912

Campbell TN, Robbins SM (2008) The Eph receptor/ephrin system: an emerging player in the invasion game. Curr Issues Mol Biol 10:61–66

Cancer Genome Atlas Research Network (2008) Comprehensive genomic characterization defines human glioblastoma genes and core pathways. Nature 455:1061–1068

Cattaneo MG, Gentilini D, Vicentini LM (2006) Deregulated human glioma cell motility: inhibitory effect of somatostatin. Mol Cell Endocrinol 256:34–39
Cecchi F, Rabe DC, Bottaro DP (2010) Targeting the HGF/Met signalling pathway in cancer. Eur J Cancer 46:1260–1270
Chow LQ, Eckhardt SG (2007) Sunitinib: from rational design to clinical efficacy. J Clin Oncol 25:884–896
Chuang YY, Tran NL, Rusk N, Nakada M, Berens ME, Symons M (2004) Role of synaptojanin 2 in glioma cell migration and invasion. Cancer Res 64:8271–8275
Citri A, Skaria KB, Yarden Y (2003) The deaf and the dumb: the biology of ErbB-2 and ErbB-3. Exp Cell Res 284:54–65
Cooper CS, Park M, Blair DG, Tainsky MA, Huebner K, Croce CM, Vande Woude GF (1984) Molecular cloning of a new transforming gene from a chemically transformed human cell line. Nature 311:29–33
Demetri GD, Von Mehren M, Blanke CD, Van Den Abbeele AD, Eisenberg B, Roberts PJ, Heinrich MC, Tuveson DA, Singer S, Janicek M, Fletcher JA, Silverman SG, Silberman SL, Capdeville R, Kiese B, Peng B, Dimitrijevic S, Druker BJ, Corless C, Fletcher CD, Joensuu H (2002) Efficacy and safety of imatinib mesylate in advanced gastrointestinal stromal tumors. N Engl J Med 347:472–480
Dresemann G, Weller M, Rosenthal MA, Wedding U, Wagner W, Engel E, Heinrich B, Mayer-Steinacker R, Karup-Hansen A, Fluge O, Nowak A, Mehdorn M, Schleyer E, Krex D, Olver IN, Steinbach JP, Hosius C, Sieder C, Sorenson G, Parker R, Nikolova Z (2010) Imatinib in combination with hydroxyurea versus hydroxyurea alone as oral therapy in patients with progressive pretreated glioblastoma resistant to standard dose temozolomide. J Neurooncol 96:393–402
Druker BJ, Talpaz M, Resta DJ, Peng B, Buchdunger E, Ford JM, Lydon NB, Kantarjian H, Capdeville R, Ohno-Jones S, Sawyers CL (2001) Efficacy and safety of a specific inhibitor of the BCR-ABL tyrosine kinase in chronic myeloid leukemia. N Engl J Med 344:1031–1037
Du J, Bernasconi P, Clauser KR, Mani DR, Finn SP, Beroukhim R, Burns M, Julian B, Peng XP, Hieronymus H, Maglathlin RL, Lewis TA, Liau LM, Nghiemphu P, Mellinghoff IK, Louis DN, Loda M, Carr SA, Kung AL, Golub TR (2009) Bead-based profiling of tyrosine kinase phosphorylation identifies SRC as a potential target for glioblastoma therapy. Nat Biotechnol 27:77–83
Edwards LA, Woolard K, Son MJ, Li A, Lee J, Ene C, Mantey SA, Maric D, Song H, Belova G, Jensen RT, Zhang W, Fine HA (2011) Effect of brain- and tumor-derived connective tissue growth factor on glioma invasion. J Natl Cancer Inst 103:1162–1178
Ekstrand AJ, Sugawa N, James CD, Collins VP (1992) Amplified and rearranged epidermal growth factor receptor genes in human glioblastomas reveal deletions of sequences encoding portions of the N- and/or C-terminal tails. Proc Natl Acad Sci U S A 89:4309–4313
Eller JL, Longo SL, Kyle MM, Bassano D, Hicklin DJ, Canute GW (2005) Anti-epidermal growth factor receptor monoclonal antibody cetuximab augments radiation effects in glioblastoma multiforme in vitro and in vivo. Neurosurgery 56:155–162, discussion 162
Emrich JG, Brady LW, Quang TS, Class R, Miyamoto C, Black P, Rodeck U (2002) Radioiodinated (I-125) monoclonal antibody 425 in the treatment of high grade glioma patients: ten-year synopsis of a novel treatment. Am J Clin Oncol 25:541–546
Engelhard HH, Wolters M, Criswell PS (1995) Analysis of c-erbB2 protein content of human glioma cells and tumor tissue. J Neurooncol 23:31–40
Erber R, Eichelsbacher U, Powajbo V, Korn T, Djonov V, Lin J, Hammes HP, Grobholz R, Ullrich A, Vajkoczy P (2006) EphB4 controls blood vascular morphogenesis during postnatal angiogenesis. EMBO J 25:628–641
Escalante M, Courtney J, Chin WG, Teng KK, Kim JI, Fajardo JE, Mayer BJ, Hempstead BL, Birge RB (2000) Phosphorylation of c-Crk II on the negative regulatory Tyr222 mediates nerve growth factor-induced cell spreading and morphogenesis. J Biol Chem 275:24787–24797
Feldkamp MM, Lala P, Lau N, Roncari L, Guha A (1999) Expression of activated epidermal growth factor receptors, Ras-guanosine triphosphate, and mitogen-activated protein kinase in human glioblastoma multiforme specimens. Neurosurgery 45:1442–1453
Fleming TP, Saxena A, Clark WC, Robertson JT, Oldfield EH, Aaronson SA, Ali IU (1992) Amplification and/or overexpression of platelet-derived growth factor receptors and epidermal growth factor receptor in human glial tumors. Cancer Res 52:4550–4553
Franceschi E, Cavallo G, Lonardi S, Magrini E, Tosoni A, Grosso D, Scopece L, Blatt V, Urbini B, Pession A, Tallini G, Crino L, Brandes AA (2007) Gefitinib in patients with progressive high-grade gliomas: a multicentre phase II study by Gruppo Italiano Cooperativo di Neuro-Oncologia (GICNO). Br J Cancer 96:1047–1051
Friedman HS, Prados MD, Wen PY, Mikkelsen T, Schiff D, Abrey LE, Yung WK, Paleologos N, Nicholas MK, Jensen R, Vredenburgh J, Huang J, Zheng M, Cloughesy T (2009) Bevacizumab alone and in combination with irinotecan in recurrent glioblastoma. J Clin Oncol 27:4733–4740
Fukai J, Yokote H, Yamanaka R, Arao T, Nishio K, Itakura T (2008) EphA4 promotes cell proliferation and migration through a novel EphA4-FGFR1 signaling pathway in the human glioma U251 cell line. Mol Cancer Ther 7:2768–2778
Gale NW, Holland SJ, Valenzuela DM, Flenniken A, Pan L, Ryan TE, Henkemeyer M, Strebhardt K, Hirai H, Wilkinson DG, Pawson T, Davis S, Yancopoulos GD (1996) Eph receptors and ligands comprise two major specificity subclasses and are reciprocally compartmentalized during embryogenesis. Neuron 17:9–19

Gale NW, Thurston G, Hackett SF, Renard R, Wang Q, Mcclain J, Martin C, Witte C, Witte MH, Jackson D, Suri C, Campochiaro PA, Wiegand SJ, Yancopoulos GD (2002) Angiopoietin-2 is required for postnatal angiogenesis and lymphatic patterning, and only the latter role is rescued by Angiopoietin-1. Dev Cell 3:411–423

Gao L, Li F, Dong B, Zhang J, Rao Y, Cong Y, Mao B, Chen X (2010) Inhibition of STAT3 and ErbB2 suppresses tumor growth, enhances radiosensitivity, and induces mitochondria-dependent apoptosis in glioma cells. Int J Radiat Oncol Biol Phys 77:1223–1231

Gilbertson RJ, Perry RH, Kelly PJ, Pearson AD, Lunec J (1997) Prognostic significance of HER2 and HER4 coexpression in childhood medulloblastoma. Cancer Res 57:3272–3280

Gilbertson RJ, Bentley L, Hernan R, Junttila TT, Frank AJ, Haapasalo H, Connelly M, Wetmore C, Curran T, Elenius K, Ellison DW (2002) ERBB receptor signaling promotes ependymoma cell proliferation and represents a potential novel therapeutic target for this disease. Clin Cancer Res 8:3054–3064

Gondi CS, Dinh DH, Klopfenstein JD, Gujrati M, Rao JS (2009) MMP-2 downregulation mediates differential regulation of cell death via ErbB-2 in glioma xenografts. Int J Oncol 35:257–263

Gu J, Tamura M, Pankov R, Danen EH, Takino T, Matsumoto K, Yamada KM (1999) Shc and FAK differentially regulate cell motility and directionality modulated by PTEN. J Cell Biol 146:389–403

Guha A, Dashner K, Black PM, Wagner JA, Stiles CD (1995) Expression of PDGF and PDGF receptors in human astrocytoma operation specimens supports the existence of an autocrine loop. Int J Cancer 60:168–173

Gulati S, Ytterhus B, Granli US, Gulati M, Lydersen S, Torp SH (2010) Overexpression of c-erbB2 is a negative prognostic factor in anaplastic astrocytomas. Diagn Pathol 5:18

Haapasalo H, Hyytinen E, Sallinen P, Helin H, Kallioniemi OP, Isola J (1996) c-erbB-2 in astrocytomas: infrequent overexpression by immunohistochemistry and absence of gene amplification by fluorescence in situ hybridization. Br J Cancer 73:620–623

Haas-Kogan DA, Prados MD, Tihan T, Eberhard DA, Jelluma N, Arvold ND, Baumber R, Lamborn KR, Kapadia A, Malec M, Berger MS, Stokoe D (2005) Epidermal growth factor receptor, protein kinase B/Akt, and glioma response to erlotinib. J Natl Cancer Inst 97:880–887

Hafizi S, Dahlback B (2006) Signalling and functional diversity within the Axl subfamily of receptor tyrosine kinases. Cytokine Growth Factor Rev 17:295–304

Hainsworth JD, Ervin T, Friedman E, Priego V, Murphy PB, Clark BL, Lamar RE (2010) Concurrent radiotherapy and temozolomide followed by temozolomide and sorafenib in the first-line treatment of patients with glioblastoma multiforme. Cancer 116:3663–3669

Hawighorst T, Skobe M, Streit M, Hong YK, Velasco P, Brown LF, Riccardi L, Lange-Asschenfeldt B, Detmar M (2002) Activation of the tie2 receptor by angiopoietin-1 enhances tumor vessel maturation and impairs squamous cell carcinoma growth. Am J Pathol 160:1381–1392

Hayes AJ, Huang WQ, Yu J, Maisonpierre PC, Liu A, Kern FG, Lippman ME, Mcleskey SW, Li LY (2000) Expression and function of angiopoietin-1 in breast cancer. Br J Cancer 83:1154–1160

Hegi ME, Diserens AC, Bady P, Kamoshima Y, Kouwenhoven MC, Delorenzi M, Lambiv WL, Hamou MF, Matter MS, Koch A, Heppner FL, Yonekawa Y, Merlo A, Frei K, Mariani L, Hofer S (2011) Pathway analysis of glioblastoma tissue after preoperative treatment with the EGFR tyrosine kinase inhibitor gefitinib–a phase II trial. Mol Cancer Ther 10:1102–1112

Hermanson M, Funa K, Hartman M, Claesson-Welsh L, Heldin CH, Westermark B, Nister M (1992) Platelet-derived growth factor and its receptors in human glioma tissue: expression of messenger RNA and protein suggests the presence of autocrine and paracrine loops. Cancer Res 52:3213–3219

Heroult M, Schaffner F, Augustin HG (2006) Eph receptor and ephrin ligand-mediated interactions during angiogenesis and tumor progression. Exp Cell Res 312:642–650

Hiesiger EM, Hayes RL, Pierz DM, Budzilovich GN (1993) Prognostic relevance of epidermal growth factor receptor (EGF-R) and c-neu/erbB2 expression in glioblastomas (GBMs). J Neurooncol 16:93–104

Hoelzinger DB, Demuth T, Berens ME (2007) Autocrine factors that sustain glioma invasion and paracrine biology in the brain microenvironment. J Natl Cancer Inst 99:1583–1593

Hortobagyi GN (2005) Trastuzumab in the treatment of breast cancer. N Engl J Med 353:1734–1736

Hu B, Jarzynka MJ, Guo P, Imanishi Y, Schlaepfer DD, Cheng SY (2006) Angiopoietin 2 induces glioma cell invasion by stimulating matrix metalloprotease 2 expression through the alphavbeta1 integrin and focal adhesion kinase signaling pathway. Cancer Res 66:775–783

Huang PH, Cavenee WK, Furnari FB, White FM (2007) Uncovering therapeutic targets for glioblastoma: a systems biology approach. Cell Cycle 6:2750–2754

Hutterer M, Knyazev P, Abate A, Reschke M, Maier H, Stefanova N, Knyazeva T, Barbieri V, Reindl M, Muigg A, Kostron H, Stockhammer G, Ullrich A (2008) Axl and growth arrest-specific gene 6 are frequently overexpressed in human gliomas and predict poor prognosis in patients with glioblastoma multiforme. Clin Cancer Res 14:130–138

Janssen JW, Schulz AS, Steenvoorden AC, Schmidberger M, Strehl S, Ambros PF, Bartram CR (1991) A novel putative tyrosine kinase receptor with oncogenic potential. Oncogene 6:2113–2120

Jones RB, Gordus A, Krall JA, Macbeath G (2006) A quantitative protein interaction network for the ErbB receptors using protein microarrays. Nature 439:168–174
Jorissen RN, Walker F, Pouliot N, Garrett TP, Ward CW, Burgess AW (2003) Epidermal growth factor receptor: mechanisms of activation and signalling. Exp Cell Res 284:31–53
Khoshyomn S, Penar PL, Rossi J, Wells A, Abramson DL, Bhushan A (1999) Inhibition of phospholipase C-gamma1 activation blocks glioma cell motility and invasion of fetal rat brain aggregates. Neurosurgery 44:568–577, discussion 577–568
Kita D, Takino T, Nakada M, Takahashi T, Yamashita J, Sato H (2001) Expression of dominant-negative form of Ets-1 suppresses fibronectin-stimulated cell adhesion and migration through down-regulation of integrin alpha5 expression in U251 glioma cell line. Cancer Res 61:7985–7991
Kita D, Yonekawa Y, Weller M, Ohgaki H (2007) PIK3CA alterations in primary (de novo) and secondary glioblastomas. Acta Neuropathol 113:295–302
Koga K, Todaka T, Morioka M, Hamada J, Kai Y, Yano S, Okamura A, Takakura N, Suda T, Ushio Y (2001) Expression of angiopoietin-2 in human glioma cells and its role for angiogenesis. Cancer Res 61:6248–6254
Koochekpour S, Jeffers M, Rulong S, Taylor G, Klineberg E, Hudson EA, Resau JH, Vande Woude GF (1997) Met and hepatocyte growth factor/scatter factor expression in human gliomas. Cancer Res 57:5391–5398
Kullander K, Klein R (2002) Mechanisms and functions of Eph and ephrin signalling. Nat Rev Mol Cell Biol 3:475–486
Lam C, Bouffet E, Bartels U (2009) Nimotuzumab in pediatric glioma. Future Oncol 5:1349–1361
Lamszus K, Laterra J, Westphal M, Rosen EM (1999) Scatter factor/hepatocyte growth factor (SF/HGF) content and function in human gliomas. Int J Dev Neurosci 17:517–530
Lassman AB, Rossi MR, Raizer JJ, Abrey LE, Lieberman FS, Grefe CN, Lamborn K, Pao W, Shih AH, Kuhn JG, Wilson R, Nowak NJ, Cowell JK, Deangelis LM, Wen P, Gilbert MR, Chang S, Yung WA, Prados M, Holland EC (2005) Molecular study of malignant gliomas treated with epidermal growth factor receptor inhibitors: tissue analysis from North American Brain Tumor Consortium Trials 01–03 and 00–01. Clin Cancer Res 11:7841–7850
Lee HS, Han J, Bai HJ, Kim KW (2009) Brain angiogenesis in developmental and pathological processes: regulation, molecular and cellular communication at the neurovascular interface. FEBS J 276:4622–4635
Li L, Quang TS, Gracely EJ, Kim JH, Emrich JG, Yaeger TE, Jenrette JM, Cohen SC, Black P, Brady LW (2010) A phase II study of anti-epidermal growth factor receptor radioimmunotherapy in the treatment of glioblastoma multiforme. J Neurosurg 113:192–198
Liang ML, Ma J, Ho M, Solomon L, Bouffet E, Rutka JT, Hawkins C (2008) Tyrosine kinase expression in pediatric high grade astrocytoma. J Neurooncol 87:247–253
Libermann TA, Nusbaum HR, Razon N, Kris R, Lax I, Soreq H, Whittle N, Waterfield MD, Ullrich A, Schlessinger J (1985) Amplification and overexpression of the EGF receptor gene in primary human glioblastomas. J Cell Sci Suppl 3:161–172
Liu D, Martin V, Fueyo J, Lee OH, Xu J, Cortes-Santiago N, Alonso MM, Aldape K, Colman H, Gomez-Manzano C (2010) Tie2/TEK modulates the interaction of glioma and brain tumor stem cells with endothelial cells and promotes an invasive phenotype. Oncotarget 1:700–709
Lohi J (2001) Laminin-5 in the progression of carcinomas. Int J Cancer Journal international du cancer 94:763–767
Lokker NA, Sullivan CM, Hollenbach SJ, Israel MA, Giese NA (2002) Platelet-derived growth factor (PDGF) autocrine signaling regulates survival and mitogenic pathways in glioblastoma cells: evidence that the novel PDGF-C and PDGF-D ligands may play a role in the development of brain tumors. Cancer Res 62:3729–3735
Machein MR, Knedla A, Knoth R, Wagner S, Neuschl E, Plate KH (2004) Angiopoietin-1 promotes tumor angiogenesis in a rat glioma model. Am J Pathol 165:1557–1570
Maisonpierre PC, Suri C, Jones PF, Bartunkova S, Wiegand SJ, Radziejewski C, Compton D, Mcclain J, Aldrich TH, Papadopoulos N, Daly TJ, Davis S, Sato TN, Yancopoulos GD (1997) Angiopoietin-2, a natural antagonist for Tie2 that disrupts in vivo angiogenesis. Science 277:55–60
Majumdar K, Radotra BD, Vasishta RK, Pathak A (2009) Platelet-derived growth factor expression correlates with tumor grade and proliferative activity in human oligodendrogliomas. Surg Neurol 72:54–60
Manfioletti G, Brancolini C, Avanzi G, Schneider C (1993) The protein encoded by a growth arrest-specific gene (gas6) is a new member of the vitamin K-dependent proteins related to protein S, a negative coregulator in the blood coagulation cascade. Mol Cell Biol 13:4976–4985
Manning G, Whyte DB, Martinez R, Hunter T, Sudarsanam S (2002) The protein kinase complement of the human genome. Science 298:1912–1934
Martinho O, Longatto-Filho A, Lambros MB, Martins A, Pinheiro C, Silva A, Pardal F, Amorim J, Mackay A, Milanezi F, Tamber N, Fenwick K, Ashworth A, Reis-Filho JS, Lopes JM, Reis RM (2009) Expression, mutation and copy number analysis of platelet-derived growth factor receptor A (PDGFRA) and its ligand PDGFA in gliomas. Br J Cancer 101:973–982
Mateos ME, Lopez-Laso E, Izquierdo L, Perez-Navero JL, Garcia S, Garzas C (2011) Response to nimotuzumab in a child with a progressive diffuse intrinsic pontine glioma. Pediatr Int 53:261–263

Meakin SO, Macdonald JI, Gryz EA, Kubu CJ, Verdi JM (1999) The signaling adapter FRS-2 competes with Shc for binding to the nerve growth factor receptor TrkA. A model for discriminating proliferation and differentiation. J Biol Chem 274:9861–9870

Mellinghoff IK, Wang MY, Vivanco I, Haas-Kogan DA, Zhu S, Dia EQ, Lu KV, Yoshimoto K, Huang JH, Chute DJ, Riggs BL, Horvath S, Liau LM, Cavenee WK, Rao PN, Beroukhim R, Peck TC, Lee JC, Sellers WR, Stokoe D, Prados M, Cloughesy TF, Sawyers CL, Mischel PS (2005) Molecular determinants of the response of glioblastomas to EGFR kinase inhibitors. N Engl J Med 353:2012–2024

Miao H, Li DQ, Mukherjee A, Guo H, Petty A, Cutter J, Basilion JP, Sedor J, Wu J, Danielpour D, Sloan AE, Cohen ML, Wang B (2009) EphA2 mediates ligand-dependent inhibition and ligand-independent promotion of cell migration and invasion via a reciprocal regulatory loop with Akt. Cancer Cell 16:9–20

Mineo JF, Bordron A, Baroncini M, Maurage CA, Ramirez C, Siminski RM, Berthou C, Dam Hieu P (2007) Low HER2-expressing glioblastomas are more often secondary to anaplastic transformation of low-grade glioma. J Neurooncol 85:281–287

Miyamori H, Hasegawa K, Kim KR, Sato H (2000) Expression of metastasis-associated mts1 gene is co-induced with membrane type-1 matrix metalloproteinase (MT1-MMP) during oncogenic transformation and tubular formation of Madin Darby canine kidney (MDCK) epithelial cells. Clin Exp Metastasis 18:51–56

Moriyama T, Kataoka H, Tsubouchi H, Koono M (1995) Concomitant expression of hepatocyte growth factor (HGF), HGF activator and c-met genes in human glioma cells in vitro. FEBS Lett 372:78–82

Moriyama T, Kataoka H, Hamasuna R, Yokogami K, Uehara H, Kawano H, Goya T, Tsubouchi H, Koono M, Wakisaka S (1998a) Up-regulation of vascular endothelial growth factor induced by hepatocyte growth factor/scatter factor stimulation in human glioma cells. Biochem Biophys Res Commun 249:73–77

Moriyama T, Kataoka H, Kawano H, Yokogami K, Nakano S, Goya T, Uchino H, Koono M, Wakisaka S (1998b) Comparative analysis of expression of hepatocyte growth factor and its receptor, c-met, in gliomas, meningiomas and schwannomas in humans. Cancer Lett 124:149–155

Nagane M (2011) Neuro-oncology: continuing multidisciplinary progress. Lancet Neurol 10:18–20

Nakada M, Kita D, Futami K, Yamashita J, Fujimoto N, Sato H, Okada Y (2001) Roles of membrane type 1 matrix metalloproteinase and tissue inhibitor of metalloproteinases 2 in invasion and dissemination of human malignant glioma. J Neurosurg 94:464–473

Nakada M, Niska JA, Miyamori H, Mcdonough WS, Wu J, Sato H, Berens ME (2004) The phosphorylation of EphB2 receptor regulates migration and invasion of human glioma cells. Cancer Res 64:3179–3185

Nakada M, Niska JA, Tran NL, Mcdonough WS, Berens ME (2005) EphB2/R-Ras signaling regulates glioma cell adhesion, growth, and invasion. Am J Pathol 167:565–576

Nakada M, Drake KL, Nakada S, Niska JA, Berens ME (2006) Ephrin-B3 ligand promotes glioma invasion through activation of Rac1. Cancer Res 66:8492–8500

Nakada M, Nakada S, Demuth T, Tran NL, Hoelzinger DB, Berens ME (2007) Molecular targets of glioma invasion. Cell Mol Life Sci 64:458–478

Nakada M, Anderson EM, Demuth T, Nakada S, Reavie LB, Drake KL, Hoelzinger DB, Berens ME (2010) The phosphorylation of ephrin-B2 ligand promotes glioma cell migration and invasion. Int J Cancer 126:1155–1165

Nakada M, Hayashi Y, Hamada J (2011) Role of Eph/ephrin tyrosine kinase in malignant glioma. Neuro Oncol 13:1163–1170

Natarajan M, Stewart JE, Golemis EA, Pugacheva EN, Alexandropoulos K, Cox BD, Wang W, Grammer JR, Gladson CL (2006) HEF1 is a necessary and specific downstream effector of FAK that promotes the migration of glioblastoma cells. Oncogene 25:1721–1732

Neyns B, Sadones J, Joosens E, Bouttens F, Verbeke L, Baurain JF, D'hondt L, Strauven T, Chaskis C, In't Veld P, Michotte A, De Greve J (2009) Stratified phase II trial of cetuximab in patients with recurrent high-grade glioma. Ann Oncol 20:1596–1603

Neyns B, Sadones J, Chaskis C, Dujardin M, Everaert H, Lv S, Duerinck J, Tynninen O, Nupponen N, Michotte A, De Greve J (2011) Phase II study of sunitinib malate in patients with recurrent high-grade glioma. J Neurooncol 103:491–501

Nister M, Libermann TA, Betsholtz C, Pettersson M, Claesson-Welsh L, Heldin CH, Schlessinger J, Westermark B (1988) Expression of messenger RNAs for platelet-derived growth factor and transforming growth factor-alpha and their receptors in human malignant glioma cell lines. Cancer Res 48:3910–3918

Normanno N, De Luca A, Bianco C, Strizzi L, Mancino M, Maiello MR, Carotenuto A, De Feo G, Caponigro F, Salomon DS (2006) Epidermal growth factor receptor (EGFR) signaling in cancer. Gene 366:2–16

Ohgaki H, Kleihues P (2009) Genetic alterations and signaling pathways in the evolution of gliomas. Cancer Sci 100:2235–2241

Ohgaki H, Dessen P, Jourde B, Horstmann S, Nishikawa T, Di Patre PL, Burkhard C, Schuler D, Probst-Hensch NM, Maiorka PC, Baeza N, Pisani P, Yonekawa Y, Yasargil MG, Lutolf UM, Kleihues P (2004) Genetic pathways to glioblastoma: a population-based study. Cancer Res 64:6892–6899

Onishi M, Ichikawa T, Kurozumi K, Date I (2011) Angiogenesis and invasion in glioma. Brain Tumor Pathol 28:13–24

Ozawa T, Brennan CW, Wang L, Squatrito M, Sasayama T, Nakada M, Huse JT, Pedraza A, Utsuki S, Yasui Y, Tandon A, Fomchenko EI, Oka H, Levine RL, Fujii K, Ladanyi M, Holland EC (2010) PDGFRA gene rearrangements are frequent genetic events in PDGFRA-amplified glioblastomas. Genes Dev 24:2205–2218

Pedersen MW, Pedersen N, Ottesen LH, Poulsen HS (2005) Differential response to gefitinib of cells expressing normal EGFR and the mutant EGFRvIII. Br J Cancer 93:915–923

Peereboom DM, Shepard DR, Ahluwalia MS, Brewer CJ, Agarwal N, Stevens GH, Suh JH, Toms SA, Vogelbaum MA, Weil RJ, Elson P, Barnett GH (2010) Phase II trial of erlotinib with temozolomide and radiation in patients with newly diagnosed glioblastoma multiforme. J Neurooncol 98:93–99

Pflug BR, Colangelo AM, Tornatore C, Mocchetti I (2001) TrkA induces differentiation but not apoptosis in C6-2B glioma cells. J Neurosci Res 64:636–645

Plate KH, Breier G, Farrell CL, Risau W (1992) Platelet-derived growth factor receptor-beta is induced during tumor development and upregulated during tumor progression in endothelial cells in human gliomas. Lab Invest 67:529–534

Pollack IF, Stewart CF, Kocak M, Poussaint TY, Broniscer A, Banerjee A, Douglas JG, Kun LE, Boyett JM, Geyer JR (2011) A phase II study of gefitinib and irradiation in children with newly diagnosed brainstem gliomas: a report from the Pediatric Brain Tumor Consortium. Neuro Oncol 13:290–297

Ponzetto C, Bardelli A, Zhen Z, Maina F, Dalla Zonca P, Giordano S, Graziani A, Panayotou G, Comoglio PM (1994) A multifunctional docking site mediates signaling and transformation by the hepatocyte growth factor/scatter factor receptor family. Cell 77:261–271

Prigent SA, Gullick WJ (1994) Identification of c-erbB-3 binding sites for phosphatidylinositol 3′-kinase and SHC using an EGF receptor/c-erbB-3 chimera. EMBO J 13:2831–2841

Quang TS, Brady LW (2004) Radioimmunotherapy as a novel treatment regimen: 125I-labeled monoclonal antibody 425 in the treatment of high-grade brain gliomas. Int J Radiat Oncol Biol Phys 58:972–975

Rabin SJ, Tornatore C, Baker-Cairns B, Spiga G, Mocchetti I (1998) TrkA receptors delay C6-2B glioma cell growth in rat striatum. Brain Res Mol Brain Res 56:273–276

Raizer JJ, Abrey LE, Lassman AB, Chang SM, Lamborn KR, Kuhn JG, Yung WK, Gilbert MR, Aldape KA, Wen PY, Fine HA, Mehta M, Deangelis LM, Lieberman F, Cloughesy TF, Robins HI, Dancey J, Prados MD (2010) A phase II trial of erlotinib in patients with recurrent malignant gliomas and nonprogressive glioblastoma multiforme postradiation therapy. Neuro Oncol 12:95–103

Ram R, Lorente G, Nikolich K, Urfer R, Foehr E, Nagavarapu U (2006) Discoidin domain receptor-1a (DDR1a) promotes glioma cell invasion and adhesion in association with matrix metalloproteinase-2. J Neurooncol 76:239–248

Ramnarain DB, Park S, Lee DY, Hatanpaa KJ, Scoggin SO, Otu H, Libermann TA, Raisanen JM, Ashfaq R, Wong ET, Wu J, Elliott R, Habib AA (2006) Differential gene expression analysis reveals generation of an autocrine loop by a mutant epidermal growth factor receptor in glioma cells. Cancer Res 66:867–874

Raymond E, Brandes AA, Dittrich C, Fumoleau P, Coudert B, Clement PM, Frenay M, Rampling R, Stupp R, Kros JM, Heinrich MC, Gorlia T, Lacombe D, Van Den Bent MJ (2008) Phase II study of imatinib in patients with recurrent gliomas of various histologies: a European Organisation for Research and Treatment of Cancer Brain Tumor Group Study. J Clin Oncol 26:4659–4665

Reardon DA, Egorin MJ, Quinn JA, Rich JN, Gururangan S, Vredenburgh JJ, Desjardins A, Sathornsumetee S, Provenzale JM, Herndon JE 2nd, Dowell JM, Badruddoja MA, Mclendon RE, Lagattuta TF, Kicielinski KP, Dresemann G, Sampson JH, Friedman AH, Salvado AJ, Friedman HS (2005) Phase II study of imatinib mesylate plus hydroxyurea in adults with recurrent glioblastoma multiforme. J Clin Oncol 23:9359–9368

Reardon DA, Vredenburgh JJ, Desjardins A, Peters K, Gururangan S, Sampson JH, Marcello J, Herndon JE 2nd, Mclendon RE, Janney D, Friedman AH, Bigner DD, Friedman HS (2011) Effect of CYP3A-inducing anti-epileptics on sorafenib exposure: results of a phase II study of sorafenib plus daily temozolomide in adults with recurrent glioblastoma. J Neurooncol 101:57–66

Ribom D, Andrae J, Frielingsdorf M, Hartman M, Nister M, Smits A (2002) Prognostic value of platelet derived growth factor alpha receptor expression in grade 2 astrocytomas and oligoastrocytomas. J Neurol Neurosurg Psychiatry 72:782–787

Rich JN, Rasheed BK, Yan H (2004) EGFR mutations and sensitivity to gefitinib. N Engl J Med 351:1260–1261, author reply 1260–1261

Rosario M, Birchmeier W (2003) How to make tubes: signaling by the Met receptor tyrosine kinase. Trends Cell Biol 13:328–335

Salhia B, Rutten F, Nakada M, Beaudry C, Berens M, Kwan A, Rutka JT (2005) Inhibition of Rho-kinase affects astrocytoma morphology, motility, and invasion through activation of Rac1. Cancer Res 65:8792–8800

Sato H, Takino T, Okada Y, Cao J, Shinagawa A, Yamamoto E, Seiki M (1994) A matrix metalloproteinase expressed on the surface of invasive tumour cells. Nature 370:61–65

Saurez G, Cabanas R, Zaldivar M, Garnier T, Iglesias B, Piedra P, Castillo MR, Longchong M, Iznaga N, Lage A (2009) Clinical experience with nimotuzumab in cuban pediatric patients with brain tumors, 2005 to 2007. MEDICC Rev 11:27–33

Schlessinger J (2000) Cell signaling by receptor tyrosine kinases. Cell 103:211–225

Schwechheimer K, Laufle RM, Schmahl W, Knodlseder M, Fischer H, Hofler H (1994) Expression of neu/c-erbB-2 in human brain tumors. Hum Pathol 25:772–780
Senner V, Ratzinger S, Mertsch S, Grassel S, Paulus W (2008) Collagen XVI expression is upregulated in glioblastomas and promotes tumor cell adhesion. FEBS Lett 582:3293–3300
Shih AH, Holland EC (2006) Platelet-derived growth factor (PDGF) and glial tumorigenesis. Cancer Lett 232:139–147
Shinojima N, Tada K, Shiraishi S, Kamiryo T, Kochi M, Nakamura H, Makino K, Saya H, Hirano H, Kuratsu J, Oka K, Ishimaru Y, Ushio Y (2003) Prognostic value of epidermal growth factor receptor in patients with glioblastoma multiforme. Cancer Res 63:6962–6970
Singer HS, Hansen B, Martinie D, Karp CL (1999) Mitogenesis in glioblastoma multiforme cell lines: a role for NGF and its TrkA receptors. J Neurooncol 45:1–8
Spencer KS, Graus-Porta D, Leng J, Hynes NE, Klemke RL (2000) ErbB2 is necessary for induction of carcinoma cell invasion by ErbB family receptor tyrosine kinases. J Cell Biol 148:385–397
Stephens P, Hunter C, Bignell G, Edkins S, Davies H, Teague J, Stevens C, O'meara S, Smith R, Parker A, Barthorpe A, Blow M, Brackenbury L, Butler A, Clarke O, Cole J, Dicks E, Dike A, Drozd A, Edwards K, Forbes S, Foster R, Gray K, Greenman C, Halliday K, Hills K, Kosmidou V, Lugg R, Menzies A, Perry J, Petty R, Raine K, Ratford L, Shepherd R, Small A, Stephens Y, Tofts C, Varian J, West S, Widaa S, Yates A, Brasseur F, Cooper CS, Flanagan AM, Knowles M, Leung SY, Louis DN, Looijenga LH, Malkowicz B, Pierotti MA, Teh B, Chenevix-Trench G, Weber BL, Yuen ST, Harris G, Goldstraw P, Nicholson AG, Futreal PA, Wooster R, Stratton MR (2004) Lung cancer: intragenic ERBB2 kinase mutations in tumours. Nature 431:525–526
Stommel JM, Kimmelman AC, Ying H, Nabioullin R, Ponugoti AH, Wiedemeyer R, Stegh AH, Bradner JE, Ligon KL, Brennan C, Chin L, Depinho RA (2007) Coactivation of receptor tyrosine kinases affects the response of tumor cells to targeted therapies. Science 318:287–290
Stratmann A, Risau W, Plate KH (1998) Cell type-specific expression of angiopoietin-1 and angiopoietin-2 suggests a role in glioblastoma angiogenesis. Am J Pathol 153:1459–1466
Surawska H, Ma PC, Salgia R (2004) The role of ephrins and Eph receptors in cancer. Cytokine Growth Factor Rev 15:419–433
Tamura M, Gu J, Matsumoto K, Aota S, Parsons R, Yamada KM (1998) Inhibition of cell migration, spreading, and focal adhesions by tumor suppressor PTEN. Science 280:1614–1617
Thiessen B, Stewart C, Tsao M, Kamel-Reid S, Schaiquevich P, Mason W, Easaw J, Belanger K, Forsyth P, Mcintosh L, Eisenhauer E (2010) A phase I/II trial of GW572016 (lapatinib) in recurrent glioblastoma multiforme: clinical outcomes, pharmacokinetics and molecular correlation. Cancer Chemother Pharmacol 65:353–361
Tokarski JS, Newitt JA, Chang CY, Cheng JD, Wittekind M, Kiefer SE, Kish K, Lee FY, Borzillerri R, Lombardo LJ, Xie D, Zhang Y, Klei HE (2006) The structure of Dasatinib (BMS-354825) bound to activated ABL kinase domain elucidates its inhibitory activity against imatinib-resistant ABL mutants. Cancer Res 66:5790–5797
Uhm JH, Ballman KV, Wu W, Giannini C, Krauss JC, Buckner JC, James CD, Scheithauer BW, Behrens RJ, Flynn PJ, Schaefer PL, Dakhill SR, Jaeckle KA (2011) Phase II evaluation of gefitinib in patients with newly diagnosed Grade 4 astrocytoma: Mayo/North Central Cancer Treatment Group Study N0074. Int J Radiat Oncol Biol Phys 80:347–353
Uhrbom L, Hesselager G, Nister M, Westermark B (1998) Induction of brain tumors in mice using a recombinant platelet-derived growth factor B-chain retrovirus. Cancer Res 58:5275–5279
Vajkoczy P, Knyazev P, Kunkel A, Capelle HH, Behrndt S, Von Tengg-Kobligk H, Kiessling F, Eichelsbacher U, Essig M, Read TA, Erber R, Ullrich A (2006) Dominant-negative inhibition of the Axl receptor tyrosine kinase suppresses brain tumor cell growth and invasion and prolongs survival. Proc Natl Acad Sci U S A 103:5799–5804
Valius M, Kazlauskas A (1993) Phospholipase C-gamma 1 and phosphatidylinositol 3 kinase are the downstream mediators of the PDGF receptor's mitogenic signal. Cell 73:321–334
Van Den Bent MJ, Brandes AA, Rampling R, Kouwenhoven MC, Kros JM, Carpentier AF, Clement PM, Frenay M, Campone M, Baurain JF, Armand JP, Taphoorn MJ, Tosoni A, Kletzl H, Klughammer B, Lacombe D, Gorlia T (2009) Randomized phase II trial of erlotinib versus temozolomide or carmustine in recurrent glioblastoma: EORTC brain tumor group study 26034. J Clin Oncol 27:1268–1274
Varela M, Ranuncolo SM, Morand A, Lastiri J, De Kier Joffe EB, Puricelli LI, Pallotta MG (2004) EGF-R and PDGF-R, but not bcl-2, overexpression predict overall survival in patients with low-grade astrocytomas. J Surg Oncol 86:34–40
Verhaak RG, Hoadley KA, Purdom E, Wang V, Qi Y, Wilkerson MD, Miller CR, Ding L, Golub T, Mesirov JP, Alexe G, Lawrence M, O'kelly M, Tamayo P, Weir BA, Gabriel S, Winckler W, Gupta S, Jakkula L, Feiler HS, Hodgson JG, James CD, Sarkaria JN, Brennan C, Kahn A, Spellman PT, Wilson RK, Speed TP, Gray JW, Meyerson M, Getz G, Perou CM, Hayes DN (2010) Integrated genomic analysis identifies clinically relevant subtypes of glioblastoma characterized by abnormalities in PDGFRA, IDH1, EGFR, and NF1. Cancer Cell 17:98–110
Vogel W, Gish GD, Alves F, Pawson T (1997) The discoidin domain receptor tyrosine kinases are activated by collagen. Mol Cell 1:13–23

Vogel WF, Abdulhussein R, Ford CE (2006) Sensing extracellular matrix: an update on discoidin domain receptor function. Cell Signal 18:1108–1116

Wen PY (2010) American Society of Clinical Oncology 2010: report of selected studies from the CNS tumors section. Expert Rev Anticancer Ther 10:1367–1369

Wen PY, Yung WK, Lamborn KR, Dahia PL, Wang Y, Peng B, Abrey LE, Raizer J, Cloughesy TF, Fink K, Gilbert M, Chang S, Junck L, Schiff D, Lieberman F, Fine HA, Mehta M, Robins HI, Deangelis LM, Groves MD, Puduvalli VK, Levin V, Conrad C, Maher EA, Aldape K, Hayes M, Letvak L, Egorin MJ, Capdeville R, Kaplan R, Murgo AJ, Stiles C, Prados MD (2006) Phase I/II study of imatinib mesylate for recurrent malignant gliomas: North American Brain Tumor Consortium Study 99–08. Clin Cancer Res 12:4899–4907

Wilhelm SM, Carter C, Tang L, Wilkie D, Mcnabola A, Rong H, Chen C, Zhang X, Vincent P, Mchugh M, Cao Y, Shujath J, Gawlak S, Eveleigh D, Rowley B, Liu L, Adnane L, Lynch M, Auclair D, Taylor I, Gedrich R, Voznesensky A, Riedl B, Post LE, Bollag G, Trail PA (2004) BAY 43–9006 exhibits broad spectrum oral antitumor activity and targets the RAF/MEK/ERK pathway and receptor tyrosine kinases involved in tumor progression and angiogenesis. Cancer Res 64:7099–7109

Wimmer-Kleikamp SH, Lackmann M (2005) Eph-modulated cell morphology, adhesion and motility in carcinogenesis. IUBMB Life 57:421–431

Wong AJ, Bigner SH, Bigner DD, Kinzler KW, Hamilton SR, Vogelstein B (1987) Increased expression of the epidermal growth factor receptor gene in malignant gliomas is invariably associated with gene amplification. Proc Natl Acad Sci U S A 84:6899–6903

Wykosky J, Gibo DM, Stanton C, Debinski W (2005) EphA2 as a novel molecular marker and target in glioblastoma multiforme. Mol Cancer Res 3:541–551

Yakes FM, Chen J, Tan J, Yamaguchi K, Shi Y, Yu P, Qian F, Chu F, Bentzien F, Cancilla B, Orf J, You A, Laird AD, Engst S, Lee L, Lesch J, Chou YC, Joly A (2011) Cabozantinib (XL184), a novel MET and VEGFR2 inhibitor, simultaneously suppresses metastasis, angiogenesis, and tumor growth. Mol Cancer Ther 10:2298–2308

Yamada T, Tsubouchi H, Daikuhara Y, Prat M, Comoglio PM, Mcgeer PL, Mcgeer EG (1994) Immunohistochemistry with antibodies to hepatocyte growth factor and its receptor protein (c-MET) in human brain tissues. Brain Res 637:308–312

Yamanaka R, Arao T, Yajima N, Tsuchiya N, Homma J, Tanaka R, Sano M, Oide A, Sekijima M, Nishio K (2006) Identification of expressed genes characterizing long-term survival in malignant glioma patients. Oncogene 25:5994–6002

Yamazaki H, Fukui Y, Ueyama Y, Tamaoki N, Kawamoto T, Taniguchi S, Shibuya M (1988) Amplification of the structurally and functionally altered epidermal growth factor receptor gene (c-erbB) in human brain tumors. Mol Cell Biol 8:1816–1820

Yarden Y, Sliwkowski MX (2001) Untangling the ErbB signalling network. Nat Rev Mol Cell Biol 2:127–137

Ye K, Hurt KJ, Wu FY, Fang M, Luo HR, Hong JJ, Blackshaw S, Ferris CD, Snyder SH (2000) Pike. A nuclear gtpase that enhances PI3kinase activity and is regulated by protein 4.1 N. Cell 103:919–930

Ye K, Aghdasi B, Luo HR, Moriarity JL, Wu FY, Hong JJ, Hurt KJ, Bae SS, Suh PG, Snyder SH (2002) Phospholipase C gamma 1 is a physiological guanine nucleotide exchange factor for the nuclear GTPase PIKE. Nature 415:541–544

Yong HY, Kim IY, Kim JS, Moon A (2010) ErbB2-enhanced invasiveness of H-Ras MCF10A breast cells requires MMP-13 and uPA upregulation via p38 MAPK signaling. Int J Oncol 36:501–507

Zadeh G, Koushan K, Pillo L, Shannon P, Guha A (2004a) Role of Ang1 and its interaction with VEGF-A in astrocytomas. J Neuropathol Exp Neurol 63:978–989

Zadeh G, Qian B, Okhowat A, Sabha N, Kontos CD, Guha A (2004b) Targeting the Tie2/Tek receptor in astrocytomas. Am J Pathol 164:467–476

Zadeh G, Koushan K, Baoping Q, Shannon P, Guha A (2010) Role of angiopoietin-2 in regulating growth and vascularity of astrocytomas. J Oncol 2010:659231

Zeng Q, Chen S, You Z, Yang F, Carey TE, Saims D, Wang CY (2002) Hepatocyte growth factor inhibits anoikis in head and neck squamous cell carcinoma cells by activation of ERK and Akt signaling independent of NFkappa B. J Biol Chem 277:25203–25208

Zhang YW, Vande Woude GF (2003) HGF/SF-met signaling in the control of branching morphogenesis and invasion. J Cell Biochem 88:408–417

Zhang QK, Boast S, De Los SK, Begemann M, Goff SP (1996) Transforming activity of retroviral genomes encoding Gag-Axl fusion proteins. J Virol 70:8089–8097

Zhang Z, Yang Y, Gong A, Wang C, Liang Y, Chen Y (2005) Localization of NGF and TrkA at mitotic apparatus in human glioma cell line U251. Biochem Biophys Res Commun 337:68–74

Chapter 9
TGF Beta Signaling and Its Role in Glioma Pathogenesis

Bozena Kaminska, Marta Kocyk, and Magdalena Kijewska

Abstract Transforming growth factor beta (TGF-β) signaling is involved in the regulation of proliferation, differentiation and survival/or apoptosis of many cells, including glioma cells. TGF-β acts via specific receptors activating multiple intracellular pathways resulting in phosphorylation of receptor-regulated Smad2/3 proteins that associate with the common mediator, Smad4. Such complex translocates to the nucleus, binds to DNA and regulates transcription of many genes. Furthermore, TGF-β-activated kinase-1 (TAK1) is a component of TGF-β signaling and activates mitogen-activated protein kinase cascades. Negative regulation of TGF-β/Smad signaling may occur through the inhibitory Smad6/7. Increased expression of TGF-β1-3 correlates with a degree of malignancy of human gliomas. TGF-β may contribute to tumor pathogenesis by direct support of tumor growth, self-renewal of glioma initiating stem cells and inhibiting of anti-tumor immunity. TGF-β1,2 stimulate expression of the vascular endothelial growth factor as well as the plasminogen activator inhibitor and some metalloproteinases that are involved in vascular remodeling, angiogenesis and degradation of the extracellular matrix. Inhibitors of TGF-β signaling reduce viability and invasion of gliomas in animal models and show promises as novel, potential anti-tumor therapeutics.

Keywords TGF-β signaling • TGF-β receptors • Activin-receptor-like kinases • MAP kinases • Smad proteins • Gliomas

B. Kaminska (✉) • M. Kijewska
Laboratory of Transcription Regulation, Department of Cell Biology, Nencki Institute of Experimental Biology, Polish Academy of Sciences, 3 Pasteur St., PL 02-093 Warsaw, Poland
e-mail: bozenakk@nencki.gov.pl; mkijewska@nencki.gov.pl

M. Kocyk
Laboratory of Transcription Regulation, Department of Cell Biology, Nencki Institute of Experimental Biology, Polish Academy of Sciences, 3 Pasteur St., PL 02-093 Warsaw, Poland

Postgraduate School of Molecular Medicine, Warsaw Medical University, Warsaw, Poland
e-mail: mkocyk@nencki.gov.pl

J. Barańska (ed.), *Glioma Signaling*, Advances in Experimental Medicine and Biology 986, DOI 10.1007/978-94-007-4719-7_9, © Springer Science+Business Media Dordrecht 2013

Abbreviations

ADAMTS-1	Metalloproteinase and disintegrin-like domain
Akt	Protein kinase B/Akt kinase
ALKs	Activin-receptor-like kinases
AP-1	Activator protein
BMPs	Bone morphogenetic proteins
CTLs	Cytotoxic T lymphocytes
EGFR	Epidermal growth factor receptor
ECM	Extracellular matrix
ERK1/2	Extracellular signal-regulated kinases 1/2
GDFs	Growth and differentiation factors
JAK	Janus kinase
JNK c	Jun N-terminal kinases
LAP	Latency-associated peptide
LIF	Leukemia inhibitory factor
LTBP	Latent TGF-β binding protein
MT1-MMP	Membrane-type 1 matrix metalloproteinase
p38 MAPK	P38 mitogen-activated protein kinases
R-Smads	Receptor-Smad proteins
SARA	Smad anchor for receptor activation, STAT, signal transducer and activator of transcription
TAK1	TGF-β-activated kinase-1
TβRI	TGF-β type I receptor
TβRII	TGF-β type II receptor
TGF-β	Transforming growth factor β
TMZ	Temozolomide
TNFα	Tumor necrosis factor α
TRAF	TNF receptor associated factor
VEGF	Vascular endothelial growth factor.

9.1 Introduction

Transforming growth factor-beta (TGF-β) is a multifunctional cytokine that regulates cell proliferation, differentiation and extracellular matrix production (Jennings and Pietenpol 1998; Verrecchia and Mauviel 2002). The TGF-β belongs to a larger superfamily including: TGF-β, BMPs (bone morphogenetic proteins) and GDFs (growth and differentiation factors), sharing the characteristic fold of cystine-knot of the TGF superfamily. TGF-β and BMP/GDF form homo- and hetero-dimers that interact with combinations of type I and type II receptor dimers to produce multiple signaling complexes, leading to the activation of SMAD transcription factors (Rider and Mulloy 2010) (Fig. 9.1). The effects of distinct TGF-β isoforms depend on the type, differentiation state and physiological conditions of target cells (Bottner et al. 2000). Functional analysis of genes in the TGF-β signaling pathway in mice either lacking entire genes, or expressing dominant negative forms of particular proteins, provided new insights into the signaling cascades, their interaction and specificity (Goumans and Mummery 2000).

Deregulation of TGF-β expression or signaling has been implicated in the pathogenesis of a variety of diseases, including cancer and fibrosis. TGF-β binds and activates a membrane receptor serine/threonine kinase complex. Upon TGF-β binding, the receptor complex phosphorylates the transcription

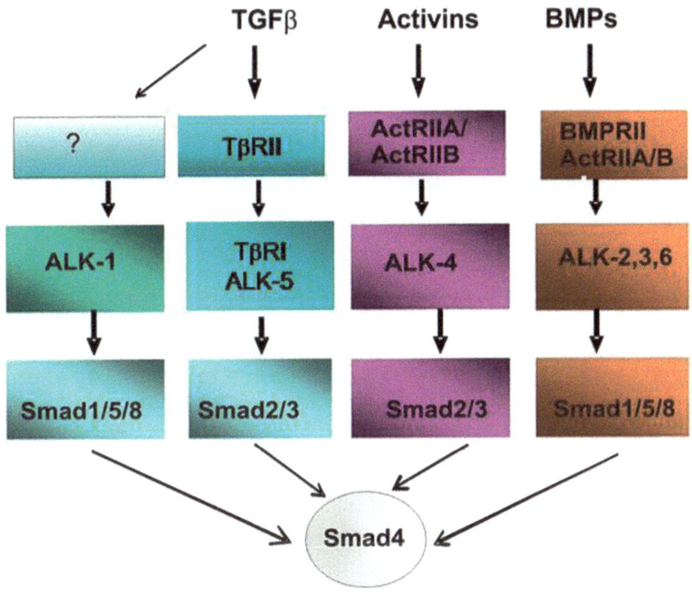

Fig. 9.1 The TGF-β superfamily signal transduction. Members of the TGF-β superfamily signal via a distinct type I and type II receptors. Ligand binding induces conformational changes and activation of kinase domains of the receptors. Smad 2/3 proteins are cytoplasmic transcription factors which are phosphorylated by a serine-threonine kinase associated with the receptor. Hetero-oligomeric complex of the R-Smad (receptor Smad proteins) associates with Smad 4, translocates to the nucleus and binds to specific DNA sequence in the promoters of target genes to regulate transcription. Four distinct type II, seven type I receptors and five R-Smads have been identified. ActR, activin receptor; ALK, activin-receptor-like-kinase; BMPR, BMP receptor; TβRI, TGF- β type I receptor; TβRII, TGF- β type II receptor

factors Smad2 and Smad3 that bind to Smad4 and accumulate in the nucleus, where they regulate transcription (Massague et al. 2005; Schmierer and Hill 2007; ten Dijke and Hill 2004). TGF-β has a dual role in oncogenesis. It is a strong inhibitor of proliferation of normal epithelial cells and astrocytes, and is considered a tumor suppressor factor. On the other hand, in some tumor types, and specifically in high-grade glioma, TGF-β becomes an oncogenic factor (Massague 2008; Seoane 2008).

Under physiological conditions TGF-β is expressed at a very low level in the brain but its expression strongly increases after injury (Lindholm et al. 1992). TGF-β inhibits proliferation of normal astrocytes, but loses its growth-inhibitory potential towards gliomas, due to alterations in the expression of cell cycle inhibitors. There is growing evidence that TGF-β is actively secreted by tumor cells in the later stages of cancer progression and contributes to cell growth, invasion, and metastasis, and decreases host immune responses against tumor. TGF-β is highly active in high-grade glioma and elevated TGF-β activity confers poor prognosis in glioma patients (Rich 2003; Penuelas et al. 2009). TGF-β induces cell proliferation and tumor progression through the induction of platelet derived growth factor-B (PDGF-B) in human gliomas with an unmethylated *PDGF-B* gene (Bruna et al. 2007). TGF-β1 produced by glioma infiltrating brain microglia/macrophages doubles glioma invasion *in vitro* and *in vivo* (Wesolowska et al. 2008). Recently, TGF-β has been shown to cooperate with leukemia inhibitory factor (LIF) in inducing the self-renewal and preventing the differentiation of glioma initiating stem cells (Ikushima et al. 2009; Penuelas et al. 2009). TGF-β induced the self-renewal capacity of those cells, but not of normal human neuroprogenitors, through the Smad-dependent induction of LIF and the subsequent activation of the JAK-STAT pathway (Penuelas et al. 2009). Despite a large number of data regarding TGF-β expression in different brain tumors, the molecular mechanisms underlying the expression, signaling and role of TGF-β in pathogenesis of glioblastomas are still unknown. We summarized here data concerning molecular mechanisms of TGF activation, its signaling and role in glioma pathogenesis. Furthermore, we discussed advances in development of molecular and pharmacological inhibitors of TGF-β signaling as cancer therapeutics.

9.2 A Brief Summary of Mechanisms of TGF-β Signaling in Normal and Malignant Cells

9.2.1 Components and Mechanisms of TGF-β Signaling

TGF-β is synthesized as 55-kDa polypeptides, which form a dimer shortly after production and is further cleaved in the Golgi apparatus by furine-like proteases to form a small latent TGF-β (Dubois et al. 1995). The mature 25-kDa TGF-β protein is non-covalently bound to the latency-associated peptide (LAP) and the latent TGF-β binding protein (LTBP). TGF-β is secreted as a latent complex. Such complexes are secreted and targeted to specific locations in the extracellular matrix by the appropriate LTBP (Hyytiainen et al. 2004). The latent state prevents the cytokine from eliciting a response until certain physiological conditions occur or until the target cell is reached. LTBPs are important for activation of the latent cytokine and as targeting molecules of the effects of TGF-β. The latency proteins also contribute to the cytokine stability: free TGF-β has a half life of about 2 min, whereas the latent form – 90 min. TGF-β is activated in an enzymatic process of LAP cleavage for example by plasmin (Grainger et al. 1995) or thrombospondin-1 which also cooperates with the plasmin- mediated process (Ribeiro et al. 1999). Other proteins participating in TGF-β activation *in vivo* include: integrins (Annes et al. 2004) and matrix metalloproteinases (membrane-type 1 matrix metalloproteinase, MT1-MMP) (Tatti et al. 2008). Activation of TGF-β is a multi-step process and represents a regulatory stage responsible for a tight control of active TGF-β formation.

TGF-β signals through binding to the type II and type I serine/threonine kinase receptors (TβRII and TβRI, respectively), inducing their hetero-oligomerization that subsequently activates different, intracellular signaling pathways (Groppe et al. 2008). Five type II and seven type I receptors, also termed activin-receptor-like kinases (ALKs) have been identified (Shi and Massague 2003) (Fig. 9.1). TGF-β first binds to TβRII and alters its conformation. Due to that TβRII can be recognized by TβRI. Constitutively active, intrinsic protein kinase of TβRII phosphorylates then the three serine and two threonine residues in the special domain of TβRI. Subsequently, the type I receptor directly phosphorylates Smad proteins on the C-terminal Ser-Ser-X-Ser motif. This phosphorylation is facilitated by other proteins: SARA (Smad anchor for receptor activation) and HRS/HGS (hepatocyte growth factor-regulated tyrosine kinase substrate) (Miura et al. 2000; Tsukazaki et al. 1998) via the phospholipids-binding FYVE domain, acting as an adaptor to aid in recruitment of R-Smad2 to the TGF-β receptor complex, and by cPML (the cytoplasmic promyelocytic leukaemia), an adaptor that promotes ALK5 mediated receptor- Smad proteins (R-Smad) phosphorylation (Lin et al. 2004). R-Smad phosphorylation is coupled to TGF-β receptor internalization (Penheiter et al. 2002).

It is generally accepted that the TGF-β/Activin/Nodal subfamily activates Smad2 and 3 whereas the BMP/GDF/MIS subfamily functions through Smad1, 5, and 8 (Goumans and Mummery 2000; Nakao et al 1997) (Fig. 9.1). The MH2 (MAD homology 2) domain is highly conserved among all Smad proteins and is responsible for receptor interaction, formation of homo- and heteromeric Smad complexes, and direct contact with the nuclear pore complex for shuttling to the nucleus. Phosphorylation of the two serine residues in the SXS motif of the MH2 domain activates R-Smad (Attisano and Wrana 2002; Piek et al. 1999; Shi and Massague 2003). Once phosphorylated, R-Smads dissociate from the receptor/SARA complex and form an oligomeric complex with the common mediator, Smad4, whereupon they translocate to the nucleus and interact with Smad binding elements (SBE: GTCT) or GC-rich sequences present in certain promoters (Fig. 9.2). Smad proteins have been shown to

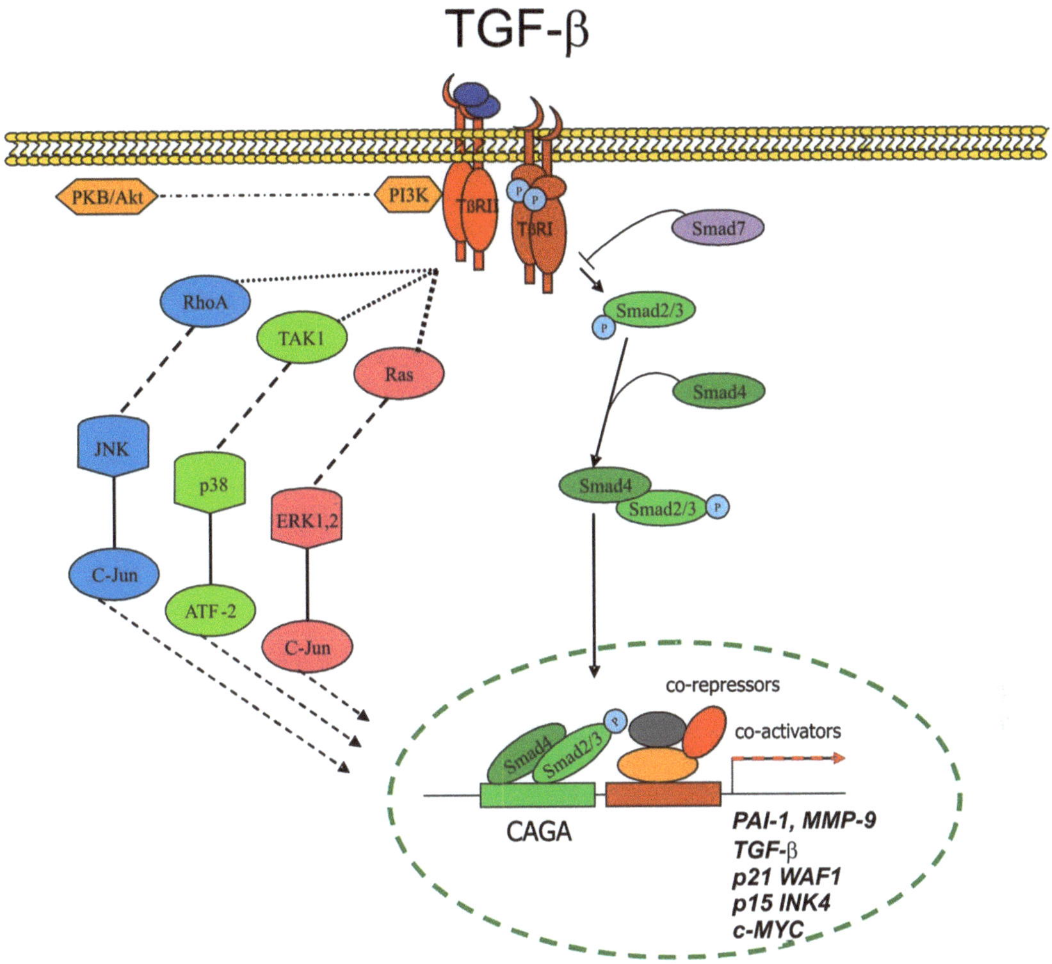

Fig. 9.2 Activation of TGF-β type I and type II receptors (TβRI, TβRII) leads to activation of receptor kinases, phoshorylation of receptor-Smad proteins (R-Smads), that form a complex with Smad4, which translocates to the nucleus. The pathway is regulated by the activity of the inhibitory Smad7. TGF-β can activate several mitogen activated protein kinases (MAPKs), including extracellular signal-regulated kinases (ERKs), c-Jun N-terminal kinases (JNKs), p38 MAPK, phosphatidylinositol 3-kinase (PI3K) and protein kinase B/Akt (PKB/Akt). The interaction with these MAP kinases may positively or negatively regulate Smad transcriptional activity

cooperate with a wide variety of transcription factors to regulate gene expression (Massague et al. 2005; Moustakas et al. 2001).

Besides the canonical TGF-β/Smad pathway, TGF-β can directly activate non-Smad signaling pathways (Moustakas and Heldin 2005; Zhang 2009), including the mitogen activated protein (MAP) kinases. TGF-β activated kinase-1 (TAK1), a member of the MEKK family and activator of JNK and p38 MAPK, is activated by TGF-β (Yamaguchi et al. 1995). ERK could be phosphorylated via TβRI-induced phosphorylation of ShcA protein which associates with adaptor Grb2 and Sos proteins, thereby initiating the well-characterised pathway linking receptor tyrosine kinases with ERK MAP kinases (Lee et al. 2007). In addition, small GTPases such as Ras, Rho, Rac and CDC42, have been implicated in non-Smad TGF-β signaling in epithelial cells (Bhowmick et al. 2001; Edlund et al. 2002; Wilkes et al. 2003). For example, activation of Ras, extracellular signal-regulated kinases 1/2 (ERK1/2), and c-Jun N-terminal kinase (JNK) by TGF-β signaling have been reported in primary intestinal epithelial cells and some breast cancer cell lines (Mulder 2000), whereas activation of protein kinase A contributes to TGF-β signaling in murine mesangial cells (Wang et al. 1998). We demonstrated that TGF-β1 which did not affect viability or proliferation of human glioblastoma T98G cells, increased transcriptional responses exemplified by the induction of *MMP-9* expression. TGF-β-induced nuclear translocation of the SMAD3/SMAD4 complex and activation of SMAD-dependent promoter was paralleled by the selective activation of p38 MAPK, and phosphorylation of its substrates: ATF2 and c-Jun proteins leading to transient activation of AP-1 transcription factor (Dziembowska et al. 2007; Kaminska 2009). Selective activation of p38 MAPK, with no apparent activation of JNK after TGF-β stimulation was reported for C2C12, Mv1Lu, and HaCaT cells (Hanafusa et al. 1999; Karsdal et al. 2003).

Mechanisms underlying non-canonical TGF-β signaling may involve the adaptor protein TRAF6, the ubiquitin ligase (E3), which interacts with a consensus motif present in TβRI and forms a complex required for TGF-β-induced autoubiquitylation of TRAF6 and subsequent activation of the TAK1-p38/JNK pathway in a receptor kinase-independent manner. Activation of Smad2 is not dependent on TRAF6 (Sorrentino et al. 2008). A recent study provided evidence that TGF-β type I receptor (TβRI) can by-pass canonical signaling pathways. It has been shown that TβRI undergoes cleavage by TACE (TNF-α converting enzyme) in cancer cells and that the liberated intracellular domain (ICD) of TβRI associates with the transcriptional regulator p300 to activate genes involved in tumor cell invasiveness, such as *Snail* and *MMP-2*. The TβRI ICD was localized in the nuclei of different types of tumor cells in tissue sections but not in normal epithelial cells (Mu et al. 2011).

9.2.2 Negative Regulators of TGF-β Signaling

A broad array of Smad interacting partners and diverse post-translational modifications of Smads have been identified (Itoh and ten Dijke 2007). The Smad signaling pathway may be negatively regulated by the inhibitory Smad6 and Smad7 (Nakao et al. 1997). Inhibitory Smad7 acts to oppose the signal mediated by R-Smads by forming stable associations with activated type I receptors, thus preventing phosphorylation of R-Smads (Shi and Massague 2003).

Other mechanisms by which I-Smads antagonize signaling include: interactions with Smad4, preventing R-Smad–Smad4 complex formation; recruitment of E3-ubiquitin ligases Smurf1 and 2 to induce type I receptor ubiquitination and subsequent receptor degradation; direct repression of Smad-induced transcriptional responses. Another important way of inhibition of the R-Smad function is through the recruitment of transcriptional co-repressors such as c-Ski and SnoN. Co-repressors Ski and SnoN can be recruited to Smad binding elements in a Smad4-dependent manner and inhibit the basal expression of TGF-β-responsive genes such as *Smad7* (ten Dijke and Hill 2004).

9.2.3 Transcriptional Responses Induced by TGF-β Signaling

The Smad binding element – the CAGAC sequence is calculated to be present on average every 1,024 bp in the genome, or about once in the regulatory region of any average size gene. Also Smad binding to the SBE lacks selectivity, as Smads 1, 3 and 4 have a similar affinity for the SBE. Therefore, additional DNA contacts are necessary for specific, high-affinity binding of a Smad complex to a target gene. Smads can achieve high-affinity, selective interactions with DNA by associating with DNA-binding partners, forming complexes of specific composition and geometry. Most of the Smad transcriptional partners identified to date: Fast1, Mixer, Jun/Fos, Runx, ATF3, E2F4/5 are highly responsive to different inputs (reviewed in Massague and Wotton 2000). FAST-1 is a member of the winged-helix family of DNA-binding proteins and interacts with Smad2–Smad4 or Smad3–Smad4 complexes, but not with BMP-activated Smad complexes (Chen et al. 1998). Smad3–Smad4 complex and an AP-1 (activator protein-1) complex synergize in the transcriptional activation from the c-*Jun* promoter by binding to separate sites located 120 bp apart from each other (Wong et al. 1999). Smads recruit the general co-activators p300 and CBP that have histone acetyltransferase activity and may increase transcription of target genes by inducing chromatin remodeling (Massague and Wotton 2000).

TGF-β1 can activate its mRNA expression and thereby its own secretion in Smad-and AP-1-dependent manner in many cell types, including glioma cells (Jachimczak et al. 1996; Jennings et al. 1991; Kim et al. 1990; Van Obberghen-Schilling et al. 1988). There are three AP-1-binding elements contributing to TGF-β1 induction and antisense against AP-1 components *c-jun* and *c-fos* blocked autocrine TGF-β1-induced expression. The TGF-β cytostatic program involves transcriptional activation of the cyclin-dependent kinase inhibitors p21WAF1/Cip1 and p15Ink4b, and repression of transcription of genes encoding transcription factors c-myc and Id1-Id3 (Alexandrow and Moses 1995; Siegel et al. 2003a, b). Negative regulators of TGF-β signaling such as I-Smads and SnoN are direct target genes of TGF-β and participate in negative feedback loops (Derynck and Zhang 2003; Massague and Wotton 2000). The canonical model has been challenged in some studies employing global gene expression profiling and showing that a subset of TGF-β induced genes does not require Smad4 (Ijichi et al. 2004; Levy and Hill 2005).

9.3 Deregulation of TGF-β Signaling in Gliomas

There is growing evidence that alterations in TGF-β signaling pathway components modify cancer risk. Approximately 14 % of the general population carry TGFBR1*6A, a variant of the TGFBR1 gene that results in decreased TGF-β-mediated growth inhibition. Recent studies show that the overall cancer risk is increased by 70 and 19 % among TGFBR1*6A homozygotes and heterozygotes, respectively (Bian et al. 2003). The expression of downstream components of TGF-β signaling has been evaluated in glioma cell lines and glioma speciments and those analyses showed alterations in many components of this pathway.

Early studies indicated that TGF-β1 induces endogenous PAI-1 protein synthesis, Smad binding element/(CAGA)12-luciferase-reporter activity, as well as mRNA expression of *SMAD6* and *SMAD7* in all tested human glioma cell lines that suggested unaffected TGF-β signaling (Piek et al. 1999). On the other hand, high-grade human gliomas secrete TGF-β, but are generally resistant to its growth inhibitory effects. Analysis of TGF-β effects on 12 human glioma cell lines showed that the cytokine mildly inhibited or had no effect on the cell growth and stimulated growth of two lines. The majority of glioma lines had homozygous deletions of the *p15(INK4B)* gene, only in three lines TGF-β slightly induced $p21^{WAF1/CIP}$ expression.

Other studies have shown that expression of the SMAD2, SMAD3 and SMAD4 proteins was lower in the glioma cell lines, while expression of the SMAD7 protein was similar to that in normal astrocytes. In particular, SMAD3 expression was low or very low in nine out of the ten malignant glioma cell lines. Seven of the ten glioma cell lines exhibited lower levels of nuclear translocation of SMAD2 and SMAD3, and two cell lines expressing very low levels of SMAD3 showed no nuclear translocation. All glioma cell lines expressed the SnoN protein and its expression was not modulated by a treatment with TGF-β1; three glioma cell lines expressed high levels of the Ski protein. The expression of the $p21^{WAF1/CIP}$, p15(INK4B), cyclin-dependent kinase-4 (CDK4), and cyclin D1 proteins was not altered by TGF-β1 treatment in a majority of glioma cell lines suggesting dysfunction of TGF-β-induced growth inhibitory signals (Zhang et al. 2006).

Up-regulation of *TGF-β2* mRNA expression was observed in EGFRvIII-positive GBM patients and in astrocytoma U87MG cells overexpressing EGFRvIII as compared with U87MG cells (Zhang et al. 2011c). This suggests that the loss of p15INK4B, a cell cycle inhibitor, may explain, in part, the selective loss of growth inhibition by TGF-β in gliomas (Held-Feindt et al. 2003; Rich et al. 1999). Most glioma lines retained other TGF-β-mediated responses, such as extracellular matrix protein and angiogenic factor secretion, which may contribute to increased malignant behavior.

9.4 Functions of TGF-β Signaling in Glioma Biology

9.4.1 TGF-β Signaling in Controlling Cell Proliferation

The TGF-β cytostatic program involves transcriptional activation of genes coding for the cyclin-dependent kinase inhibitors p21WAF1/Cip1 and p15Ink4b, and repression of c-*myc* and *Id1*-*Id3* genes (Alexandrow and Moses 1995; Siegel et al. 2003b). Cooperatively, these gene responses mediate cell cycle arrest in the G_0/G_1 phase of the cell cycle. Mechanisms for Smad-mediated repression of c-*myc* and *Id* have been elucidated. A TGFβ inhibitory element (TIE) located between positions –92 and –63 relative to the c-*myc* P2 transcription start site mediates repression in response to TGFβ in human skin keratinocytes and mammary epithelial cells (Chen et al. 2002; Siegel et al. 2003a, b). c-Myc, which binds to the *p21Cip1* and *p15Ink4b* promoters in mitogen-stimulated cells, must be down-regulated before the *p21Cip1* and *p15Ink4b* genes could be activated by TGF-β. c-Myc down-regulation by TGF-β renders those promoters competent for activation that requires another transactivation complex (Seoane et al. 2001). FoxO proteins (Fox – members of the Forkhead box) have been identified as key partners of SMAD3 and SMAD4 in the TGF-β-dependent generation of the *p21Cip1* activation complex. FoxO-Smad complexes are also inhibited by FoxG1, a distinct Forkhead family member, and the combined actions of FoxG1 and phosphatidylinositol 3-kinase (PI3K) signaling, (negatively regulating Forkhead proteins) mediate resistance of human glioblastoma cells to TGF-β-induced growth inhibition. Three pathways—the SMAD, PI3K, and FoxG1 -converge on FoxO factors in the control of glioblastoma cell proliferation (Seoane et al. 2004).

9.4.2 TGF-β Signaling in the Regulation of Invasion

The invasion of neoplastic cells into brain tissue is a pathologic hallmark of gliomas and contributes to the failure of current therapies (surgery, radiation and chemotherapy). Gliomas are highly invasive, partly due to a unique composition of the brain parenchyma, composed mainly of hyaluronan and devoid of rigid protein barriers made up of collagen, fibronectin and laminin. Proteases secreted during glioma progression degrade extracellular matrix (ECM) allowing tumor cells to spread and diffusely infiltrate the brain parenchyma (Rao 2003).

TGF-β1 is an important mediator of invasion in malignant gliomas. The effects of TGF-β1 on mobility and migration are associated with changes in ECM components, including Tenascin C, fibronectin, laminin, vitronectin, MMP-2 and MMP-9 (Hau et al. 2006; Platten et al. 2001). Exogenous TGF-β1 directly increased the motility of glioma cells by enhancing expression of subunits of $\alpha_{2,5}$, β_3 integrin (Platten et al. 2000) as well as by up-regulating the activity of metalloproteinases: MMP-2, 9 and MT1-MMP (Wick et al. 2001). Increased enzymatic degradation of extracellular matrix proteins may facilitate tumor spread (Platten et al. 2001; Wick et al. 2001). TGF-*β*2-specific phosphorothioate antisense molecules inhibited glioma migration and down-regulated *versican* expression as determined by gene arrays (Arslan et al. 2007). Versican, a large multi-domain chondroitin sulfate proteoglycan, is a major component of the extracellular matrix. The largest splice variants of versican, V0 and V1, are the predominant forms present in most glioma cell lines. TGF-*β*2 has been shown to up-regulate expression the versican V0/V1 isoforms and to modulate glioma migration (Arslan et al. 2007). TGF-*β*2 was capable of increasing MMP-2 activity and induced the degradation of versican V1. In T98G glioblastoma cells TGF-β1-induced migration and invasiveness were blocked by exposure to an ADAM17 inhibitor, TAPI-2. TGF-β1 can up-regulate *ADAM17* mRNA and protein expression and the ADAMTS-1 (metalloproteinase and disintegrin-like domain) proteases are known to cleave versican (Lu et al. 2011). A different study demonstrated that cleavage of brevican, another member of the lectican family by ADAMTS-5 is functionally involved in glioma invasion *in vivo* (Nakada et al. 2005). It is likely that TGF-β induces migration and invasiveness via up-regulation of ADAMs expression and their activity that leads to degradation of extracellular matrix proteoglycans (e.g. aggrecan, brevican and versican).

9.4.3 TGF-β1 as Pro-Angiogenic Factor

Genetic studies have revealed a role for TGF-β1 and its receptors in embryonic vascular assembly, the establishment and maintenance of vessel wall integrity (Pepper 1997). In gliomas TGF-β1 may act as an angiogenic factor promoting neovascularization. TGF-β signaling stimulates production of vascular endothelial growth factor (VEGF), which is a major stimulus in the promotion of angiogenesis, as well as plasminogen activator inhibitor (PAI-I) involved in maturation of blood vessels (Goumans et al. 2002). Some studies demonstrated that hypoxia and TGF-β signaling pathways can synergize in the regulation of *VEGF* gene expression at the transcriptional level. This cooperation has been mapped on the human *VEGF* gene promoter within a region at −1006 to −954 that contains functional DNA-binding sequences for HIF-1 (hypoxia -inducible factor) and SMADs (Sanchez-Elsner et al. 2001).

Furthermore, integrin-mediated activation of TGF-β by astrocytes may influence endothelial cell function. The integrin α(v)β8 on human astrocytes is a major cell surface receptor for a latent TGF-β and acts as a central regulator of brain vessel differentiation and stabilization through regulation of TGF-β activation and expression of TGF-β-responsive genes, most notably *PAI-1* and *THROMBOSPONDIN-1* (Cambier et al. 2005; Tchaicha et al. 2011).

Insulin-like growth factor-binding protein 7 (IGFBP7) is a selective biomarker of glioblastoma (GBM) vessels, strongly expressed in tumor endothelial cells and a vascular basement membrane. Conditioned media from human U87MG glioma cells up-regulated of *IGFBP7* mRNA and protein in human brain endothelial cells. Addition of pan-TGF-beta-neutralizing antibody or the ALK5 antagonist, SB431542, blocked *IGFBP7* expression, indicating that TGF-β1 is a tumor-secreted factor capable of IGFBP7 induction in endothelial cells (Pen et al. 2008).

9.4.4 A Role of TGF-β Signaling in Glioma Cancer Initiating Cells

Autocrine TGF-β signaling has be shown to play an essential role in maintenance of stem-like phenotype of glioma-initiating cells (GICs). TGF-β induced expression of *Sox2*, a stemness gene, and this induction was mediated by *Sox4*, a direct TGF-β target gene. Inhibitors of TGF-β signaling promoted GIC differentiation, and reduced their tumorigenicity in intracranial transplantation assay (Ikushima et al. 2009). Further studies demonstrated that TGF-β has to cooperate with leukemia inhibitory factor (LIF) in inducing the self-renewal and preventing the differentiation of glioma initiating stem cells (Ikushima et al. 2009; Penuelas et al. 2009). TGF-β induced the self-renewal capacity of those cells, but not of normal human neuroprogenitors, through the Smad-dependent induction of LIF and the subsequent activation of the JAK-STAT pathway (Penuelas et al. 2009).

TGF-β inhibitors can target the glioma initiating cell population in human glioblastoma patients. A GIC-enriched cell population expresses high levels of CD44 and Id1 (inhibitor of DNA-binding protein), and tend to be located in a perivascular niche. The inhibition of the TGF-β pathway (blockade of TβIR) decreased the $CD44^{high}/Id1^{high}$ GIC pop-ulation through reduction of Id1 and Id3 transcription factors levels, inhibiting capacity of those cells to initiate tumors. High CD44 and Id1 levels confer poor prognosis in GBM patients (Anido et al. 2010). A recent study showed that temozolomide (TMZ), an alkylating agent with anti-tumor efficacy for malignant gliomas, inhibits the invasion of glioma-initiating cells. TMZ reduced the TGF-β2-mediated invasion, and down-regulated TGF-β2 expression at the mRNA and protein levels (Zhang et al. 2011a, b, c).

9.4.5 *TGF-β Signaling in Tumor-Mediated Immunosuppression*

TGF-β plays a crucial role in the escape of gliomas from host immunity. The anti–tumor response in patients with glioma may be ineffective because of loss of a specific tumor antigen and professional antigen presenting cells in the brain. TGF-β1 enhances this effect by inhibition of MHC class II expression on glioma cells, macrophages and microglia (Zagzag et al. 2005). TGF-β1 exerts an immunosuppressive effect on all cells of the immune system (Beck et al. 2001; Chen et al. 2005; Jachimczak et al. 1993) by blocking differentiation into cytotoxic T lymphocytes (CTLs) and $CD4^+$ cells, and their maturation to Th1 or Th2 phenotype (Gorelik and Flavell 2001). TGF-β1 inhibits generation of cytotoxic $CD8^+$ T cell subpopulation and directly suppresses the expression of cytotoxic molecules such as granzyme B and perforin that are crucial for the cytolytic action of lymphocytes (Smyth et al. 1991).

TGF-β1 can also suppress activation of macrophages by down-regulation of TNFα, H_2O_2 and NO production, enhancing at the same time the production of the immunosuppressive cytokine IL-10 by macrophages (Maeda et al. 1995). TGF-β1 decreased the activating receptor NKG2D on the surface of NK cells and $CD8^+$ T cells in glioma patients, rendering them less efficient at tumor cell killing (Crane et al. 2010). Interference with TGF-β1 orTGF-β2 expression by a small RNA interference (RNAi) technology prevented down-regulation of NKG2D on immune cells mediated by LNT-229 glioma conditioned medium and strongly promoted their recognition and lysis by CD8+ T and NK cells. NK cells isolated from mice bearing LNT-229 glioma cells deficient in TGF-β1 showed an activated phenotype (Friese et al. 2004).

Blocking of TGF-β1 signaling in the immune system cells led to enhanced anti-tumor response. Both thymoma and melanoma tumors were eradicated in animals expressing a dominant negative TβRII under the control of a T- specific promoter (Gorelik and Flavell 2001). CTLs transduced with a vector expressing dominant negative TβRII were resistant to the anti-proliferative and anti-cytotoxic effects of exogenous TGF-β1 in EBV-positive Hodgkin disease (Bollard et al. 2002). C6 glioma cells depleted of TβRII by expression of the specific TβRII shRNA formed significantly smaller subcutanous tumors in mice than control glioma cells (Wesolowska et al. 2008). Taken together, efforts to bypass TGF-β-mediated immunosuppression represent an attractive therapeutic strategy for the treatment of gliomas.

9.5 Molecular and Pharmacological Strategies to Interfere with TGF-β Signaling for Potential Therapeutic Intervention in Gliomas

TGF-β signaling pathway is emerging as an attractive target in cancer and inhibitors of this pathway are tested in cancer clinical trials, showing reduction of tumor progression and improvement in overall survival. Table 9.1 summarizes recent therapeutic approaches targeting the TGF-β pathway. Some anti-TGF-β strategies are based on blocking the interaction between the cytokine and its receptor, e.g. by an application of a soluble TGFIIR or human alpha 2-macroglobulin plasma protein that bind TGF-β (Won et al. 1999). Two humanized monoclonal antibodies: CAT-192 specific to TGF-β1 and CAT-152 against TGF-β2 were promising in preclinical trials for a treatment of fibrosis associated with nephropathy (Benigni et al. 2003) or as anti-scarring agents in glaucoma surgery (Mead et al. 2003), but CAT-152 mAb was not effective in preventing the progression of fibrosis in patients undergoing glaucoma surgery in a phase III study (Khaw et al. 2007). A systemic inhibition of TGF-β signaling with a

Table 9.1 Development of inhibitors of TGF-β signaling for glioma therapy

Agent	Type	Development status	Company	References
Lerdelimumb CAT-152	TGF-β2 mAb	Phase III	Cambridge Antibody Technology	Mead et al. (2003)
Metelimumab CAT-192	TGF-β1 mAb	Phase II	Cambridge Antibody Technology	Benigni et al. (2003)
AP-12009	Oligonucleotide anti TGF-β2	Phase II	Antisense Pharma	Hau et al. (2007)
AP-11014	Oligonucleotide anti TGF-β2	Preclinical	Antisense Pharma	Schlingensiepen et al. (2004)
LY550410 LY580276	Small molecule TβRI inhibitor	Preclinical	Eli Lilly Pharmaceuticals	Sawyer et al. (2004)
LY2109761	TβR I/II dual inhibitor	Preclinical	Eli Lilly Pharmaceuticals	Melisi et al. (2008)
SB505124	Small molecule TβRI inhibitor	Preclinical	GlaxoSmithKline	DaCosta Byfield et al. (2004)
SB-431542	Small molecule TβRI inhibitor	Preclinical	GlaxoSmithKline	Hjelmeland et al. (2004)
SD-208	Small molecule TßRI inhibitor	Preclinical	Academic Institution	Uhl et al. (2004)

TGF-β-neutralizing monoclonal antibody – 1D11 improved the therapeutic efficacy of glioma- associated antigen peptide vaccines in mice bearing orthotopic GL261 gliomas (Ueda et al. 2009).

The phosphorothioate antisense oligodeoxynucleotides specific to TGF-β2 were effective in blocking TGF-β2 expression in malignant glioma and reversing cytokine effects on lymphocyte proliferation and autologous tumor cytotoxicity (Jachimczak et al. 1993). The antisense oligonucleotide trabedersen (AP 12009), specifically blocking *TGF-β2* mRNA was tested in I/II studies and a randomized, active-controlled dose-finding phase II study, and led to long-lasting tumor responses and promising survival data in high-grade glioma patients with recurrent or refractory tumor disease (Hau et al. 2007, 2011; Schlingensiepen et al. 2006, 2008). Survival of modified C6 glioma cells transfected with the TGF-β2 antisense vector was improved (Liau et al. 1998) leading to phase I clinical trial of a TGF-beta antisense-modified tumor cell vaccine in patients with advanced glioma (Fakhrai et al. 2006). Intracranial administration of antisense TGF-β2 ODNs with a systemic tumor vaccine improved survival of 9L glioma-bearing Fisher 344 rats (Liu et al. 2007).

Strategies based on RNA interference are used to block TGF-β signaling in glioma cells. Blockade of cytokine expression using siRNA against TGF-β inhibited tumor cell migration, invasiveness and restored anti-tumor immune response in a mouse model of glioma (Friese et al. 2004). We have developed vectors coding for small hairpin RNA (shRNA) that effectively silenced the expression of TGF-β receptor Type II gene expression, abolished TGF-activated SMAD signaling and reduced activation of the PAI-1 promoter in rat and human glioblastoma cells (Wesolowska et al. 2008).

Many small molecules inhibiting the catalytic activity of TGF-β receptor kinase have been developed. A group of competitive inhibitors of the ATP binding site of TGF receptor Type I kinase, such as LY550410, LY580276 and SB-505124 has been designed. Such compounds consist of a domain with a hydrogen-bond acceptor (which may be imidazole core, pyrozole ring or quinoline scaffold) essential for blocking of kinase activity. Several studies demonstrated physiological efficacy of such molecules, as well as their kinase inhibitory activity (DaCosta Byfield et al. 2004; Sawyer et al. 2004). A small-molecule inhibitor SD-208 (an antagonist of TGF-β receptor) significantly prolonged survival of glioma-bearing mice, enhancing the immunogenicity of glioma cells, while their migratory and invasion properties were diminished (Uhl et al. 2004).

A small molecule inhibitor of the type I TGF-β receptor – SB-431542 blocked phosphorylation and nuclear translocation of SMAD proteins, abolished TGF-β-mediated up-regulation of critical genes, inhibited proliferation and motility of human glioma cells (Hjelmeland et al. 2004). SX-007, an orally active, small-molecule TGF-β RI kinase inhibitor reduced TGF-β-mediated invasion in cell culture and reversed immune suppression, and improved the median survival in an in vivo SMA-560 glioma model (Tran et al. 2007). LY2109761, a novel TGF-β receptor type I and type II dual inhibitor, has shown a SMAD2-selective inhibitory profile with antitumor activity in various tumor models (Melisi et al. 2008). LY2109761 reduced clonogenic survival of U87 and T98 glioma cells, had anti-migratory and anti-angiogenic effects in Matrigel migration and tube formation assays. Furthermore, LY2109761 alone and in combination with fractionated radiation and temozolomide delayed tumor growth in human xenograft tumors growing subcutaneously on BALB/c nu/nu mice (Zhang et al. 2011a, b, c).

A recent study evaluated a panel of small molecule inhibitors of type 1 receptor serine threonine kinases (ALKs1-7): inhibitors of the TGF-β pathway (SB-431542, SB-505124, LY-364947 and A-83-01) and the BMP pathway (Dorsomorphin and LDN-193189) against 123 protein kinases. The inhibitors of the TGF-β pathway were found to be more selective than the inhibitors of the BMP pathway and SB-505124 was recommended as the best inhibitor of ALKs 4, 5 and 7 (Vogt et al. 2011).

A large–molecule antagonist of TGF-β signaling are considered to be more selective and having a broader action than small-molecule inhibitors. Despite earlier predictions of severe toxicity, neutralizing antibodies to TGF-β are well tolerated and have potent anti-metastatic activity. Si/shRNAs are recognized as a new class of potential therapeutics against a wide range of diseases, however delivering siRNA or miRNA specifically and efficiently into tumor cells *in vivo* remains a great challenge (Tiemann and Rossi 2009). Lentiviral or adeno-associated vectors expressing shRNA in cultured mammalian cells and in whole animals may be a promising approach for a specific, efficient, and stable knockdown of various genes.

Acknowledgments The work in the Author's Laboratory is supported by a grant NN301 786240 from the Ministry of Science and Higher Education. M.K. is a recipient of a scholarship from Postgraduate School of Molecular Medicine.

References

Alexandrow MG, Moses HL (1995) Transforming growth factor beta and cell cycle regulation. Cancer Res 55:1452–1457

Anido J, Saez-Borderias A, Gonzalez-Junca A, Rodon L, Folch G, Carmona MA, Prieto-Sanchez RM, Barba I, Martinez-Saez E, Prudkin L, Cuartas I, Raventos C, Martinez-Ricarte F, Poca MA, Garcia-Dorado D, Lahn MM, Yingling JM, Rodon J, Sahuquillo J, Baselga J, Seoane J (2010) TGF-beta receptor inhibitors target the CD44(high)/Id1(high) glioma-initiating cell population in human glioblastoma. Cancer Cell 18:655–668

Annes JP, Chen Y, Munger JS, Rifkin DB (2004) Integrin alphaVbeta6-mediated activation of latent TGF-beta requires the latent TGF-beta binding protein-1. J Cell Biol 165:723–734

Arslan F, Bosserhoff AK, Nickl-Jockschat T, Doerfelt A, Bogdahn U, Hau P (2007) The role of versican isoforms V0/V1 in glioma migration mediated by transforming growth factor-beta2. Br J Cancer 96:1560–1568

Attisano L, Wrana JL (2002) Signal transduction by the TGF-beta superfamily. Science 296:1646–1647

Beck C, Schreiber H, Rowley D (2001) Role of TGF-beta in immune-evasion of cancer. Microsc Res Tech 52:387–395

Benigni A, Zoja C, Corna D, Zatelli C, Conti S, Campana M, Gagliardini E, Rottoli D, Zanchi C, Abbate M, Ledbetter S, Remuzzi G (2003) Add-on anti-TGF-beta antibody to ACE inhibitor arrests progressive diabetic nephropathy in the rat. J Am Soc Nephrol 14:1816–1824

Bhowmick NA, Zent R, Ghiassi M, McDonnell M, Moses HL (2001) Integrin beta 1 signaling is necessary for transforming growth factor-beta activation of p38MAPK and epithelial plasticity. J Biol Chem 276:46707–46713

Bian Y, Kaklamani V, Reich J, Pasche B (2003) TGF-beta signaling alterations in cancer. Cancer Treat Res 115:73–94

Bollard CM, Rossig C, Calonge MJ, Huls MH, Wagner HJ, Massague J, Brenner MK, Heslop HE, Rooney CM (2002) Adapting a transforming growth factor beta-related tumor protection strategy to enhance antitumor immunity. Blood 99:3179–3187

Bottner M, Krieglstein K, Unsicker K (2000) The transforming growth factor-betas: structure, signaling, and roles in nervous system development and functions. J Neurochem 75:2227–2240

Bruna A, Darken RS, Rojo F, Ocana A, Penuelas S, Arias A, Paris R, Tortosa A, Mora J, Baselga J, Seoane J (2007) High TGFbeta-Smad activity confers poor prognosis in glioma patients and promotes cell proliferation depending on the methylation of the PDGF-B gene. Cancer Cell 11:147–160

Cambier S, Gline S, Mu D, Collins R, Araya J, Dolganov G, Einheber S, Boudreau N, Nishimura SL (2005) Integrin alpha(v)beta8-mediated activation of transforming growth factor-beta by perivascular astrocytes: an angiogenic control switch. Am J Pathol 166:1883–1894

Chen YG, Hata A, Lo RS, Wotton D, Shi Y, Pavletich N, Massague J (1998) Determinants of specificity in TGF-beta signal transduction. Genes Dev 12:2144–2152

Chen CR, Kang Y, Siegel PM, Massague J (2002) E2F4/5 and p107 as Smad cofactors linking the TGFbeta receptor to c-myc repression. Cell 110:19–32

Chen ML, Pittet MJ, Gorelik L, Flavell RA, Weissleder R, von Boehmer H, Khazaie K (2005) Regulatory T cells suppress tumor-specific CD8 T cell cytotoxicity through TGF-beta signals in vivo. Proc Natl Acad Sci U S A 102:419–424

Crane CA, Han SJ, Barry JJ, Ahn BJ, Lanier LL, Parsa AT (2010) TGF-beta downregulates the activating receptor NKG2D on NK cells and CD8+ T cells in glioma patients. Neuro Oncol 12:7–13

DaCosta Byfield S, Major C, Laping NJ, Roberts AB (2004) SB-505124 is a selective inhibitor of transforming growth factor-beta type I receptors ALK4, ALK5, and ALK7. Mol Pharmacol 65:744–752

Derynck R, Zhang YE (2003) Smad-dependent and Smad-independent pathways in TGF-beta family signalling. Nature 425:577–584

Dubois CM, Laprise MH, Blanchette F, Gentry LE, Leduc R (1995) Processing of transforming growth factor beta 1 precursor by human furin convertase. J Biol Chem 270:10618–10624

Dziembowska M, Danilkiewicz M, Wesolowska A, Zupanska A, Chouaib S, Kaminska B (2007) Cross-talk between Smad and p38 MAPK signalling in transforming growth factor beta signal transduction in human glioblastoma cells. Biochem Biophys Res Commun 354:1101–1106

Edlund S, Landstrom M, Heldin CH, Aspenstrom P (2002) Transforming growth factor-beta-induced mobilization of actin cytoskeleton requires signaling by small GTPases Cdc42 and RhoA. Mol Biol Cell 13:902–914

Fakhrai H, Mantil JC, Liu L, Nicholson GL, Murphy-Satter CS, Ruppert J, Shawler DL (2006) Phase I clinical trial of a TGF-beta antisense-modified tumor cell vaccine in patients with advanced glioma. Cancer Gene Ther 13:1052–1060

Friese MA, Wischhusen J, Wick W, Weiler M, Eisele G, Steinle A, Weller M (2004) RNA interference targeting transforming growth factor-beta enhances NKG2D-mediated antiglioma immune response, inhibits glioma cell migration and invasiveness, and abrogates tumorigenicity in vivo. Cancer Res 64:7596–7603

Gorelik L, Flavell RA (2001) Immune-mediated eradication of tumors through the blockade of transforming growth factor beta signaling in T cells. Nat Med 7:1118–1122

Goumans MJ, Mummery C (2000) Functional analysis of the TGFbeta receptor/Smad pathway through gene ablation in mice. Int J Dev Biol 44:253–265

Goumans MJ, Valdimarsdottir G, Itoh S, Rosendahl A, Sideras P, ten Dijke P (2002) Balancing the activation state of the endothelium via two distinct TGF-beta type I receptors. EMBO J 21:1743–1753

Grainger DJ, Wakefield L, Bethell HW, Farndale RW, Metcalfe JC (1995) Release and activation of platelet latent TGF-beta in blood clots during dissolution with plasmin. Nat Med 1:932–937

Groppe J, Hinck CS, Samavarchi-Tehrani P, Zubieta C, Schuermann JP, Taylor AB, Schwarz PM, Wrana JL, Hinck AP (2008) Cooperative assembly of TGF-beta superfamily signaling complexes is mediated by two disparate mechanisms and distinct modes of receptor binding. Mol Cell 29:157–168

Hanafusa H, Ninomiya-Tsuji J, Masuyama N, Nishita M, Fujisawa J, Shibuya H, Matsumoto K, Nishida E (1999) Involvement of the p38 mitogen-activated protein kinase pathway in transforming growth factor-beta-induced gene expression. J Biol Chem 274:27161–27167

Hau P, Kunz-Schughart LA, Rummele P, Arslan F, Dorfelt A, Koch H, Lohmeier A, Hirschmann B, Muller A, Bogdahn U, Bosserhoff AK (2006) Tenascin-C protein is induced by transforming growth factor-beta1 but does not correlate with time to tumor progression in high-grade gliomas. J Neurooncol 77:1–7

Hau P, Jachimczak P, Schlingensiepen R, Schulmeyer F, Jauch T, Steinbrecher A, Brawanski A, Proescholdt M, Schlaier J, Buchroithner J, Pichler J, Wurm G, Mehdorn M, Strege R, Schuierer G, Villarrubia V, Fellner F, Jansen O, Straube T, Nohria V, Goldbrunner M, Kunst M, Schmaus S, Stauder G, Bogdahn U, Schlingensiepen KH (2007) Inhibition of TGF-beta2 with AP 12009 in recurrent malignant gliomas: from preclinical to phase I/II studies. Oligonucleotides 17:201–212

Hau P, Jachimczak P, Schlaier J, Bogdahn U (2011) TGF-beta2 signaling in high-grade gliomas. Curr Pharm Biotechnol 12:2150–2157

Held-Feindt J, Lutjohann B, Ungefroren H, Mehdorn HM, Mentlein R (2003) Interaction of transforming growth factor-beta (TGF-beta) and epidermal growth factor (EGF) in human glioma cells. J Neurooncol 63:117–127

Hjelmeland MD, Hjelmeland AB, Sathornsumetee S, Reese ED, Herbstreith MH, Laping NJ, Friedman HS, Bigner DD, Wang XF, Rich JN (2004) SB-431542, a small molecule transforming growth factor-beta-receptor antagonist, inhibits human glioma cell line proliferation and motility. Mol Cancer Ther 3:737–745

Hyytiainen M, Penttinen C, Keski-Oja J (2004) Latent TGF-beta binding proteins: extracellular matrix association and roles in TGF-beta activation. Crit Rev Clin Lab Sci 41:233–264
Ijichi H, Otsuka M, Tateishi K, Ikenoue T, Kawakami T, Kanai F, Arakawa Y, Seki N, Shimizu K, Miyazono K, Kawabe T, Omata M (2004) Smad4-independent regulation of p21/WAF1 by transforming growth factor-beta. Oncogene 23:1043–1051
Ikushima H, Todo T, Ino Y, Takahashi M, Miyazawa K, Miyazono K (2009) Autocrine TGF-beta signaling maintains tumorigenicity of glioma-initiating cells through Sry-related HMG-box factors. Cell Stem Cell 5:504–514
Itoh S, ten Dijke P (2007) Negative regulation of TGF-beta receptor/Smad signal transduction. Curr Opin Cell Biol 19:176–184
Jachimczak P, Bogdahn U, Schneider J, Behl C, Meixensberger J, Apfel R, Dorries R, Schlingensiepen KH, Brysch W (1993) The effect of transforming growth factor-beta 2-specific phosphorothioate-anti-sense oligodeoxynucleotides in reversing cellular immunosuppression in malignant glioma. J Neurosurg 78:944–951
Jachimczak P, Hessdorfer B, Fabel-Schulte K, Wismeth C, Brysch W, Schlingensiepen KH, Bauer A, Blesch A, Bogdahn U (1996) Transforming growth factor-beta-mediated autocrine growth regulation of gliomas as detected with phosphorothioate antisense oligonucleotides. Int J Cancer 65:332–337
Jennings MT, Pietenpol JA (1998) The role of transforming growth factor beta in glioma progression. J Neurooncol 36:123–140
Jennings MT, Maciunas RJ, Carver R, Bascom CC, Juneau P, Misulis K, Moses HL (1991) TGF beta 1 and TGF beta 2 are potential growth regulators for low-grade and malignant gliomas in vitro: evidence in support of an autocrine hypothesis. Int J Cancer 49:129–139
Kaminska B (2009) Molecular characterization of inflammation-induced JNK/c-Jun signaling pathway in connection with tumorigenesis. Methods Mol Biol 512:249–264
Karsdal MA, Hjorth P, Henriksen K, Kirkegaard T, Nielsen KL, Lou H, Delaisse JM, Foged NT (2003) Transforming growth factor-beta controls human osteoclastogenesis through the p38 MAPK and regulation of RANK expression. J Biol Chem 278:44975–44987
Khaw P, Grehn F, Hollo G, Overton B, Wilson R, Vogel R, Smith Z (2007) A phase III study of subconjunctival human anti-transforming growth factor beta(2) monoclonal antibody (CAT-152) to prevent scarring after first-time trabeculectomy. Ophthalmology 114:1822–1830
Kim SJ, Angel P, Lafyatis R, Hattori K, Kim KY, Sporn MB, Karin M, Roberts AB (1990) Autoinduction of transforming growth factor beta 1 is mediated by the AP-1 complex. Mol Cell Biol 10:1492–1497
Lee MK, Pardoux C, Hall MC, Lee PS, Warburton D, Qing J, Smith SM, Derynck R (2007) TGF-beta activates Erk MAP kinase signalling through direct phosphorylation of ShcA. EMBO J 26:3957–3967
Levy L, Hill CS (2005) Smad4 dependency defines two classes of transforming growth factor beta (TGF-{beta}) target genes and distinguishes TGF-{beta}-induced epithelial-mesenchymal transition from its antiproliferative and migratory responses. Mol Cell Biol 25:8108–8125
Liau LM, Fakhrai H, Black KL (1998) Prolonged survival of rats with intracranial C6 gliomas by treatment with TGF-beta antisense gene. Neurol Res 20:742–747
Lin HK, Bergmann S, Pandolfi PP (2004) Cytoplasmic PML function in TGF-beta signalling. Nature 431:205–211
Lindholm D, Castren E, Kiefer R, Zafra F, Thoenen H (1992) Transforming growth factor-beta 1 in the rat brain: increase after injury and inhibition of astrocyte proliferation. J Cell Biol 117:395–400
Liu Y, Wang Q, Kleinschmidt-DeMasters BK, Franzusoff A, Ng KY, Lillehei KO (2007) TGF-beta2 inhibition augments the effect of tumor vaccine and improves the survival of animals with pre-established brain tumors. J Neurooncol 81:149–162
Lu Y, Jiang F, Zheng X, Katakowski M, Buller B, To SS, Chopp M (2011) TGF-beta1 promotes motility and invasiveness of glioma cells through activation of ADAM17. Oncol Rep 25:1329–1335
Maeda H, Kuwahara H, Ichimura Y, Ohtsuki M, Kurakata S, Shiraishi A (1995) TGF-beta enhances macrophage ability to produce IL-10 in normal and tumor-bearing mice. J Immunol 155:4926–4932
Massague J (2008) TGFbeta in cancer. Cell 134:215–230
Massague J, Wotton D (2000) Transcriptional control by the TGF-beta/Smad signaling system. EMBO J 19:1745–1754
Massague J, Seoane J, Wotton D (2005) Smad transcription factors. Genes Dev 19:2783–2810
Mead AL, Wong TT, Cordeiro MF, Anderson IK, Khaw PT (2003) Evaluation of anti-TGF-beta2 antibody as a new postoperative anti-scarring agent in glaucoma surgery. Invest Ophthalmol Vis Sci 44:3394–3401
Melisi D, Ishiyama S, Sclabas GM, Fleming JB, Xia Q, Tortora G, Abbruzzese JL, Chiao PJ (2008) LY2109761, a novel transforming growth factor beta receptor type I and type II dual inhibitor, as a therapeutic approach to suppressing pancreatic cancer metastasis. Mol Cancer Ther 7:829–840
Miura S, Takeshita T, Asao H, Kimura Y, Murata K, Sasaki Y, Hanai JI, Beppu H, Tsukazaki T, Wrana JL, Miyazono K, Sugamura K (2000) Hgs (Hrs), a FYVE domain protein, is involved in Smad signaling through cooperation with SARA. Mol Cell Biol 20:9346–9355
Moustakas A, Heldin CH (2005) Non-Smad TGF-beta signals. J Cell Sci 118:3573–3584

Moustakas A, Souchelnytskyi S, Heldin CH (2001) Smad regulation in TGF-beta signal transduction. J Cell Sci 114:4359–4369
Mu Y, Sundar R, Thakur N, Ekman M, Gudey SK, Yakymovych M, Hermansson A, Dimitriou H, Bengoechea-Alonso MT, Ericsson J, Heldin CH, Landstrom M (2011) TRAF6 ubiquitinates TGFbeta type I receptor to promote its cleavage and nuclear translocation in cancer. Nat Commun 2:330
Mulder KM (2000) Role of Ras and Mapks in TGFbeta signaling. Cytokine Growth Factor Rev 11:23–35
Nakada M, Miyamori H, Kita D, Takahashi T, Yamashita J, Sato H, Miura R, Yamaguchi Y, Okada Y (2005) Human glioblastomas overexpress ADAMTS-5 that degrades brevican. Acta Neuropathol 110:239–246
Nakao A, Imamura T, Souchelnytskyi S, Kawabata M, Ishisaki A, Oeda E, Tamaki K, Hanai J, Heldin CH, Miyazono K, ten Dijke P (1997) TGF-beta receptor-mediated signalling through Smad2, Smad3 and Smad4. EMBO J 16:5353–5362
Pen A, Moreno MJ, Durocher Y, Deb-Rinker P, Stanimirovic DB (2008) Glioblastoma-secreted factors induce IGFBP7 and angiogenesis by modulating Smad-2-dependent TGF-beta signaling. Oncogene 27:6834–6844
Penheiter SG, Mitchell H, Garamszegi N, Edens M, Dore JJ Jr, Leof EB (2002) Internalization-dependent and -independent requirements for transforming growth factor beta receptor signaling via the Smad pathway. Mol Cell Biol 22:4750–4759
Penuelas S, Anido J, Prieto-Sanchez RM, Folch G, Barba I, Cuartas I, Garcia-Dorado D, Poca MA, Sahuquillo J, Baselga J, Seoane J (2009) TGF-beta increases glioma-initiating cell self-renewal through the induction of LIF in human glioblastoma. Cancer Cell 15:315–327
Pepper MS (1997) Transforming growth factor-beta: vasculogenesis, angiogenesis, and vessel wall integrity. Cytokine Growth Factor Rev 8:21–43
Piek E, Westermark U, Kastemar M, Heldin CH, van Zoelen EJ, Nister M, Ten Dijke P (1999) Expression of transforming-growth-factor (TGF)-beta receptors and Smad proteins in glioblastoma cell lines with distinct responses to TGF-beta1. Int J Cancer 80:756–763
Platten M, Wick W, Wild-Bode C, Aulwurm S, Dichgans J, Weller M (2000) Transforming growth factors beta(1) (TGF-beta(1)) and TGF-beta(2) promote glioma cell migration via Up-regulation of alpha(V)beta(3) integrin expression. Biochem Biophys Res Commun 268:607–611
Platten M, Wick W, Weller M (2001) Malignant glioma biology: role for TGF-beta in growth, motility, angiogenesis, and immune escape. Microsc Res Tech 52:401–410
Rao JS (2003) Molecular mechanisms of glioma invasiveness: the role of proteases. Nat Rev Cancer 3:489–501
Ribeiro SM, Poczatek M, Schultz-Cherry S, Villain M, Murphy-Ullrich JE (1999) The activation sequence of thrombospondin-1 interacts with the latency-associated peptide to regulate activation of latent transforming growth factor-beta. J Biol Chem 274:13586–13593
Rich JN (2003) The role of transforming growth factor-beta in primary brain tumors. Front Biosci 8:e245–e260
Rich JN, Zhang M, Datto MB, Bigner DD, Wang XF (1999) Transforming growth factor-beta-mediated p15(INK4B) induction and growth inhibition in astrocytes is SMAD3-dependent and a pathway prominently altered in human glioma cell lines. J Biol Chem 274:35053–35058
Rider CC, Mulloy B (2010) Bone morphogenetic protein and growth differentiation factor cytokine families and their protein antagonists. Biochem J 429:1–12
Sanchez-Elsner T, Botella LM, Velasco B, Corbi A, Attisano L, Bernabeu C (2001) Synergistic cooperation between hypoxia and transforming growth factor-beta pathways on human vascular endothelial growth factor gene expression. J Biol Chem 276:38527–38535
Sawyer JS, Beight DW, Britt KS, Anderson BD, Campbell RM, Goodson T Jr, Herron DK, Li HY, McMillen WT, Mort N, Parsons S, Smith EC, Wagner JR, Yan L, Zhang F, Yingling JM (2004) Synthesis and activity of new aryl- and heteroaryl-substituted 5,6-dihydro-4H-pyrrolo[1,2-b]pyrazole inhibitors of the transforming growth factor-beta type I receptor kinase domain. Bioorg Med Chem Lett 14:3581–3584
Schlingensiepen KH, Bischof A, Egger T, Hafner M, Herrmuth H, Jachimczak P, Kielmanowicz M, Niewel M, Zavadova E, Stauder G (2004) The TGF-beta1 antisense oligonucleotide AP 11014 for the treatment of non-small cell lung, colorectal and prostate cancer: preclinical studies. J Clin Oncol, 2004 ASCO annual meeting proceedings, 22, 14S:3132
Schlingensiepen KH, Schlingensiepen R, Steinbrecher A, Hau P, Bogdahn U, Fischer-Blass B, Jachimczak P (2006) Targeted tumor therapy with the TGF-beta 2 antisense compound AP 12009. Cytokine Growth Factor Rev 17:129–139
Schlingensiepen KH, Fischer-Blass B, Schmaus S, Ludwig S (2008) Antisense therapeutics for tumor treatment: the TGF-beta2 inhibitor AP 12009 in clinical development against malignant tumors. Cancer Res 177:137–150
Schmierer B, Hill CS (2007) TGFbeta-SMAD signal transduction: molecular specificity and functional flexibility. Nat Rev Mol Cell Biol 8:970–982
Seoane J (2008) The TGFBeta pathway as a therapeutic target in cancer. Clin Transl Oncol 10:14–19
Seoane J, Pouponnot C, Staller P, Schader M, Eilers M, Massague J (2001) TGFbeta influences Myc, Miz-1 and Smad to control the CDK inhibitor p15INK4b. Nat Cell Biol 3:400–408
Seoane J, Le HV, Shen L, Anderson SA, Massague J (2004) Integration of Smad and forkhead pathways in the control of neuroepithelial and glioblastoma cell proliferation. Cell 117:211–223
Shi Y, Massague J (2003) Mechanisms of TGF-beta signaling from cell membrane to the nucleus. Cell 113:685–700

Siegel PM, Shu W, Cardiff RD, Muller WJ, Massague J (2003a) Transforming growth factor beta signaling impairs Neu-induced mammary tumorigenesis while promoting pulmonary metastasis. Proc Natl Acad Sci U S A 100:8430–8435
Siegel PM, Shu W, Massague J (2003b) Mad upregulation and Id2 repression accompany transforming growth factor (TGF)-beta-mediated epithelial cell growth suppression. J Biol Chem 278:35444–35450
Smyth MJ, Strobl SL, Young HA, Ortaldo JR, Ochoa AC (1991) Regulation of lymphokine-activated killer activity and pore-forming protein gene expression in human peripheral blood CD8+ T lymphocytes. Inhibition by transforming growth factor-beta. J Immunol 146:3289–3297
Sorrentino A, Thakur N, Grimsby S, Marcusson A, von Bulow V, Schuster N, Zhang S, Heldin CH, Landstrom M (2008) The type I TGF-beta receptor engages TRAF6 to activate TAK1 in a receptor kinase-independent manner. Nat Cell Biol 10:1199–1207
Tatti O, Vehvilainen P, Lehti K, Keski-Oja J (2008) MT1-MMP releases latent TGF-beta1 from endothelial cell extracellular matrix via proteolytic processing of LTBP-1. Exp Cell Res 314:2501–2514
Tchaicha JH, Reyes SB, Shin J, Hossain MG, Lang FF, McCarty JH (2011) Glioblastoma angiogenesis and tumor cell invasiveness are differentially regulated by {beta}8 integrin. Cancer Res 71:6371–6381
ten Dijke P, Hill CS (2004) New insights into TGF-beta-Smad signalling. Trends Biochem Sci 29:265–273
Tiemann K, Rossi JJ (2009) RNAi-based therapeutics-current status, challenges and prospects. EMBO Mol Med 1:142–151
Tran TT, Uhl M, Ma JY, Janssen L, Sriram V, Aulwurm S, Kerr I, Lam A, Webb HK, Kapoun AM, Kizer DE, McEnroe G, Hart B, Axon J, Murphy A, Chakravarty S, Dugar S, Protter AA, Higgins LS, Wick W, Weller M, Wong DH (2007) Inhibiting TGF-beta signaling restores immune surveillance in the SMA-560 glioma model. Neuro Oncol 9:259–270
Tsukazaki T, Chiang TA, Davison AF, Attisano L, Wrana JL (1998) SARA, a FYVE domain protein that recruits Smad2 to the TGFbeta receptor. Cell 95:779–791
Ueda R, Fujita M, Zhu X, Sasaki K, Kastenhuber ER, Kohanbash G, McDonald HA, Harper J, Lonning S, Okada H (2009) Systemic inhibition of transforming growth factor-beta in glioma-bearing mice improves the therapeutic efficacy of glioma-associated antigen peptide vaccines. Clin Cancer Res 15:6551–6559
Uhl M, Aulwurm S, Wischhusen J, Weiler M, Ma JY, Almirez R, Mangadu R, Liu YW, Platten M, Herrlinger U, Murphy A, Wong DH, Wick W, Higgins LS, Weller M (2004) SD-208, a novel transforming growth factor beta receptor I kinase inhibitor, inhibits growth and invasiveness and enhances immunogenicity of murine and human glioma cells in vitro and in vivo. Cancer Res 64:7954–7961
Van Obberghen-Schilling E, Roche NS, Flanders KC, Sporn MB, Roberts AB (1988) Transforming growth factor beta 1 positively regulates its own expression in normal and transformed cells. J Biol Chem 263:7741–7746
Verrecchia F, Mauviel A (2002) Transforming growth factor-beta signaling through the Smad pathway: role in extracellular matrix gene expression and regulation. J Invest Dermatol 118:211–215
Vogt J, Traynor R, Sapkota GP (2011) The specificities of small molecule inhibitors of the TGFss and BMP pathways. Cell Signal 23:1831–1842
Wang L, Zhu Y, Sharma K (1998) Transforming growth factor-beta1 stimulates protein kinase A in mesangial cells. J Biol Chem 273:8522–8527
Wesolowska A, Kwiatkowska A, Slomnicki L, Dembinski M, Master A, Sliwa M, Franciszkiewicz K, Chouaib S, Kaminska B (2008) Microglia-derived TGF-beta as an important regulator of glioblastoma invasion–an inhibition of TGF-beta-dependent effects by shRNA against human TGF-beta type II receptor. Oncogene 27:918–930
Wick W, Platten M, Weller M (2001) Glioma cell invasion: regulation of metalloproteinase activity by TGF-beta. J Neurooncol 53:177–185
Wilkes MC, Murphy SJ, Garamszegi N, Leof EB (2003) Cell-type-specific activation of PAK2 by transforming growth factor beta independent of Smad2 and Smad3. Mol Cell Biol 23:8878–8889
Won J, Kim H, Park EJ, Hong Y, Kim SJ, Yun Y (1999) Tumorigenicity of mouse thymoma is suppressed by soluble type II transforming growth factor beta receptor therapy. Cancer Res 59:1273–1277
Wong C, Rougier-Chapman EM, Frederick JP, Datto MB, Liberati NT, Li JM, Wang XF (1999) Smad3-Smad4 and AP-1 complexes synergize in transcriptional activation of the c-Jun promoter by transforming growth factor beta. Mol Cell Biol 19:1821–1830
Yamaguchi K, Shirakabe K, Shibuya H, Irie K, Oishi I, Ueno N, Taniguchi T, Nishida E, Matsumoto K (1995) Identification of a member of the MAPKKK family as a potential mediator of TGF-beta signal transduction. Science 270:2008–2011
Zagzag D, Salnikow K, Chiriboga L, Yee H, Lan L, Ali MA, Garcia R, Demaria S, Newcomb EW (2005) Downregulation of major histocompatibility complex antigens in invading glioma cells: stealth invasion of the brain. Lab Invest 85:328–341
Zhang YE (2009) Non-Smad pathways in TGF-beta signaling. Cell Res 19:128–139

Zhang L, Sato E, Amagasaki K, Nakao A, Naganuma H (2006) Participation of an abnormality in the transforming growth factor-beta signaling pathway in resistance of malignant glioma cells to growth inhibition induced by that factor. J Neurosurg 105:119–128
Zhang D, Jing Z, Qiu B, Wu A, Wang Y (2011a) Temozolomide decreases invasion of glioma stem cells by down-regulating TGF-beta2. Oncol Rep 26:901–908
Zhang M, Herion TW, Timke C, Han N, Hauser K, Weber KJ, Peschke P, Wirkner U, Lahn M, Huber PE (2011b) Trimodal glioblastoma treatment consisting of concurrent radiotherapy, temozolomide, and the novel TGF-beta receptor I kinase inhibitor LY2109761. Neoplasia 13:537–549
Zhang X, Wu A, Fan Y, Wang Y (2011c) Increased transforming growth factor-beta2 in epidermal growth factor receptor variant III-positive glioblastoma. J Clin Neurosci 18:821–826

Chapter 10
STAT Signaling in Glioma Cells

Karolina Swiatek-Machado and Bozena Kaminska

Abstract STAT (signal transducers and activators of transcription) are latent cytoplasmic transcription factors that function as downstream effectors of cytokine and growth factor receptor signaling. The canonical JAK/STAT signaling pathway involves the activation of Janus kinases (JAK) or growth factors receptor kinases, phosphorylation of STAT proteins, their dimerization and translocation into the nucleus where STATs act as transcription factors with pleiotropic downstream effects. STAT signaling is tightly controlled with restricted kinetics due to action of its negative regulators. While STAT1 is believed to play an important role in growth arrest and apoptosis, and to act as a tumor suppressor, STAT3 and 5 are involved in promoting cell cycle progression, cellular transformation, and preventing apoptosis. Aberrant activation of STATs, in particular STAT3 and STAT5, have been found in a large number of human tumors, including gliomas and may contribute to oncogenesis. In this chapter, we have (1) summarized the mechanisms of STAT activation in normal and malignant signaling; (2) discussed evidence for the critical role of constitutively activated STAT3 and STAT5 in glioma pathobiology; (3) disclosed molecular and pharmacological strategies to interfere with STAT signaling for potential therapeutic intervention in gliomas.

Keywords Cytokine and growth factor receptor signaling • Protein tyrosine kinases • STAT proteins • Gliomas • Transcription regulation

Abbreviations

Bcl-2	B-cell lymphoma 2
Bcl-x_L	B-cell lymphoma-extra large
BRG1	Brahma-related gene 1
EGFR	Epidermal growth factor receptor

K. Swiatek-Machado • B. Kaminska (✉)
Laboratory of Transcription Regulation, Department of Cell Biology,
Nencki Institute of Experimental Biology, Polish Academy of Sciences,
3 Pasteur St, PL 02-093 Warsaw, Poland
e-mail: k.swiatek@nencki.gov.pl; b.kaminska@nencki.gov.pl

J. Barańska (ed.), *Glioma Signaling*, Advances in Experimental Medicine and Biology 986,
DOI 10.1007/978-94-007-4719-7_10,

GAS	IFNγ-activated sequence
GBM	*Glioblastoma multiforme*
GSC	Glioma stem cells
IFNγ	Interferon γ
IRF	IFN regulatory factor
ISRE	IFN-α/β–stimulated response element
JAK	Janus kinase
Mcl-1	Induced myeloid leukemia cell differentiation protein
MMP	Metalloproteinase
ODN	Oligodeoxynucleotide
PDGFR	Platelet-derived growth factor receptor
PIAS	Protein inhibitors of activated STAT
PTEN	Phosphatase and tensin homolog
SH2	Src homology domain 2
SOCS	Suppressors of cytokine signaling
STAT	Signal transducers and activators of transcription
TMZ	Temozolomide
VEGF	Vascular endothelial growth factor
VEGFR	Vascular endothelial growth factor receptor

10.1 Introduction

STATs (signal transducers and activators of transcription) are a family of latent transcription factors that are activated in response to intracellular stimuli. Seven STAT proteins have been identified (STATs 1, 2, 3, 4, 5a, 5b, 6), and the corresponding gene products share a high degree of similarity. Each STAT protein has a DNA binding domain, a src-homology 2 (SH2) domain necessary for homo- or heterodimerization and a conserved tyrosine residue 705, phosphorylated by tyrosine kinases (Kisseleva et al. 2002).

Over 40 different cytokines or growth factors can activate STAT signaling pathway. Upon binding of cytokines to cognate receptors on the surface of cells, receptors dimerize and thereby activate receptor-associated tyrosine kinases such as Janus kinases (JAKs) that phosphorylate the receptor cytoplasmic portion. STATs are also activated by growth factor receptors with the intrinsic tyrosine kinase activity exemplified by epidermal growth factor receptor (EGFR) or platelet-derived growth factor receptor (PDGFR). These kinases phosphorylate specific tyrosine residues in the cytoplasmic tail of the receptor, providing docking sites for the SH2 domain of inactive STAT monomers that are recruited to the activated receptors. Phosphorylated STAT dimers translocate to the nucleus and bind to a consensus DNA element upstream of regulated genes (Bromberg 2001; Bromberg and Darnell 2000). Genes that are known to be regulated by STAT are involved in many fundamental biological processes, such as proliferation, apoptosis, angiogenesis and immune response (Yu and Jove 2004).

The activation duration of individual STATs is temporary and usually lasts from a few minutes to several hours under normal physiological conditions. However, numerous studies have demonstrated constitutive activation of STATs, in particular STAT1, STAT3, and STAT5, in a large number of diverse human tumor cell lines and tumors, including gliomas. Aberrant activation of STATs is a consequence of constitutively activated cytokine receptor, most common by autocrine or paracrine expression of their respective ligands, mutation or overexpression of tyrosine kinase encoding genes or loss of endogenous inhibitors. Aberrations in these pathways, as those caused by the recently identified $JAK2^{V617F}$ mutation and translocations of the *JAK2* gene, are underlying causes of leukemias and other myeloproliferative disorders (Jatiani et al. 2010).

Significant progress in defining normal physiological functions of individual STAT proteins was derived from Stat knockout mice and/or by tissue-specific deletions. Stat3-deficient mice die during early embryogenesis in contrast to other Stat knockout mice. Conditional *Stat3* gene targeting using Cre-loxP system (Takeda et al. 1998) allowed for analysis of tissue specific Stat3-deficient mice demonstrating its crucial role in a variety of biological processes such as wound healing, T-cell development, mammary gland development, cell growth, apoptosis or cell motility. All those studies demonstrated that STAT signaling is involved in a broad spectrum of fundamental processes such as embryonic development, organogenesis, innate and adaptive immunity, cell differentiation, growth, and apoptosis (Akira 2000). Constitutive activation of the STAT family members such as STAT3 and STAT5, and/or loss of STAT1 signaling, are found in a large group of diverse tumors.

10.2 A Brief Summary of Mechanisms of STAT Activation in Normal and Malignant Signaling

10.2.1 Mechanisms of STAT Activation

Upon binding of cytokines to cognate surface receptors, receptors dimerize and thereby activate receptor-associated tyrosine kinases, such as JAKs that phosphorylate the receptor cytoplasmic portion. Receptors with intrinsic tyrosine kinase activity, such as PDGFR or EGFR, autophosphorylate the receptor cytoplasmic tail (Reddy et al. 2000). Tyrosine-phosphorylated receptors provide docking sites for the recruitment of cytoplasmic monomeric STAT proteins via their SH2 domains. In some cases, non-receptor tyrosine kinases of the Src family can participate in STAT activation. Oncogenic derivatives of cytoplasmic tyrosine kinases such as v-Src or Bcr-Abl can phosphorylate STATs independently of receptor engagement (Danial and Rothman 2000). The critical tyrosine (Y) residue in all STATs is required for SH2-phosphotyrosine interaction. Tyrosine phosphorylation regulates the dimerization of STATs which is prerequisite for the establishment of a classical JAK/STAT signaling pathway (Fig. 10.1). Other routes of STAT activation occurring through G-protein signaling and JAK activation have been reported following angiotensin binding to its 7-transmembrane receptor (Marrero et al. 1995; Vila-Coro et al. 1999) or RANTES (regulated on activation, normal T cell expressed and secreted) and MIP (macrophage inflammatory protein)-1α binding to shared chemokine receptors (Wong and Fish 1998).

Most STATs (except STAT2 and 6) were found phosphorylated also on serine residues in a stimulus-regulated manner. The C-terminal part of STAT1 and STAT3, contains a serine (S) residue 727 phosphorylation site that enhances transcriptional activity (Decker and Kovarik 2000; Wen et al. 1995). Studies on murine fibroblasts revealed that candidate serine kinases for the phosphorylation of Stat3 include the various mitogen-activated protein kinase family members (Decker and Kovarik 2000; Lo et al. 2003; Turkson et al. 1999). Inhibition of p38 and JNK (c-jun N-terminal kinase) activities suppresses constitutive Stat3 serine phosphorylation and Stat3-mediated gene regulation in Src-transformed fibroblasts (Turkson et al. 1999). Recent studies demonstrated that PKCε (protein kinase C), a phosphatidylserine (PS)-dependent serine/threonine kinase, interacts with STAT3, integrates with MAPK (mitogen-activated protein kinase) cascade to phosphorylate Ser727, and increases both DNA-binding and transcriptional activity of STAT3 in skin, prostate cancers, T98G and MO59K glioma cells (Aziz et al. 2010).

An active STAT dimer is formed through reciprocal interactions between the SH2 domain of one monomer and the phosphorylated tyrosine of other (Heim et al. 1995). Phosphorylated STATs dimerize via reciprocal phosphotyrosine-SH2 domain interactions as homodimers, as seen with all STATs (except STAT2), or as heterodimers, as seen with STAT1/2, STAT1/3, and STAT5a/5b. Within minutes, the dimers translocate to the nucleus, interact with other transcriptional modulators bound to specific

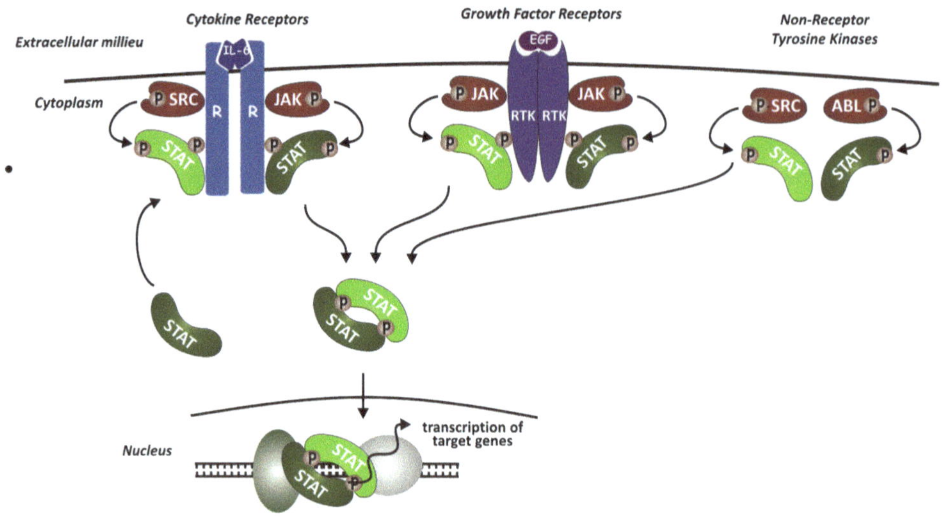

Fig. 10.1 A scheme of STAT signaling pathway. Stimulation of cells with growth factors or cytokines results in dimerization of their cognate receptors and activation of intrinsic receptor tyrosine kinase or receptor associated kinases such as JAKs or Src. These kinases phosphorylate specific tyrosine residues of the cytoplasmic tail of the receptor, providing docking sites for the SH2 domain of inactive, cytoplasmic STAT monomers that are recruited to the activated receptors. Oncogenic kinases such as Src or Abl can also phosphorylate STATs independently of receptor engagement. Phosphorylated STAT dimers translocate to the nucleus where through their DNA-binding domain bind with a consensus DNA element upstream of regulated genes. Genes that are known to be regulated by STAT are involved in many fundamental biological processes, such as proliferation, apoptosis, angiogenesis and immune response

promoter sequences and induce gene expression. Dimer-dimer interactions can occur via the NH_2-terminal portion of STATs to form tetrameric STAT molecules. Tetramerization of STATs contributes to stabilized DNA-binding activity on weak promoters (John et al. 1999). The DNA-binding domain in the center of the molecule determines DNA sequence specificity of individual STATs.

10.2.2 Negative Regulators of STAT Signaling

The activation of individual STAT proteins in normal physiological conditions is tightly controlled and usually lasts from a few minutes to several hours. STAT signaling is assumed to be terminated by dephosphorylation through nuclear tyrosine phosphatases (Lehmann et al. 2003) and/or through proteolytic degradation (Haspel and Darnell 1999). Because STAT1, STAT3, and STAT5 are often activated by the same ligand and/or intracellular tyrosine kinase, it has been suggested that cytoplasmic and nuclear proteins interact with common and unique elements to modulate STAT-specific responses.

A number of modulators of STAT signaling pathways have been described. The suppressors of cytokine signaling (SOCS) protein family can suppress STAT signaling by binding to and inhibiting JAKs (Croker et al. 2008). Eight proteins, cytokine-inducible SH2-domain-containing protein (Fuh et al. 2009) and suppressors of cytokine signaling 1–7 (SOCS-1-7), have been identified in the SOCS family so far. Expression of *SOCS-1-3* and *CIS* is induced by cytokine or growth factor stimulation, and some of these proteins are transcriptionally regulated by STATs themselves, which directly antagonizes STAT activation as part of a classic feedback loop. In particular, SOCS1 is strongly involved in the IFNγ signaling and can associate with all known JAKs directly inhibiting their catalytic activity. SOCS3 in contrast do not directly interact with JAK kinases but needs to be recruited to phosphotyrosine residues of activated receptors, in particular gp130, leptin, growth hormone and erythropoietin receptors (Nicholson et al. 2000). In addition, all SOCS members are thought to act as E3 ubiquitin ligases and to mediate proteasomal degradation of associated proteins (Yoshimura 2005). SOCS3 is frequently silenced by hypermethylation in human cancers (He et al. 2003).

The PIAS (protein inhibitors of activated STAT) represent another group of proteins blocking STAT signaling. Unlike SOCS proteins, which are expressed upon cytokine stimulation, PIAS1 is constitutively expressed in a number of cell lines. PIAS bind only to activated STAT dimers and inhibit their DNA binding or their transactivating capacity by multiple molecular mechanisms, including the recruitment of histone deacetylases, the promotion of the sumoylation of STATs, the induction of the dissociation of dimers, or the sequestration of transcription factors to subnuclear structures (Schmidt and Muller 2003; Shuai and Liu 2005). Overexpression of PIAS1 and PIAS3, specific nuclear inhibitors of STAT1 and STAT3, respectively, suppressed gene transcription mediated by these STATs (Chung et al. 1997; Liu et al. 1998). The constitutive expression of these molecules implies that their physiological function differs from that of SOCS proteins, which are induced in a classical negative feedback loop upon cytokine stimulation (Greenhalgh and Hilton 2001).

10.2.3 Transcriptional Targets of STATs

Differences in physiological properties of distinct STATs are largely attributed to their preferential activation by specific ligands, their binding to and activating specific genes. All of the STAT proteins bind a palindromic consensus sequence TTC$(N)_{2-4}$GAA. The optimal binding sites for STAT1 and STAT3 have the same consensus TTCC(C/G)GGAA (Horvath et al. 1995), and the experimentally verified binding sites in genes share a consensus TTC$(N)_3$GAA (Ehret et al. 2001). All STAT homodimers (with the exception of STAT2) differentially bind more than ten related IFNγ-activated sequence (Letimier et al.) elements that are characterized by the consensus sequence, TTNCNNNAA. A complex comprised of STAT1, STAT2, and IFN regulatory factor (IRF) nine binds to the IFN-α/β–stimulated response element (ISRE) (AGTTN$_3$TTTC) (O'Shea et al. 2002).

For a few genes, including *Socs3* regulation by both Stats *via* the same binding site was demonstrated (Ehret et al. 2001). However, the expression of a larger number of genes, including *Myc* (Ramana et al. 2000; Zhang et al. 2003), *Bcl2l1* (Catlett-Falcone et al. 1999; Fujio et al. 1997), *Icam1* and *Ccl2* (Fujio et al. 1997; Naik et al. 1997; Valente et al. 1998) was differentially affected by treatments that activate specific Stat. Binding sites specific for a particular Stat or a subset of them have been experimentally demonstrated. For example, the IFNγ-activated sequences(Letimier et al. 2007) of mouse *Ly6E* gene (Khan et al. 1993) was shown to bind preferentially to Stat1 homodimer, but not Stat3 homodimer or Stat1-Stat3 heterodimer (Horvath et al. 1995). The SIE (sis-inducible element) in the human *FOS* (Wagner et al. 1990) was shown to bind STAT1 and STAT3 (Horvath et al. 1995), but not STAT4 or STAT5 (Leong et al. 2003).

Studies on STAT null mice and transcriptional profiling has helped to clarify genes whose expression is controlled by STATs, and provided evidence that these transcription factors are clearly important for regulating a wide array of genes (Chen et al. 2003; Hoey et al. 2003; Lund et al. 2004). Immune, growth and apoptosis regulatory activities associated with STAT1-dependent gene transcription are mostly associated with the transcriptional regulation of *IRF-1, MHC, FcγRI, Fas and FasL, TRAIL, cyclin-dependent kinase inhibitors, p21waf1* and *caspase* genes (Battle and Frank 2002; Ivashkiv and Hu 2004). Many studies have shown that the anti-apoptotic gene encoding Bcl-x_L protein is a downstream target of Stat3 (Bromberg et al. 1999) and Stat5 (Gesbert and Griffin 2000; Socolovsky et al. 1999). The cell cycle control gene *c-Myc* has been shown to be induced in response to Stat3 signaling in *v-Src*-transformed NIH3T3 fibroblasts as well as through Stat5 activation (Bowman et al. 2001; Lord et al. 2000).

Recent results challenge a simple view that distinct STATs regulate discrete and non-overlapping sets of genes. In non-stimulated rat C6 glioma cells Stat1 and Stat3 proteins are expressed and phosphorylated to some extent (Fig. 10.2). Both proteins are strongly phosphorylated after a treatment with either IFNγ or IL-6. The IFN-α/β–stimulated response element (ISRE)-driven transcription was inhibited by the ISRE oligodeoxynucleotide decoys (ODNs) (Fig. 10.2). Furthermore, our studies employing ODNs and siRNA interfering with either Stat1 or Stat3 activity/expression suggest that Stat1 supports the basal expression of Stat3 target genes *Bcl2l1* and possibly *c-Myc* in proliferating C6 cells (Adach-Kilon et al. 2011). Also, studies on STAT3–/– cells showed that the immediate early genes *Fos* and *Egr1* become STAT1 transcriptional targets in the absence of STAT3 (Schiavone et al. 2011). It suggests that the abrogation of STAT3 expression directs the STAT1 to transcribe new target genes, known to drive mitogen responses and tumor transformation.

Until recently, a technology did not permit the discrimination of genes that are regulated directly compared to indirect regulation. Recent analyses employing DNA chromatin immunoprecipitation followed by microarray technique (Chip-chip) brought new insight into the accurate determination of Stat1-DNA binding sites. Those studies revealed that Stat1 binds to many sites on chromosome 22 and uncovered many new candidate target genes, not associated previously with IFN-responsive genes which are induced only in certain cell types and under specific conditions (Hartman et al. 2005; Wormald et al. 2006). Chromatin immunoprecipitation studies demonstrated that in the case of STAT4, known targets include *Ifng*, *Il18r1*, *Il12rb2*, *Il2ra* and *Furin* in Th1 lymphocytes (Letimier et al. 2007; O'Sullivan et al. 2004; Pesu et al. 2006; Thieu et al. 2008), whereas STAT6-bound genes include *Il4, Gata3*, and *Ccl17* (Kubo et al. 1997; Wirnsberger et al. 2006).

Interactions with basic transcription co-regulators may modulate the functional specificity of individual STAT. Induction of a subset of IL-6-responsive genes was found to be dependent on *Brg1* (*Brahma-related gene 1*). BRG1 protein is the ATPase subunits of the SWI/SNF complexes, chromatin remodeling proteins involved in altering local chromatin structure and facilitation of recruitment of essential transcription factors. BRG1 was required for STAT3 recruitment to the IFN regulatory factor (IRF) 1 promoter, downstream histone modifications, and IL-6-induced chromatin remodeling. Therefore, it has been suggested that BRG1 plays a role in mediating STAT accessibility at multiple cytokine-responsive promoters and may influence access of different STAT proteins to the same target (Ni and Bremner 2007). Interestingly, while both hBrm (human Brahma) and BRG1 interact with STAT1 *in vitro*, under normal conditions only hBrm is recruited by STAT1 to IFNγ-activated sequences of individual genes; while phosphorylated STAT1 mainly binds to BRG1 under stress conditions. Under basal conditions, hBrm exists in a mSin3/HDAC co-repressor complex associated with a compact chromatin structure. Upon heat-shock, the phosphorylated STAT1 binds and recruits BRG1 to the GAS, leading to induction of gene expression. This hBrm/BRG1 switch occurs at the GAS in specific cell types upon exposure to IFNγ (Zhang et al. 2010).

Furthermore, a wide variety of factors interacting with STATs exemplified by NF-κB, SMADs, c-Jun (Baker et al. 2008), Sp1 (Look et al. 1995), BRCA1 (Ouchi et al. 2000), etc. may antagonistically or synergistically influence the functional specificity of individual STAT.

10.3 Dysfunction of STAT Signaling in Gliomas

10.3.1 Constitutive Activation of STAT3 in Gliomas

Clinical and experimental studies show various expressions of STATs in gliomas. Immunohistological studies showed STAT1 expression in a majority of glioblastomas, but in tumor tissue the signal was mostly localized in the cytoplasm suggesting the predominant presence of an inactive form of STAT1. Within the infiltration area strong STAT1 expression was found in reactive astrocytes and in microglial components (Haybaeck et al. 2007). STAT3 was shown to be constitutively active, as assessed by tyrosine

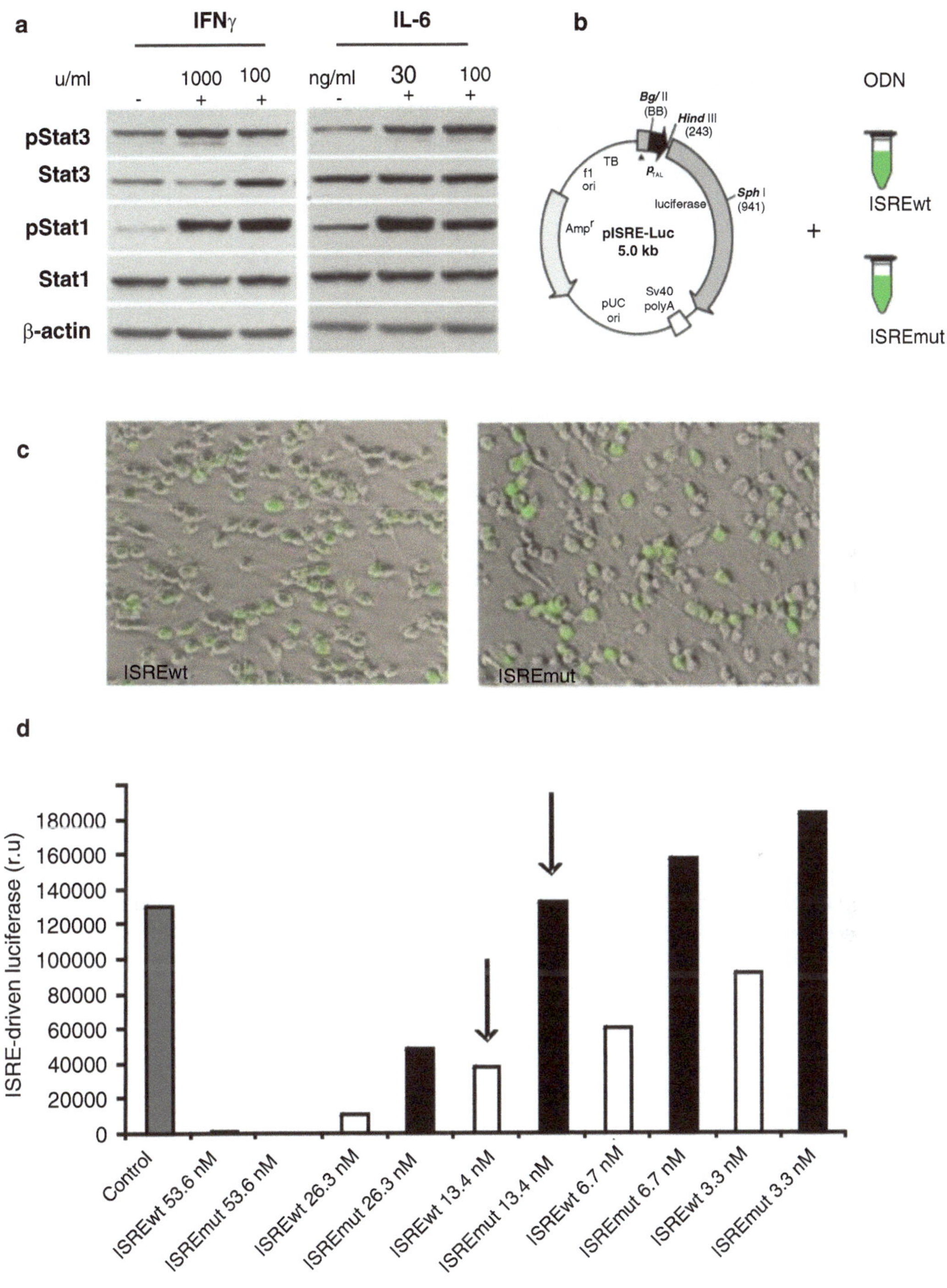

Fig. 10.2 STAT signaling in rat C6 glioma cells. (**a**). The level of total and phosphorylated Stat1 and Stat3 in rat C6 glioma cells untreated or exposed to interferon (IFN)-γ or IL-6 30 min after stimulation. Immunoblots were re-probed with an antibody recognizing β-actin to ensure equal loading. (**b**). A scheme of the experiment with ODN decoys. (**c**). Representative pictures show cells 4 h after transfection with the FITC-labelled ISRE decoys (13.4 nM) visualized by fluorescence microscopy. (**d**). The luciferase activity was measured in untreated cells or 24 h after co-transfection pISRE-luc with the indicated decoy at various concentrations. The bars indicate mean values of luciferase activity in a representative experiment (in triplicate). *Arrows* mark the ODN concentration (13.4 nM) selected for experiments. ISRE, IFN-α/β-stimulated response element

phosphorylation status in malignant gliomas tumors and cell lines when compared with normal human astrocytes, white matter, and normal tissue adjacent to tumor (Mizoguchi et al. 2006; Rahaman et al. 2002; Schaefer et al. 2002). Experimental mouse gliomas express constitutively activated STAT3 (Weissenberger et al. 2004). Many studies reported constitutive STAT3 activation, corresponding to high expression of STAT3 target genes such as *Bcl2l1, Bcl-2* or *Mcl-1* in glioblastoma cell lines (Iwamaru et al. 2007; Rahaman et al. 2002).

There is a controversy concerning potential correlation between activation STAT3 and histological glioma grade. While a nuclear staining for pSTAT3 reflecting its activation was detected by immunohistochemistry in 259 glioma samples of different grades, the positive rate was <9 % in high grade gliomas (Wang et al. 2004). On the other hand, immunohistochemical analysis of 55 glioma samples of different grades showed that STAT3 was constitutively activated in 28 % of low (I-II) and 60 % in high (III-IV) grade gliomas (Lo et al. 2008). Activation of STAT3 was essentially identical in 82 malignant astrocytic gliomas (55 glioblastomas and 27 anaplastic astrocytomas) (Mizoguchi et al. 2006). Measurements of the amounts of pSTAT3 in 15 human gliomas, five diffuse astrocytomas (WHO grade II), five anaplastic astrocytomas (WHO grade III), five *glioblastoma multiforme* (GBMs) (WHO grade IV) and in normal human brain by Western blot analysis showed the level of pSTAT3 increased with malignancy being highest in GBM (Weissenberger et al. 2004). Interestingly, in low-grade human gliomas pSTAT3 was located in the nucleus, while in GBM activated STAT3 was localized to the nucleus and to plasma membrane (Weissenberger et al. 2004).

Recent investigation of prognostic relevance of activated STAT3 in a larger collection of patients with glioblastoma provided evidence that activation of STAT3 is linked with clinically more aggressive behavior. Glioblastoma patients with high or very high numbers of pSTAT3-positive tumor cells had significantly shorter overall survival than those with no or low numbers (Birner et al. 2010). The proportion of grade III glioma cases with high or very high numbers of pSTAT3-positive tumor cells was similar to that in grade IV glioblastoma suggesting lack of association of STAT3 activation with tumor grade III glioma progression (Birner et al. 2010).

Immunohistochemical staining revealed markedly increased expression of STAT5b in GBM (57.1 %) compared with that in normal cortex (22.2 %) and diffuse astrocytoma (27.3 %) (Liang et al. 2009). Phosphorylated STAT5 was detected in primary gliomas, predominantly in the nucleus (Cao et al. 2010).

Studies using multiparameter flow cytometric cell sorting of *ex vivo* tumor specimens demonstrate the up-regulation of STAT3 and STAT5 in microglia sorted from GBM tumor specimens but not from meningiomas (Kostianovsky et al. 2008). Up-regulation of STATs in immune cells infiltrating gliomas may be involved in a complex network of inhibitory pathways, responsible for the GBM-mediated suppression of monocyte/microglial function (Hussain et al. 2007).

10.3.2 STAT Activation in Gliomas Results from Dysfunction in Control Mechanisms

While mutations of cytokine receptors and JAKs are rare, gain-of-function mutations of growth factor receptor kinases are quite common in human cancers including gliomas. The gene encoding EGFR and its constitutively activated variant, EGFRvIII, are often amplified and overexpressed in human adult gliomas. EGFRvIII is a product of rearrangement with an in-frame deletion of 801 bp of the coding sequence of the extracellular domain, resulting in a deletion of residues 6–273 and a glycine insertion as residue 6. EGFRvIII deletion results in a ligand-independent, constitutively active, and cell surface–retained receptor. Both EGFR and EGFRvIII are tumorigenic for gliomas and major targets for glioma therapy (Friedman and Bigner 2005; Nishikawa et al. 1994). Blocking of EGFR, but not PDGFR or SRC kinase activity, by pharmacological compounds markedly reduced the constitutive activation of STAT3 in U251 glioma cells suggesting EGFR activation contributes to the constitutive activation of STAT3 in those cells (Rahaman et al. 2002). Several studies demonstrated that three STAT3-activating kinases, JAK2, EGFR, and EGFRvIII, contribute to STAT3 activation. Constitutive STAT3 activation coexisted with EGFR expression in 27.2 % of primary high-grade gliomas. Combination of an anti-EGFR agent – Iressa and a JAK2/STAT3 inhibitor – JSI-124, synergistically suppressed STAT3 activation and potently killed glioblastoma cell lines expressing EGFR or EGFRvIII (Lo et al. 2008).

In addition to the tyrosine kinase EGFR, other tyrosine kinases known to activate STAT3 may be important in activating signaling pathways relevant to gliomagenesis. These proteins include SRC (Yu et al. 1995) and the endothelial receptor vascular endothelial growth factor receptor-2 (VEGFR2) (Korpelainen et al. 1999). Unlike STAT activation by PDGFR and EGFR, which activate both STAT1 and STAT3, VEGFR2 did not activate STAT1.

GBM samples contain significantly higher levels of IL6 protein compared to those of control brains (Weissenberger et al. 2004). In U251glioblastoma cells STAT3 activation was in part caused by autocrine IL-6 , as neutralizing IL-6 antibodies reduced STAT3 activation by 70 % (Rahaman et al. 2002). A recent study demonstrates expression of the interleukin 6 receptor α (IL6Rα) and glycoprotein 130 (gp130) in glioma stem cells (GSCs). Targeting IL6Rα or IL6 ligand expression in those cells with short hairpin RNAs (shRNAs) attenuated STAT3 activation indicating that STAT3 is a downstream mediator of pro-survival IL6 signals in GSCs (Wang et al. 2009).

Altered expression of negative regulators of STAT may contribute to their constitutive activation. SOCS1 and SOCS3 are aberrantly expressed in GBM cell lines and primary tissues. The promoter of SOCS1-2-3 was methylated in 24, 6.5 and 35 % of GBM, respectively, that resulted in reduced SOCSs expression (SOCS1-2-3 mRNA was reduced by 5, 3 and 7-folds, respectively) when compared with unmethylated GBM. Hypermethylation of SOCS3 promoter was significantly associated with an unfavorable clinical outcome (Martini et al. 2008). The other study demonstrated that 10 tested GBM cell lines lacked SOCS1 expression, whereas GBM cell lines and primary GBM tumor samples constitutively expressed SOCS3. SOCS1 gene repression was linked to hypermethylation of the SOCS1 genetic locus in GBM cells. Reintroduction of SOCS1 or blocking SOCS3 expression sensitized cells to radiation and decreased the levels of activated ERK in GBM cells (Zhou et al. 2007). Interestingly, assessment of the relationship between SOCS3 and EGFR aberrations revealed that SOCS3 promoter hypermethylation was inversely related to both the EGFR gene dosage as well as the EGFR protein expression (Lindemann et al. 2011). About 89 % of glioblastoma samples were found to be PIAS3 negative and pSTAT3 positive. The ectopic expression of PIAS3 in a glioblastoma cell line caused the inhibition of the transcriptional activity of STAT3 (Brantley et al. 2008).

10.4 Functions of STAT3 in Gliomas

10.4.1 STAT3 as Oncogene

There is an accepted concept that STAT1 and STAT3, despite their similar structures, have antagonistic effects on cellular proliferation and apoptosis, with STAT3 acting like an oncogene and STAT1 playing a role of tumor suppressor (Battle and Frank 2002; Bromberg and Darnell 2000; Stephanou and Latchman 2005). A first evidence for a potential oncogenic function of STAT3 came from findings showing its constitutive activation in Src-transformed cell lines (Bromberg et al. 1998; Turkson et al. 1998). Further studies showed that STAT3C, a constitutively active mutant of STAT3, can transform cultured fibroblasts which form tumors when injected into mice (Bromberg et al. 1999). Therefore, STAT3 is considered to be one of the major mediators of tumorigenesis and numerous studies show how interfering with STAT3 signaling affects growth, survival and tumorigenicity of many tumors, including gliomas.

U87-derived cell lines stably expressing a dominant negative mutant DN-STAT3 in hypoxia inducible manner failed to grow in mice due to impaired cell proliferation, survival and reduced angiogenesis. Mice implanted with DN-STAT3 expressing clones survived significantly longer than control mice (Dasgupta et al. 2009). Knockdown of STAT3 expression by RNAi suppressed growth, induced apoptosis and differentiation in glioblastoma stem cells (Li et al. 2010). Effects of STAT3 on cell cycle and proliferation were mediated through its ability to regulate the expression of *Cyclin D1* and *c-Myc*. Double-stranded decoy oligodeoxynucleotides which correspond closely to the STAT3 response element within the *c-Fos* promoter blocked STAT3 signaling and subsequently inhibited cell proliferation by inducing apoptosis and cell-cycle arrest in two glioma cell lines U251 and A172 (Gu et al. 2005; Iwamaru et al. 2007; Lo et al. 2008). On the other hand, we found no reduction in viability of C6 glioma cells after a treatment with the STAT3 decoys or STAT3 silencing by siRNA (Adach-Kilon et al. 2011).

Suppression of apoptosis by STAT3 is mediated through expression of various survival genes that are regulated by STAT3 such as *Bcl2l1*, *Bcl-2*, *survivin*, *Mcl-1*. Pharmacological inhibition of STAT3 activation in glioma cells leads to down-regulation of survival-related genes and apoptosis (Iwamaru et al. 2007; Konnikova et al. 2003; Rahaman et al. 2002).

STAT3 plays also an important role in regulating tumor cell invasion. STAT3 is known to directly up-regulate the expression of matrix metalloproteinases: MMP-2 and MMP-9, involved in the degradation of extracellular matrix. Inhibition of STAT3 activity impaired migratory and invasive potential of numerous glioma cell lines and decreased *MMP-2* and *MMP-9* transcription, and their proteolytic activity (Chen et al. 2010; Senft et al. 2010).

STAT3 was shown to be a direct transcriptional activator of the VEGF gene which is the most potent angiogenesis inducing signal. In gliomas VEGF is expressed in cells with activated STAT3 (Lo et al. 2008) and activated STAT3 dramatically increases the transcription from the *VEGF* gene promoter (Schaefer et al. 2002). As enhanced expression of VEGFR leads to activation of STAT3, it suggests its participation in a VEGF/VEGFR autocrine loop facilitating angiogenesis in malignant gliomas (Schaefer et al. 2002).

Infiltration of immune cells into tumors, their switch to the alternative, pro-tumorigenic phenotype and formation of an inflammatory microenvironment support glioma progression (Gabrusiewicz et al. 2011). Cancer-associated inflammation is marked by the presence of specific inflammatory mediators, including numerous cytokines and chemokines. Recent evidence suggest a crucial role of STATs, in particular STAT3, in tumor-induced immunosuppression. Interference with STAT3 expression/activity in tumor-infiltrating immune cells reduced tumor progression in animal models (Yu et al. 2007, 2009). STAT3 signaling in innate immune cells is required for the immunsuppressive and tumor promoting effects of myeloid-derived suppressor cells and tumor-associated macrophages in several experimental models. STAT3 also modulates expansion of T regulatory cells in tumors and is necessary for the development of TH17 T cells (Yu et al. 2009). Inhibition of STAT3 activity by JSI-124 (a STAT3 inhibitor) promoted maturation of tumor infiltrating CD11c+dendritic cells and activation of tumor-conditioned cytotoxic T cells. Brain infiltrating lymphocytes isolated from JSI-124-treated mice exhibited enhanced expression of surface maturation markers such as MHC class II, CD40, CD80, CD86 on tumor infiltrating CD11c+dendritic cells. Moreover, *in vivo* JSI-124 treatment reduced CD11b+/Gr1+ myeloid suppressor cells and CD4+/CD25+ regulatory T cells in a murine GL261 intracranial glioma model (Fujita et al. 2008). These data further support the hypothesis that systemic inhibition of STAT3 signaling can reverse the immunosuppressive environment in gliomas.

Furthermore, *ex vivo* STAT3 inhibition has been reported to activate T cell, monocytes and microglia isolated from glioma patients. Glioma-infiltrating microglia/macrophages expressed MHCII but lacked expression of the co-stimulatory molecules CD80, CD86 and CD40 critical for T-cell activation and were unable to stimulate T lymphocytes. Inhibition of STAT3 by WP1066 (a JAK inhibitor) resulted in up-regulation of CD80 and CD86 on both normal donor PBMCs and on tumor infiltrating microglia/macrophages isolated from GBM patients. Moreover, the treatment stimulated the production of immune-stimulatory cytokines IL2, IL4, IL12, IL15 and induced proliferation of effector T cells (Hussain et al. 2007). Altogether, STAT3 targeting in tumor microenvironment seems to be an effective strategy to overcome glioma-induced immunosuppression and to induce anti-tumor immunity in gliomas.

10.4.2 STAT3 as a Tumor Suppressor in Gliomas

Although most studies have highlighted the oncogenic function of STAT3 in non-brain tumors, recent data suggest that STAT3 may play both tumor suppressive and oncogenic function in glioma pathogenesis, depending on the genetic background of the tumor (de la Iglesia et al. 2008a, b, 2009). STAT3 has been shown to promote cell differentiation along the astrocytic lineage (Bonni et al. 1997; Rajan and McKay 1998). *Stat3*$^{-/-}$ mouse astrocytes displayed increased proliferation and invasiveness as compared to control astrocytes, indicating that STAT3 inhibits astrocyte proliferation and invasiveness. Although loss of *Stat3* gene was not sufficient to transform astrocytes, combined with knockdown of the tumor suppressor PTEN led to malignant transformation (de la Iglesia et al. 2008b). Deficiency of PTEN triggered the cascade of events that inhibits STAT3 signaling in murine astrocytes and human glioblastoma tumors (de la Iglesia et al. 2008b). Reactivation of STAT3 in PTEN-deficient but not in PTEN-expressing glioblastoma cells inhibited their proliferation, invasiveness and spreading on myelin (de la Iglesia et al. 2008a, b). In a panel of human brain tumors, PTEN loss correlates tightly with down-regulation of LIFRβ and low levels of phosphorylated STAT3. These findings provide correlative evidence that PTEN loss and inhibition of LIFRβ-STAT3 signaling are linked in gliomas (de la Iglesia et al. 2008b).

We have performed a global analysis of phospho-STAT3 binding sites in rat C6 glioma cells by chIP-on chip and crossed putative STAT3 target genes with the profiles of gene expression in control and JAK/STAT3 inhibitor treated cells. The distribution of genes containing peaks for phospho-STAT3, which demonstrates STAT3 binding to promoters, correlated ($p < 10^{-10}$ in the Kolmogorov-Smirnov test) with changes in gene expression induced by inhibition of STAT3 phosphorylation (mostly with expression increases). It suggests that STAT3 binding contributes mostly to negative regulation of target gene expression in proliferating C6 glioma cells (unpublished).

In contrast to STAT3 acting as a tumor suppressor in PTEN deficient cells, STAT3 acts as an oncogene in EGFRvIII-expressing cells. STAT3 was required for the malignant transformation of astrocytes that are both PTEN-deficient and express EGFRvIII, and physically associates with EGFRvIII within the nucleus. It has been suggested that nuclear EGFRvIII acts as a switch to convert STAT3 from the tumor suppressive to pro-oncogenic protein (de la Iglesia et al. 2008a). Thus, the role of STAT3 as a tumor suppressive and oncogenic protein depends on the genetic background of the tumor.

10.4.3 STAT3 in Glioma Cancer Initiating Cells

A subpopulation of cells with stem-like features, glioma-initiating cells or glioma stem cells were identified in GBM (Singh et al. 2003, 2004; Yuan et al. 2004). Those cells are characterized by their ability to undergo self-renewal and to differentiate into neuronal, astroglial and oligodendroglial cells, are highly tumorigenic (Galli et al. 2004; Vescovi et al. 2006). STAT3 mediates self-renewal of pluripotent embryonic stem cells (Niwa et al. 1998; Raz et al. 1999) and is implicated in neurogenesis and gliogenesis in neural stem cells (Gu et al. 2005). Isolated glioma-initiating cells grow as anchorage-independent spheres and express STAT3 phosphorylated on activating tyrosine and serine residues. Inhibition of STAT3 signaling in those cells with either small molecule inhibitors or RNAi inhibited

Table 10.1 STAT3 inhibitors tested in preclinical/clinical trials in glioblastomas

Inhibitor	Postulated mechanism	References
LLL3	Inhibition of DNA binding	Fuh et al. (2009)
LLL12	Inhibition of DNA binding	Lin et al. (2010)
Peptides aptamers	Disruption of STAT3 dimerization	Borghouts et al. (2008)
ODN	Inhibition of DNA binding	Gu et al. (2008), Shen et al. (2009)
AG490	JAK1/2 inhibition	Rahaman et al. (2002)
WP1066	JAK2 inhibition	Iwamaru et al. (2007)
JSI-124	JAK2 inhibition	Lo et al. (2008), Su et al. (2008)
Curcumin	JAK1/2 inhibition	Weissenberger et al. (2010)
ZD6474	VEGFR, EGFR inhibition	Yiin et al. (2010)
Sorafenib	Multikinase inhibition	Yang et al. (2010)

cell growth and spheres formation (Sherry et al. 2009; Villalva et al. 2010), and sensitized cells to an anti-tumor drug – Temozolomide (Villalva et al. 2010). STAT3 has emerged as the important regulator of immunosuppressive pathway in glioma initiating cells (Hatiboglu et al. 2010). The inhibition of STAT3 did not alter the immunological phenotype of those cells, but reduced Treg induction, restored T-cell function and induced T-cell proliferation (Wei et al. 2010).

10.5 Molecular and Pharmacological Strategies to Interfere with STAT Signaling for Potential Therapeutic Intervention in Gliomas

A number of studies have focused on development of STAT3 inhibitors due to their expected therapeutic potential in cancer therapy. Therapeutic strategies for blocking STAT3 activity are based either on direct targeting of STAT3 protein or indirect targeting of the upstream components of the STAT3 signaling pathway (Table 10.1). Due to difficulties in designing specific inhibitors of particular kinases, problems with drug specificities and side effects, recent investigator's efforts focused on developing direct STAT3 inhibitors. These include dominant negative STAT3 expression vectors, oligonucleotides decoys, small interfering RNA, peptides, peptidomimetics and small molecules inhibitors. Mechanism of action of these compounds is usually based on disruption of STAT3 dimerization or STAT3 DNA binding activity (Turkson 2004).

Recently several direct STAT3 inhibitors have been successfully applied in glioblastoma. Two non-peptide, cell-permeable, small molecules, termed LLL3 and LLL12 were developed using structure-based drug design (Fuh et al. 2009; Lin et al. 2010). Computer modeling with docking simulation showed that these compounds bind directly to the phosphoryl tyrosine 705 binding sites of STAT3 monomer. Both compounds inhibit STAT3 phosphorylation and downstream STAT3 target genes that results in reduced cell viability and induction of apoptosis (Fuh et al. 2009; Lin et al. 2010). LLL12 is a more potent inhibitor of cell viability than previously described JAK2 inhibitor WP1066 (Iwamaru et al. 2007), with half maximal inhibitory concentration values 0.21 and 0.86 μM for U87-MG and U373-MG, respectively (Lin et al. 2010). Furthermore, both inhibitors demonstrated very potent activity *in vivo*. The U87 glioblastoma tumor-bearing mice treated with LLL3 or LLL12 had smaller intracranial tumors and LLL3 treated mice exhibited prolonged survival relative to vehicle treated mice (28.5 vs 16 days) (Fuh et al. 2009).

A dimerization-disrupting phosphopeptide sequence derived from the SH2 domain-binding region of STAT3, PY*LKTK (where Y* represents phosphotyrosine) and its tripeptide derivatives PY*L and AY*L, were developed as inhibitors of STAT3 activation and biological function. Specific peptidomimetics selectively disrupt STAT3 DNA-binding activity *in vitro* and interfere with STAT3 DNA binding and a reporter gene activity in Src-transformed fibroblasts (Turkson et al. 2001, 2004).

Dimerization of STAT3 proteins can be blocked by peptide aptamers, short, usually 12 to 20 amino acids in length, peptides that specifically bind to a target protein. Peptide aptamers specifically interacting with the STAT3 dimerization domain were selected from a peptide library by an adaptation of the yeast two-hybrid procedure (Nagel-Wolfrum et al. 2004). Purified recombinant peptide aptamer, tagged with a protein transduction motif of nine arginine and fused with thioredoxin as a scaffold protein, was cell-permeable and selectively induced growth inhibition and apoptosis of glioblastoma cells (Borghouts et al. 2008). Currently, peptides aptamers represent one of the most effective approaches to disrupt STAT3 function *in vitro* with half maximal inhibitory concentration values for glioblastoma cells <1 μM. No prior knowledge of the structure of the target protein is required and their binding ability is not limited to preexisting small molecular weight compound binding pocket (Borghouts et al. 2008). However, if the peptide aptamers are considered for *in vivo* usage, limitations such as stability and permeability of the fusion protein must be optimized.

Another approach developed for direct STAT3 targeting are oligodeoxynucleotides decoys, a short, synthetic DNA carrying the cognate DNA-binding sites of transcription factor (Yu and Jove 2004). STAT3 decoy-ODN inhibited the binding of STAT3 to DNA and altered the downstream gene expression in U251 and A172 glioma cell lines (Gu et al. 2008). Intratumorally administrated STAT3 decoy-ODN significantly suppressed the growth of glioma by inhibiting proliferation and promoting apoptosis in xenografts (Shen et al. 2009). Our studies on different tumor cells, including rat C6 and human T98 glioma cells demonstrated lack of significant STAT3 decoy-ODN effects on the expression of endogenous, STAT dependent genes and cell survival (Adach-Kilon et al. 2011 and unpublished). Although such nucleic acids based strategy works well in cell culture and in animal model, it might be limited in terms of their clinical development as therapeutic agents.

Indirect inhibition of STAT3 can be achieved by disruption of the ligand-receptor interaction at the extracellular surface or by blocking upstream tyrosine kinases that are responsible for its activation. A plethora of small molecule inhibitors of JAK, Src or EGFR has been shown to effectively block STAT3 signaling, inhibit glioma proliferation and induce apoptosis (Heimberger and Priebe 2008). These STAT3 inhibitors display marked efficacy in murine glioma models, including intracerebral tumors. The mechanism of this *in vivo* efficacy of the STAT3 blockade agents is a combination of direct tumor cytotoxicity and immune cytotoxic clearance (Hussain et al. 2007; Iwamaru et al. 2007).

One of the first reports showing that STAT3 is a valid target for therapy of glioblastoma came from a study with AG490, the JAK1/2 inhibitor. A treatment with AG490 reduced a constitutive STAT3 activation, inhibited proliferation of cultured U251 GBM cells and induced apoptosis by reducing the steady-state level of Bcl-x_L, Bcl-2 and Mcl-1 anti-apoptotic proteins (Rahaman et al. 2002). Inhibition of the JAK/STAT3 signaling pathway impedes the migratory and invasive potential of human glioblastoma cells. Treatment with AG490 reduced migratory and invasive potential of five different glioblastoma cell lines that was paralleled by a decrease in transcription of *MMP-2* and *MMP-9*, and their reduced proteolytic activity (Senft et al. 2010).

WP1066 was designed by modifying the structure of AG490 and inhibited STAT3 activation by blocking JAK2. It showed selective cytotoxicity toward cultured malignant glioma U87-MG and U373-MG cells in lower doses than AG490 and significantly inhibited the growth of subcutaneous tumors generated from U87-MG in mice (Iwamaru et al. 2007). Although the subcutaneous glioma model was used, the concentration of the compound in the brain was 10 times higher than in the plasma after intraperitoneal treatment, indicating a good penetrance of the blood–brain barrier. Furthermore, WP1066 reversed immune tolerance in immune cells isolated from GBM patients (Hussain et al. 2007).

Interestingly, the traditional herbal medicine cucrubitacin I (JSI-124) is a selective inhibitor of JAK kinase and STAT3 signaling. JSI-124 induced G2/M arrest and apoptosis of GBM cell lines via down-regulation of *Cyclin B1* and *Cdc2* expression (Lo et al. 2008; Su et al. 2008). Combination of JSI-124 and dasatinib (Src family kinase inhibitor) synergistically decreased cell proliferation and viability, and had a significant effect on cell migration exceeding those observed with either drug alone (Premkumar et al. 2010). Another naturally occurring, dietary compound, displaying the inhibitory potential toward JAK signaling is curcumin. Curcumin was shown to suppress malignant glioma growth *in vitro* with induction of G2/M arrest and attenuation of migratory and invasive behavior. *In vivo*, curcumin reduced growth of intracranial gliomas and proliferation of tumors cells but failed to increase survival of glioma-bearing mice (Weissenberger et al. 2010).

The multikinase inhibitors such as sorafenib, sunitib or vandetanib that are already in clinical trials were shown to mediate the antitumor effects by reducing STAT3 activity. ZD6474 (ZACTIMA, vandetanib) is an inhibitor of various receptor tyrosine kinases, in particular VEGFR2 and EGFR. ZD6474 inhibited growth and survival through attenuation of STAT3 phosphorylation, Akt, and Bcl-x_L expression in glioma cells expressing EGFRvIII but not in cells with non-detectable EGFRvIII (Yiin et al. 2010). Vandetanib was used in the phase I of clinical trial in children with newly diagnosed diffuse intrinsic pontine glioma (Broniscer et al. 2010). Sorafenib (BAY43-9006, Nexavar) is an oral multikinase inhibitor originally developed to Raf and receptor tyrosine kinase signaling. Sorafenib (≤10 μmol/L) inhibited cell proliferation and induced apoptosis of U87 and U251 glioma cell lines, and two primary cultures (PBT015 and PBT022) from human glioblastomas. The effects of sorafenib were associated with inhibition of STAT3 phosphorylation. Overexpression of a constitutively activated STAT3 partially blocked the effects of sorafenib. The level of phosphorylated JAK1 was reduced in U87 and U251 cells, whereas phosphorylated JAK2 was down-regulated in primary cultures (Yang et al. 2010). Glioma pathogenesis involves abnormalities in many cellular pathways, thus molecularly targeted therapies with multikinase inhibitors may provide clinical benefits in the treatment of glioblastomas.

Currently, the major approaches to treating glioma are surgical resection, radiotherapy, and adjuvant chemotherapy. The combination of oral cytotoxic chemotherapy with concomitant radiotherapy has been shown to improve survival of patients with glioblastoma (Stupp et al. 2005). However, GBM cells often develop resistance to ionizing radiation (IR) and chemotherapeutics, rendering therapy ineffective. Therefore, increasing cell sensitivity to radiation and chemotherapy could significantly increase therapeutic outcome. Attempts to sensitize GBM cells to radiation have focused on the use of broadly acting inhibitors of kinases known to be mutated or amplified in gliomas. In particular, targeting EGFR has been shown to sensitize GBM cells to IR (Stea et al. 2003; Zhou et al. 2007). Expression of DN-STAT3 sensitized U87 cells to the cytotoxic effects of IR (Zhou et al. 2007). Moreover, suppression of STAT3 with siRNA enhanced radiation-induced growth inhibition in a U251 glioma model. Simultaneous inhibition of STAT3 and ErbB2 combined with radiotherapy led to the most significant reduction of tumor growth (Gao et al. 2010). Inhibition of STAT3 with JSI-124 sensitized malignant glioma cells to TMZ (temozolomide), the most commonly used agent in the therapy of GBM and alykalating agents (cisplatin or 1,3 bis(2chlorylethyl)-1-nitrosourea) (Lo et al. 2008). When STAT3 expression/activity was blocked by either shRNA or Sttatic (a direct STAT3 inhibitor) in glioma initiating stem cells, the treatment with IC20 concentration of TMZ decreased proliferation rate by 50 % in comparison to untreated cells (Villalva et al. 2010).

While an array of STAT3 inhibitors that induce antitumor effects cultured cells and animal models have been identified, most of the STAT3 inhibitors reported to date have not undergone an *in vivo* efficacy, pharmacology or toxicity testing (Yue and Turkson 2009). Overall, there is a need for re-examination of the ongoing strategies to target STAT3 intended not only for refinement, but also to incorporate newest technologies to transform current compounds into clinically useful anticancer therapeutics.

Acknowledgements We thank Kavita Ramji for a critical reading of the manuscript. Studies were supported by a grant N N405621938 from the Ministry of Science and Higher Education.

References

Adach-Kilon A, Swiatek-Machado K, Kaminska B, Dabrowski M (2011) Signal transducer and activator of transcription 1 (Stat1) maintains basal mRNA expression of pro-survival Stat3-target genes in glioma C6 cells. J Cell Biochem 112:3685–3694

Akira S (2000) Roles of STAT3 defined by tissue-specific gene targeting. Oncogene 19:2607–2611

Aziz MH, Hafeez BB, Sand JM, Pierce DB, Aziz SW, Dreckschmidt NE, Verma AK (2010) Protein kinase Cvarepsilon mediates Stat3Ser727 phosphorylation, Stat3-regulated gene expression, and cell invasion in various human cancer cell lines through integration with MAPK cascade (RAF-1, MEK1/2, and ERK1/2). Oncogene 29:3100–3109

Baker BJ, Qin H, Benveniste EN (2008) Molecular basis of oncostatin M-induced SOCS-3 expression in astrocytes. Glia 56:1250–1262

Battle TE, Frank DA (2002) The role of STATs in apoptosis. Curr Mol Med 2:381–392

Birner P, Toumangelova-Uzeir K, Natchev S, Guentchev M (2010) STAT3 tyrosine phosphorylation influences survival in glioblastoma. J Neurooncol 100:339–343

Bonni A, Sun Y, Nadal-Vicens M, Bhatt A, Frank DA, Rozovsky I, Stahl N, Yancopoulos GD, Greenberg ME (1997) Regulation of gliogenesis in the central nervous system by the JAK-STAT signaling pathway. Science 278:477–483

Borghouts C, Kunz C, Delis N, Groner B (2008) Monomeric recombinant peptide aptamers are required for efficient intracellular uptake and target inhibition. Mol Cancer Res 6:267–281

Bowman T, Broome MA, Sinibaldi D, Wharton W, Pledger WJ, Sedivy JM, Irby R, Yeatman T, Courtneidge SA, Jove R (2001) Stat3-mediated Myc expression is required for Src transformation and PDGF-induced mitogenesis. Proc Natl Acad Sci U S A 98:7319–7324

Brantley EC, Nabors LB, Gillespie GY, Choi YH, Palmer CA, Harrison K, Roarty K, Benveniste EN (2008) Loss of protein inhibitors of activated STAT-3 expression in glioblastoma multiforme tumors: implications for STAT-3 activation and gene expression. Clin Cancer Res 14:4694–4704

Bromberg JF (2001) Activation of STAT proteins and growth control. Bioessays 23:161–169

Bromberg J, Darnell JE Jr (2000) The role of STATs in transcriptional control and their impact on cellular function. Oncogene 19:2468–2473

Bromberg JF, Horvath CM, Besser D, Lathem WW, Darnell JE Jr (1998) Stat3 activation is required for cellular transformation by v-src. Mol Cell Biol 18:2553–2558

Bromberg JF, Wrzeszczynska MH, Devgan G, Zhao Y, Pestell RG, Albanese C, Darnell JE Jr (1999) Stat3 as an oncogene. Cell 98:295–303

Broniscer A, Baker JN, Tagen M, Onar-Thomas A, Gilbertson RJ, Davidoff AM, Pai Panandiker AS, Leung W, Chin TK, Stewart CF, Kocak M, Rowland C, Merchant TE, Kaste SC, Gajjar A (2010) Phase I study of vandetanib during and after radiotherapy in children with diffuse intrinsic pontine glioma. J Clin Oncol 28:4762–4768

Cao S, Wang C, Zheng Q, Qiao Y, Xu K, Jiang T, Wu A (2010) STAT5 regulates glioma cell invasion by pathways dependent and independent of STAT5 DNA binding. Neurosci Lett 487:228–233

Catlett-Falcone R, Landowski TH, Oshiro MM, Turkson J, Levitzki A, Savino R, Ciliberto G, Moscinski L, Fernandez-Luna JL, Nunez G, Dalton WS, Jove R (1999) Constitutive activation of Stat3 signaling confers resistance to apoptosis in human U266 myeloma cells. Immunity 10:105–115

Chen Z, Lund R, Aittokallio T, Kosonen M, Nevalainen O, Lahesmaa R (2003) Identification of novel IL-4/Stat6-regulated genes in T lymphocytes. J Immunol 171:3627–3635

Chen F, Xu Y, Luo Y, Zheng D, Song Y, Yu K, Li H, Zhang L, Zhong W, Ji Y (2010) Down-regulation of Stat3 decreases invasion activity and induces apoptosis of human glioma cells. J Mol Neurosci 40:353–359

Chung CD, Liao J, Liu B, Rao X, Jay P, Berta P, Shuai K (1997) Specific inhibition of Stat3 signal transduction by PIAS3. Science 278:1803–1805

Croker BA, Kiu H, Nicholson SE (2008) SOCS regulation of the JAK/STAT signalling pathway. Semin Cell Dev Biol 19:414–422

Danial NN, Rothman P (2000) JAK-STAT signaling activated by Abl oncogenes. Oncogene 19:2523–2531

Dasgupta A, Raychaudhuri B, Haqqi T, Prayson R, Van Meir EG, Vogelbaum M, Haque SJ (2009) Stat3 activation is required for the growth of U87 cell-derived tumours in mice. Eur J Cancer 45:677–684

de la Iglesia N, Konopka G, Lim KL, Nutt CL, Bromberg JF, Frank DA, Mischel PS, Louis DN, Bonni A (2008a) Deregulation of a STAT3-interleukin 8 signaling pathway promotes human glioblastoma cell proliferation and invasiveness. J Neurosci 28:5870–5878

de la Iglesia N, Konopka G, Puram SV, Chan JA, Bachoo RM, You MJ, Levy DE, Depinho RA, Bonni A (2008b) Identification of a PTEN-regulated STAT3 brain tumor suppressor pathway. Genes Dev 22:449–462

de la Iglesia N, Puram SV, Bonni A (2009) STAT3 regulation of glioblastoma pathogenesis. Curr Mol Med 9:580–590

Decker T, Kovarik P (2000) Serine phosphorylation of STATs. Oncogene 19:2628–2637

Ehret GB, Reichenbach P, Schindler U, Horvath CM, Fritz S, Nabholz M, Bucher P (2001) DNA binding specificity of different STAT proteins. Comparison of in vitro specificity with natural target sites. J Biol Chem 276:6675–6688

Friedman HS, Bigner DD (2005) Glioblastoma multiforme and the epidermal growth factor receptor. N Engl J Med 353:1997–1999
Fuh B, Sobo M, Cen L, Josiah D, Hutzen B, Cisek K, Bhasin D, Regan N, Lin L, Chan C, Caldas H, DeAngelis S, Li C, Li PK, Lin J (2009) LLL-3 inhibits STAT3 activity, suppresses glioblastoma cell growth and prolongs survival in a mouse glioblastoma model. Br J Cancer 100:106–112
Fujio Y, Kunisada K, Hirota H, Yamauchi-Takihara K, Kishimoto T (1997) Signals through gp130 upregulate bcl-x gene expression via STAT1-binding cis-element in cardiac myocytes. J Clin Invest 99:2898–2905
Fujita M, Zhu X, Sasaki K, Ueda R, Low KL, Pollack IF, Okada H (2008) Inhibition of STAT3 promotes the efficacy of adoptive transfer therapy using type-1 CTLs by modulation of the immunological microenvironment in a murine intracranial glioma. J Immunol 180:2089–2098
Gabrusiewicz K, Ellert-Miklaszewska A, Lipko M, Sielska M, Frankowska M, Kaminska B (2011) Characteristics of the alternative phenotype of microglia/macrophages and its modulation in experimental gliomas. PLoS One 6:e23902
Galli R, Binda E, Orfanelli U, Cipelletti B, Gritti A, De Vitis S, Fiocco R, Foroni C, Dimeco F, Vescovi A (2004) Isolation and characterization of tumorigenic, stem-like neural precursors from human glioblastoma. Cancer Res 64:7011–7021
Gao L, Li F, Dong B, Zhang J, Rao Y, Cong Y, Mao B, Chen X (2010) Inhibition of STAT3 and ErbB2 suppresses tumor growth, enhances radiosensitivity, and induces mitochondria-dependent apoptosis in glioma cells. Int J Radiat Oncol Biol Phys 77:1223–1231
Gesbert F, Griffin JD (2000) Bcr/Abl activates transcription of the Bcl-X gene through STAT5. Blood 96:2269–2276
Greenhalgh CJ, Hilton DJ (2001) Negative regulation of cytokine signaling. J Leukoc Biol 70:348–356
Gu F, Hata R, Ma YJ, Tanaka J, Mitsuda N, Kumon Y, Hanakawa Y, Hashimoto K, Nakajima K, Sakanaka M (2005) Suppression of Stat3 promotes neurogenesis in cultured neural stem cells. J Neurosci Res 81:163–171
Gu J, Li G, Sun T, Su Y, Zhang X, Shen J, Tian Z, Zhang J (2008) Blockage of the STAT3 signaling pathway with a decoy oligonucleotide suppresses growth of human malignant glioma cells. J Neurooncol 89:9–17
Hartman SE, Bertone P, Nath AK, Royce TE, Gerstein M, Weissman S, Snyder M (2005) Global changes in STAT target selection and transcription regulation upon interferon treatments. Genes Dev 19:2953–2968
Haspel RL, Darnell JE Jr (1999) A nuclear protein tyrosine phosphatase is required for the inactivation of Stat1. Proc Natl Acad Sci U S A 96:10188–10193
Hatiboglu MA, Wei J, Wu AS, Heimberger AB (2010) Immune therapeutic targeting of glioma cancer stem cells. Target Oncol 5:217–227
Haybaeck J, Obrist P, Schindler CU, Spizzo G, Doppler W (2007) STAT-1 expression in human glioblastoma and peritumoral tissue. Anticancer Res 27:3829–3835
He B, You L, Uematsu K, Zang K, Xu Z, Lee AY, Costello JF, McCormick F, Jablons DM (2003) SOCS-3 is frequently silenced by hypermethylation and suppresses cell growth in human lung cancer. Proc Natl Acad Sci U S A 100:14133–14138
Heim MH, Kerr IM, Stark GR, Darnell JE Jr (1995) Contribution of STAT SH2 groups to specific interferon signaling by the Jak-STAT pathway. Science 267:1347–1349
Heimberger AB, Priebe W (2008) Small molecular inhibitors of p-STAT3: novel agents for treatment of primary and metastatic CNS cancers. Recent Pat CNS Drug Discov 3:179–188
Hoey T, Zhang S, Schmidt N, Yu Q, Ramchandani S, Xu X, Naeger LK, Sun YL, Kaplan MH (2003) Distinct requirements for the naturally occurring splice forms Stat4alpha and Stat4beta in IL-12 responses. EMBO J 22:4237–4248
Horvath CM, Wen Z, Darnell JE Jr (1995) A STAT protein domain that determines DNA sequence recognition suggests a novel DNA-binding domain. Genes Dev 9:984–994
Hussain SF, Kong LY, Jordan J, Conrad C, Madden T, Fokt I, Priebe W, Heimberger AB (2007) A novel small molecule inhibitor of signal transducers and activators of transcription 3 reverses immune tolerance in malignant glioma patients. Cancer Res 67:9630–9636
Ivashkiv LB, Hu X (2004) Signaling by STATs. Arthritis Res Ther 6:159–168
Iwamaru A, Szymanski S, Iwado E, Aoki H, Yokoyama T, Fokt I, Hess K, Conrad C, Madden T, Sawaya R, Kondo S, Priebe W, Kondo Y (2007) A novel inhibitor of the STAT3 pathway induces apoptosis in malignant glioma cells both in vitro and in vivo. Oncogene 26:2435–2444
Jatiani SS, Cosenza SC, Reddy MV, Ha JH, Baker SJ, Samanta AK, Olnes MJ, Pfannes L, Sloand EM, Arlinghaus RB, Reddy EP (2010) A non-ATP-competitive dual inhibitor of JAK2 and BCR-ABL kinases: elucidation of a novel therapeutic spectrum based on substrate competitive inhibition. Genes Cancer 1:331–345
John S, Vinkemeier U, Soldaini E, Darnell JE Jr, Leonard WJ (1999) The significance of tetramerization in promoter recruitment by Stat5. Mol Cell Biol 19:1910–1918
Khan KD, Shuai K, Lindwall G, Maher SE, Darnell JE Jr, Bothwell AL (1993) Induction of the Ly-6A/E gene by interferon alpha/beta and gamma requires a DNA element to which a tyrosine-phosphorylated 91-kDa protein binds. Proc Natl Acad Sci U S A 90:6806–6810
Kisseleva T, Bhattacharya S, Braunstein J, Schindler CW (2002) Signaling through the JAK/STAT pathway, recent advances and future challenges. Gene 285:1–24

Konnikova L, Kotecki M, Kruger MM, Cochran BH (2003) Knockdown of STAT3 expression by RNAi induces apoptosis in astrocytoma cells. BMC Cancer 3:23

Korpelainen EI, Karkkainen M, Gunji Y, Vikkula M, Alitalo K (1999) Endothelial receptor tyrosine kinases activate the STAT signaling pathway: mutant Tie-2 causing venous malformations signals a distinct STAT activation response. Oncogene 18:1–8

Kostianovsky AM, Maier LM, Anderson RC, Bruce JN, Anderson DE (2008) Astrocytic regulation of human monocytic/microglial activation. J Immunol 181:5425–5432

Kubo M, Ransom J, Webb D, Hashimoto Y, Tada T, Nakayama T (1997) T-cell subset-specific expression of the IL-4 gene is regulated by a silencer element and STAT6. EMBO J 16:4007–4020

Lehmann U, Schmitz J, Weissenbach M, Sobota RM, Hortner M, Friederichs K, Behrmann I, Tsiaris W, Sasaki A, Schneider-Mergener J, Yoshimura A, Neel BG, Heinrich PC, Schaper F (2003) SHP2 and SOCS3 contribute to Tyr-759-dependent attenuation of interleukin-6 signaling through gp130. J Biol Chem 278:661–671

Leong PL, Andrews GA, Johnson DE, Dyer KF, Xi S, Mai JC, Robbins PD, Gadiparthi S, Burke NA, Watkins SF, Grandis JR (2003) Targeted inhibition of Stat3 with a decoy oligonucleotide abrogates head and neck cancer cell growth. Proc Natl Acad Sci U S A 100:4138–4143

Letimier FA, Passini N, Gasparian S, Bianchi E, Rogge L (2007) Chromatin remodeling by the SWI/SNF-like BAF complex and STAT4 activation synergistically induce IL-12Rbeta2 expression during human Th1 cell differentiation. EMBO J 26:1292–1302

Li GH, Wei H, Lv SQ, Ji H, Wang DL (2010) Knockdown of STAT3 expression by RNAi suppresses growth and induces apoptosis and differentiation in glioblastoma stem cells. Int J Oncol 37:103–110

Liang QC, Xiong H, Zhao ZW, Jia D, Li WX, Qin HZ, Deng JP, Gao L, Zhang H, Gao GD (2009) Inhibition of transcription factor STAT5b suppresses proliferation, induces G1 cell cycle arrest and reduces tumor cell invasion in human glioblastoma multiforme cells. Cancer Lett 273:164–171

Lin L, Hutzen B, Li PK, Ball S, Zuo M, DeAngelis S, Foust E, Sobo M, Friedman L, Bhasin D, Cen L, Li C, Lin J (2010) A novel small molecule, LLL12, inhibits STAT3 phosphorylation and activities and exhibits potent growth-suppressive activity in human cancer cells. Neoplasia 12:39–50

Lindemann C, Hackmann O, Delic S, Schmidt N, Reifenberger G, Riemenschneider MJ (2011) SOCS3 promoter methylation is mutually exclusive to EGFR amplification in gliomas and promotes glioma cell invasion through STAT3 and FAK activation. Acta Neuropathol 122:241–251

Liu B, Liao J, Rao X, Kushner SA, Chung CD, Chang DD, Shuai K (1998) Inhibition of Stat1-mediated gene activation by PIAS1. Proc Natl Acad Sci U S A 95:10626–10631

Lo RK, Cheung H, Wong YH (2003) Constitutively active Galpha16 stimulates STAT3 via a c-Src/JAK- and ERK-dependent mechanism. J Biol Chem 278:52154–52165

Lo HW, Cao X, Zhu H, Ali-Osman F (2008) Constitutively activated STAT3 frequently coexpresses with epidermal growth factor receptor in high-grade gliomas and targeting STAT3 sensitizes them to Iressa and alkylators. Clin Cancer Res 14:6042–6054

Look DC, Pelletier MR, Tidwell RM, Roswit WT, Holtzman MJ (1995) Stat1 depends on transcriptional synergy with Sp1. J Biol Chem 270:30264–30267

Lord JD, McIntosh BC, Greenberg PD, Nelson BH (2000) The IL-2 receptor promotes lymphocyte proliferation and induction of the c-myc, bcl-2, and bcl-x genes through the trans-activation domain of Stat5. J Immunol 164:2533–2541

Lund RJ, Chen Z, Scheinin J, Lahesmaa R (2004) Early target genes of IL-12 and STAT4 signaling in th cells. J Immunol 172:6775–6782

Marrero MB, Schieffer B, Paxton WG, Heerdt L, Berk BC, Delafontaine P, Bernstein KE (1995) Direct stimulation of Jak/STAT pathway by the angiotensin II AT1 receptor. Nature 375:247–250

Martini M, Pallini R, Luongo G, Cenci T, Lucantoni C, Larocca LM (2008) Prognostic relevance of SOCS3 hypermethylation in patients with glioblastoma multiforme. Int J Cancer 123:2955–2960

Mizoguchi M, Betensky RA, Batchelor TT, Bernay DC, Louis DN, Nutt CL (2006) Activation of STAT3, MAPK, and AKT in malignant astrocytic gliomas: correlation with EGFR status, tumor grade, and survival. J Neuropathol Exp Neurol 65:1181–1188

Nagel-Wolfrum K, Buerger C, Wittig I, Butz K, Hoppe-Seyler F, Groner B (2004) The interaction of specific peptide aptamers with the DNA binding domain and the dimerization domain of the transcription factor Stat3 inhibits transactivation and induces apoptosis in tumor cells. Mol Cancer Res 2:170–182

Naik SM, Shibagaki N, Li LJ, Quinlan KL, Paxton LL, Caughman SW (1997) Interferon gamma-dependent induction of human intercellular adhesion molecule-1 gene expression involves activation of a distinct STAT protein complex. J Biol Chem 272:1283–1290

Ni Z, Bremner R (2007) Brahma-related gene 1-dependent STAT3 recruitment at IL-6-inducible genes. J Immunol 178:345–351

Nicholson SE, De Souza D, Fabri LJ, Corbin J, Willson TA, Zhang JG, Silva A, Asimakis M, Farley A, Nash AD, Metcalf D, Hilton DJ, Nicola NA, Baca M (2000) Suppressor of cytokine signaling-3 preferentially binds to the SHP-2-binding site on the shared cytokine receptor subunit gp130. Proc Natl Acad Sci U S A 97:6493–6498

Nishikawa R, Ji XD, Harmon RC, Lazar CS, Gill GN, Cavenee WK, Huang HJ (1994) A mutant epidermal growth factor receptor common in human glioma confers enhanced tumorigenicity. Proc Natl Acad Sci U S A 91:7727–7731

Niwa H, Burdon T, Chambers I, Smith A (1998) Self-renewal of pluripotent embryonic stem cells is mediated via activation of STAT3. Genes Dev 12:2048–2060

O'Shea JJ, Gadina M, Schreiber RD (2002) Cytokine signaling in 2002: new surprises in the Jak/Stat pathway. Cell 109(Suppl):S121–S131

O'Sullivan A, Chang HC, Yu Q, Kaplan MH (2004) STAT4 is required for interleukin-12-induced chromatin remodeling of the CD25 locus. J Biol Chem 279:7339–7345

Ouchi T, Lee SW, Ouchi M, Aaronson SA, Horvath CM (2000) Collaboration of signal transducer and activator of transcription 1 (STAT1) and BRCA1 in differential regulation of IFN-gamma target genes. Proc Natl Acad Sci U S A 97:5208–5213

Pesu M, Muul L, Kanno Y, O'Shea JJ (2006) Proprotein convertase furin is preferentially expressed in T helper 1 cells and regulates interferon gamma. Blood 108:983–985

Premkumar DR, Jane EP, Agostino NR, Scialabba JL, Pollack IF (2010) Dasatinib synergizes with JSI-124 to inhibit growth and migration and induce apoptosis of malignant human glioma cells. J Carcinog 9:7

Rahaman SO, Harbor PC, Chernova O, Barnett GH, Vogelbaum MA, Haque SJ (2002) Inhibition of constitutively active Stat3 suppresses proliferation and induces apoptosis in glioblastoma multiforme cells. Oncogene 21:8404–8413

Rajan P, McKay RD (1998) Multiple routes to astrocytic differentiation in the CNS. J Neurosci 18:3620–3629

Ramana CV, Chatterjee-Kishore M, Nguyen H, Stark GR (2000) Complex roles of Stat1 in regulating gene expression. Oncogene 19:2619–2627

Raz R, Lee CK, Cannizzaro LA, D'Eustachio P, Levy DE (1999) Essential role of STAT3 for embryonic stem cell pluripotency. Proc Natl Acad Sci U S A 96:2846–2851

Reddy EP, Korapati A, Chaturvedi P, Rane S (2000) IL-3 signaling and the role of Src kinases, JAKs and STATs: a covert liaison unveiled. Oncogene 19:2532–2547

Schaefer LK, Ren Z, Fuller GN, Schaefer TS (2002) Constitutive activation of Stat3alpha in brain tumors: localization to tumor endothelial cells and activation by the endothelial tyrosine kinase receptor (VEGFR-2). Oncogene 21:2058–2065

Schiavone D, Avalle L, Dewilde S, Poli V (2011) The immediate early genes Fos and Egr1 become STAT1 transcriptional targets in the absence of STAT3. FEBS Lett 585:2455–2460

Schmidt D, Muller S (2003) PIAS/SUMO: new partners in transcriptional regulation. Cell Mol Life Sci 60:2561–2574

Senft C, Priester M, Polacin M, Schroder K, Seifert V, Kogel D, Weissenberger J (2010) Inhibition of the JAK-2/STAT3 signaling pathway impedes the migratory and invasive potential of human glioblastoma cells. J Neurooncol 101:393–403

Shen J, Li R, Li G (2009) Inhibitory effects of decoy-ODN targeting activated STAT3 on human glioma growth in vivo. In Vivo 23:237–243

Sherry MM, Reeves A, Wu JK, Cochran BH (2009) STAT3 is required for proliferation and maintenance of multipotency in glioblastoma stem cells. Stem Cells 27:2383–2392

Shuai K, Liu B (2005) Regulation of gene-activation pathways by PIAS proteins in the immune system. Nat Rev Immunol 5:593–605

Singh SK, Clarke ID, Terasaki M, Bonn VE, Hawkins C, Squire J, Dirks PB (2003) Identification of a cancer stem cell in human brain tumors. Cancer Res 63:5821–5828

Singh SK, Hawkins C, Clarke ID, Squire JA, Bayani J, Hide T, Henkelman RM, Cusimano MD, Dirks PB (2004) Identification of human brain tumour initiating cells. Nature 432:396–401

Socolovsky M, Fallon AE, Wang S, Brugnara C, Lodish HF (1999) Fetal anemia and apoptosis of red cell progenitors in Stat5a–/–5b–/– mice: a direct role for Stat5 in Bcl-X(L) induction. Cell 98:181–191

Stea B, Falsey R, Kislin K, Patel J, Glanzberg H, Carey S, Ambrad AA, Meuillet EJ, Martinez JD (2003) Time and dose-dependent radiosensitization of the glioblastoma multiforme U251 cells by the EGF receptor tyrosine kinase inhibitor ZD1839 ('Iressa'). Cancer Lett 202:43–51

Stephanou A, Latchman DS (2005) Opposing actions of STAT-1 and STAT-3. Growth Factors 23:177–182

Stupp R, Mason WP, van den Bent MJ, Weller M, Fisher B, Taphoorn MJ, Belanger K, Brandes AA, Marosi C, Bogdahn U, Curschmann J, Janzer RC, Ludwin SK, Gorlia T, Allgeier A, Lacombe D, Cairncross JG, Eisenhauer E, Mirimanoff RO (2005) Radiotherapy plus concomitant and adjuvant temozolomide for glioblastoma. N Engl J Med 352:987–996

Su Y, Li G, Zhang X, Gu J, Zhang C, Tian Z, Zhang J (2008) JSI-124 inhibits glioblastoma multiforme cell proliferation through G(2)/M cell cycle arrest and apoptosis augment. Cancer Biol Ther 7:1243–1249

Takeda K, Kaisho T, Yoshida N, Takeda J, Kishimoto T, Akira S (1998) Stat3 activation is responsible for IL-6-dependent T cell proliferation through preventing apoptosis: generation and characterization of T cell-specific Stat3-deficient mice. J Immunol 161:4652–4660

Thieu VT, Yu Q, Chang HC, Yeh N, Nguyen ET, Sehra S, Kaplan MH (2008) Signal transducer and activator of transcription 4 is required for the transcription factor T-bet to promote T helper 1 cell-fate determination. Immunity 29:679–690

Turkson J (2004) STAT proteins as novel targets for cancer drug discovery. Expert Opin Ther Targets 8:409–422
Turkson J, Bowman T, Garcia R, Caldenhoven E, De Groot RP, Jove R (1998) Stat3 activation by Src induces specific gene regulation and is required for cell transformation. Mol Cell Biol 18:2545–2552
Turkson J, Bowman T, Adnane J, Zhang Y, Djeu JY, Sekharam M, Frank DA, Holzman LB, Wu J, Sebti S, Jove R (1999) Requirement for Ras/Rac1-mediated p38 and c-Jun N-terminal kinase signaling in Stat3 transcriptional activity induced by the Src oncoprotein. Mol Cell Biol 19:7519–7528
Turkson J, Ryan D, Kim JS, Zhang Y, Chen Z, Haura E, Laudano A, Sebti S, Hamilton AD, Jove R (2001) Phosphotyrosyl peptides block Stat3-mediated DNA binding activity, gene regulation, and cell transformation. J Biol Chem 276:45443–45455
Turkson J, Kim JS, Zhang S, Yuan J, Huang M, Glenn M, Haura E, Sebti S, Hamilton AD, Jove R (2004) Novel peptidomimetic inhibitors of signal transducer and activator of transcription 3 dimerization and biological activity. Mol Cancer Ther 3:261–269
Valente AJ, Xie JF, Abramova MA, Wenzel UO, Abboud HE, Graves DT (1998) A complex element regulates IFN-gamma-stimulated monocyte chemoattractant protein-1 gene transcription. J Immunol 161:3719–3728
Vescovi AL, Galli R, Reynolds BA (2006) Brain tumour stem cells. Nat Rev Cancer 6:425–436
Vila-Coro AJ, Rodriguez-Frade JM, Martin De Ana A, Moreno-Ortiz MC, Martinez AC, Mellado M (1999) The chemokine SDF-1alpha triggers CXCR4 receptor dimerization and activates the JAK/STAT pathway. FASEB J 13:1699–1710
Villalva C, Martin-Lanneree S, Cortes U, Dkhissi F, Wager M, Le Corf A, Tourani JM, Dusanter-Fourt I, Turhan AG, Karayan-Tapon L (2010) STAT3 is essential for the maintenance of neurosphere-initiating tumor cells in patients with glioblastomas: a potential for targeted therapy? Int J Cancer 128:826–838
Wagner BJ, Hayes TE, Hoban CJ, Cochran BH (1990) The SIF binding element confers sis/PDGF inducibility onto the c-fos promoter. EMBO J 9:4477–4484
Wang H, Zhang W, Huang HJ, Liao WS, Fuller GN (2004) Analysis of the activation status of Akt, NFkappaB, and Stat3 in human diffuse gliomas. Lab Invest 84:941–951
Wang H, Lathia JD, Wu Q, Wang J, Li Z, Heddleston JM, Eyler CE, Elderbroom J, Gallagher J, Schuschu J, MacSwords J, Cao Y, McLendon RE, Wang XF, Hjelmeland AB, Rich JN (2009) Targeting interleukin 6 signaling suppresses glioma stem cell survival and tumor growth. Stem Cells 27:2393–2404
Wei J, Barr J, Kong LY, Wang Y, Wu A, Sharma AK, Gumin J, Henry V, Colman H, Sawaya R, Lang FF, Heimberger AB (2010) Glioma-associated cancer-initiating cells induce immunosuppression. Clin Cancer Res 16:461–473
Weissenberger J, Loeffler S, Kappeler A, Kopf M, Lukes A, Afanasieva TA, Aguzzi A, Weis J (2004) IL-6 is required for glioma development in a mouse model. Oncogene 23:3308–3316
Weissenberger J, Priester M, Bernreuther C, Rakel S, Glatzel M, Seifert V, Kogel D (2010) Dietary curcumin attenuates glioma growth in a syngeneic mouse model by inhibition of the JAK1,2/STAT3 signaling pathway. Clin Cancer Res 16:5781–5795
Wen Z, Zhong Z, Darnell JE Jr (1995) Maximal activation of transcription by Stat1 and Stat3 requires both tyrosine and serine phosphorylation. Cell 82:241–250
Wirnsberger G, Hebenstreit D, Posselt G, Horejs-Hoeck J, Duschl A (2006) IL-4 induces expression of TARC/CCL17 via two STAT6 binding sites. Eur J Immunol 36:1882–1891
Wong M, Fish EN (1998) RANTES and MIP-1alpha activate stats in T cells. J Biol Chem 273:309–314
Wormald S, Hilton DJ, Smyth GK, Speed TP (2006) Proximal genomic localization of STAT1 binding and regulated transcriptional activity. BMC Genomics 7:254
Yang F, Brown C, Buettner R, Hedvat M, Starr R, Scuto A, Schroeder A, Jensen M, Jove R (2010) Sorafenib induces growth arrest and apoptosis of human glioblastoma cells through the dephosphorylation of signal transducers and activators of transcription 3. Mol Cancer Ther 9:953–962
Yiin JJ, Hu B, Schornack PA, Sengar RS, Liu KW, Feng H, Lieberman FS, Chiou SH, Sarkaria JN, Wiener EC, Ma HI, Cheng SY (2010) ZD6474, a multitargeted inhibitor for receptor tyrosine kinases, suppresses growth of gliomas expressing an epidermal growth factor receptor mutant, EGFRvIII, in the brain. Mol Cancer Ther 9:929–941
Yoshimura A (2005) Negative regulation of cytokine signaling. Clin Rev Allergy Immunol 28:205–220
Yu H, Jove R (2004) The STATs of cancer–new molecular targets come of age. Nat Rev Cancer 4:97–105
Yu CL, Meyer DJ, Campbell GS, Larner AC, Carter-Su C, Schwartz J, Jove R (1995) Enhanced DNA-binding activity of a Stat3-related protein in cells transformed by the Src oncoprotein. Science 269:81–83
Yu H, Kortylewski M, Pardoll D (2007) Crosstalk between cancer and immune cells: role of STAT3 in the tumour microenvironment. Nat Rev Immunol 7:41–51
Yu H, Pardoll D, Jove R (2009) STATs in cancer inflammation and immunity: a leading role for STAT3. Nat Rev Cancer 9:798–809
Yuan X, Curtin J, Xiong Y, Liu G, Waschsmann-Hogiu S, Farkas DL, Black KL, Yu JS (2004) Isolation of cancer stem cells from adult glioblastoma multiforme. Oncogene 23:9392–9400
Yue P, Turkson J (2009) Targeting STAT3 in cancer: how successful are we? Expert Opin Investig Drugs 18:45–56

Zhang F, Li C, Halfter H, Liu J (2003) Delineating an oncostatin M-activated STAT3 signaling pathway that coordinates the expression of genes involved in cell cycle regulation and extracellular matrix deposition of MCF-7 cells. Oncogene 22:894–905

Zhang Y, Cheng MB, Zhang YJ, Zhong X, Dai H, Yan L, Wu NH, Shen YF (2010) A switch from hBrm to Brg1 at IFNgamma-activated sequences mediates the activation of human genes. Cell Res 20:1345–1360

Zhou H, Miki R, Eeva M, Fike FM, Seligson D, Yang L, Yoshimura A, Teitell MA, Jamieson CA, Cacalano NA (2007) Reciprocal regulation of SOCS 1 and SOCS3 enhances resistance to ionizing radiation in glioblastoma multiforme. Clin Cancer Res 13:2344–2353

Chapter 11
Cannabinoid Signaling in Glioma Cells

Aleksandra Ellert-Miklaszewska, Iwona Ciechomska, and Bozena Kaminska

Abstract Cannabinoids are a group of structurally heterogeneous but pharmacologically related compounds, including plant-derived cannabinoids, synthetic substances and endogenous cannabinoids, such as anandamide and 2-arachidonoylglycerol. Cannabinoids elicit a wide range of central and peripheral effects mostly mediated through cannabinoid receptors. There are two types of specific $G_{i/o}$-protein-coupled receptors cloned so far, called CB1 and CB2, although an existence of additional cannabinoid-binding receptors has been suggested. CB1 and CB2 differ in their predicted amino acid sequence, tissue distribution, physiological role and signaling mechanisms. Significant alterations of a balance in the cannabinoid system between the levels of endogenous ligands and their receptors occur during malignant transformation in various types of cancer, including gliomas. Cannabinoids exert anti-proliferative action in tumor cells. Induction of cell death by cannabinoid treatment relies on the generation of a pro-apoptotic sphingolipid ceramide and disruption of signaling pathways crucial for regulation of cellular proliferation, differentiation or apoptosis. Increased ceramide levels lead also to ER-stress and autophagy in drug-treated glioblastoma cells.

Keywords Cannabinoids • Apoptosis • Autophagy • ER-stress • Gliomas

Abbreviations

AEA	Anandamide, arachidonoylethanolamide
2-AG	2-Arachidonoylglycerol
Akt	Protein kinase B/Akt
ATF4	Activating transcription factor 4
cAMP	Cylic adenosine monophosphate
CB1	Cannabinoid receptor type 1
CB2	Cannabinoid receptor type 2
CHOP	The C/EBP-homologous protein

A. Ellert-Miklaszewska (✉) • I. Ciechomska • B. Kaminska
Laboratory of Transcription Regulation, Department of Cell Biology,
Nencki Institute of Experimental Biology, Polish Academy of Sciences,
3, Pasteur Str., PL 02-093 Warsaw, Poland
e-mail: aellert@nencki.gov.pl; jcjech@nencki.gov.pl; bozenakk@nencki.gov.pl

J. Barańska (ed.), *Glioma Signaling*, Advances in Experimental Medicine and Biology 986,
DOI 10.1007/978-94-007-4719-7_11, © Springer Science+Business Media Dordrecht 2013

DAG	Diacylglycerol
eIF2α	Eukaryotic translation initiation factor 2α
ER	Endoplasmic reticulum
ERK1/2	Extracellular signal-regulated kinase 1/2
FAAH	Fatty acid amide hydrolase
IP_3	Inositol 1,4,5-trisphosphate
LPS	Lipopolysacharide
MAPK	Mitogen-activated protein kinase
MEK	MAP kinase-ERK kinase
MGL	Monoacylglycerol lipase
mTORC1	Mammalian target of rapamycin, complex 1
NAPE	N-Arachidonylphosphatidylethanolamide
PI3K	Phosphatidylinositol 3-kinase
PIP_2	Phosphatidylinositol 4,5-bisphosphate
PKA	Protein kinase A
PKC	Protein kinase C
PLC	Phospholipase C
Δ^9-THC	(−)-*trans*-Δ^9-Tetrahydrocannabinol
TRB3	Pseudo-kinase tribbles homologue 3
TRPV1	Transient receptor potential cation channel subfamily V member 1, capsaicin or vanilloid receptor
WIN55,212-2	Synthetic cannabinoid

11.1 Introduction

Preparations from *Cannabis sativa*, the hemp plant, have been used for centuries for both medicinal and recreational purposes (Howlett et al. 2002; Mackie 2006). Isolation of the active components of the plant, called cannabinoids, in 1960s, as much as subsequent cloning of cannabinoid receptors, discovery of their endogenous ligands and development of synthetic cannabinoids contributed to an intensive burst of cannabinoid research. Along with our expanding comprehension of mechanisms of cannabinoids action, targeting cannabinoid signaling for therapeutic purposes has inevitably emerged as an interesting area of scientific and clinical investigations.

One of the most extensively studied applications of cannabinoids is their potential use as anti-cancer agents. Anti-proliferative effects of cannabinoids have been reported in various cultured cancer cells, including neural, breast, prostate, skin, thyroid cancer cells and leukemia cells. Several studies demonstrated anti-tumor activity of cannabinoids in animal models (Guzman 2003). Since the first publication by Sanchez and co-workers in 1998, providing evidence that cannabinoids are effective in inducing glioma cell death, a growing interest of several groups, including ours, has been focused on understanding of molecular mechanisms of cannabinoid signaling in glioma cells and therapeutic potential of cannabinoids in glial tumors (Sanchez et al. 1998).

Signaling pathways and intracellular processes underlying cannabinoid action on glioma cells are reviewed here. Due to the number of studies carried out in the recent years on mechanisms of anti-proliferative effects of cannabinoids, these mechanisms are described in details. A brief overview of the endocannabinoid system, canonical signal transduction pathways coupled to the activation of cannabinoid receptors and current scientific data relevant to the use of cannabinoids in treatments of glioblastomas are also included.

11.2 Cannabinoids and Their Receptors

Cannabinoids are a group of structurally heterogeneous but pharmacologically related compounds classified into three subtypes: plant-derived, synthetic and endogenous cannabinoids (Fig. 11.1). Plant-derived cannabinoids (phytocannabinoids) are uniquely found in the cannabis plant. Although the pharmacology of the majority of them is unknown, (−)-*trans*-Δ^9-tetrahydrocannabinol (Δ^9-THC) is recognized as the most potent out of approximately 70 identified phytocannabinoids. Various modifications of the chemical structure of natural cannabinoids led to generation of a still growing set of synthetic cannabinoids. Exogenous cannabinoids mimic the action of endogenous compounds, known as endocannabinoids, ubiquitously produced in both vertebrate and invertebrate tissues.

Two arachidonic acid derivatives, arachidonoylethanolamide (anandamide, AEA) and 2-arachidonoylglycerol (2-AG) are the best characterized endogenous cannabinoids, although some other amides and esters of long chain polyunsaturated fatty acids also exhibit cannabimimetic properties. Endocannabinoids are produced as rapidly inactivated lipid mediators and their levels are strictly controlled by a transporter system and hydrolyzing enzymes (Fig. 11.2). Biosynthesis of endocannabinoids is activated "on demand". AEA is produced from lipid membrane precursors upon the stimuli that increase intracellular concentrations of calcium. Increased calcium levels activate N-acyltransferase, the enzyme generating an AEA precursor N-arachidonylphosphatidylethanolamide (NAPE), which is then hydrolyzed by NAPE phospholipase D. Separate mechanisms have been described for 2-AG, which is most likely produced through the phospholipase C/diacylglycerol lipase pathway.

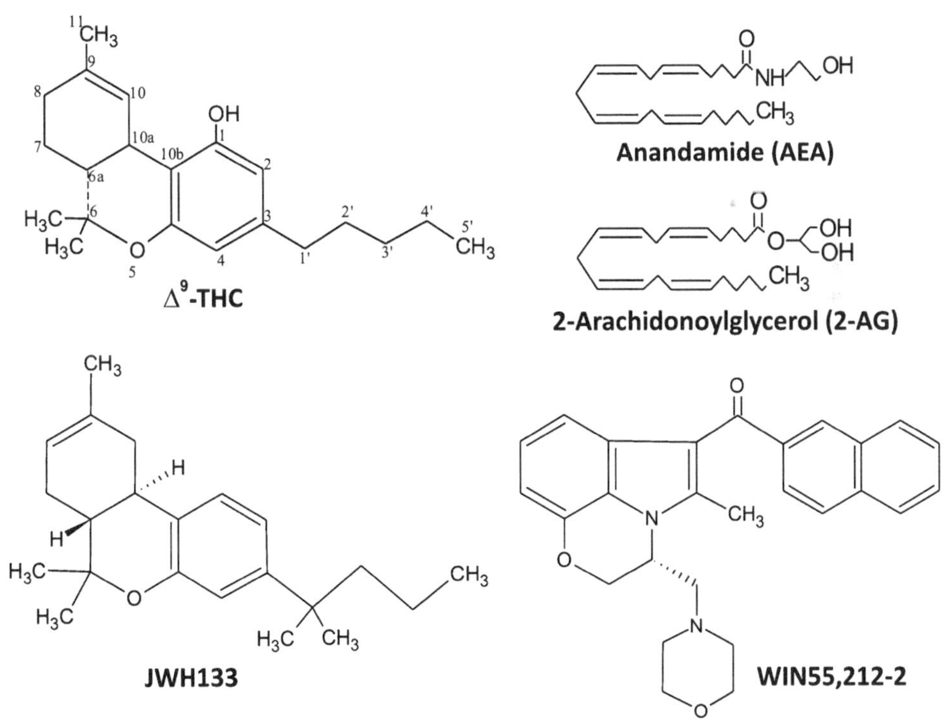

Fig. 11.1 Chemical structures of cannabinoids. Plant-derived Δ^9-THC ((−)-*trans*-Δ^9-tetrahydrocannabinol), endocannabinoids: anandamide (N-arachidonoylethanolamine) and 2-arachidonoylglycerol, synthetic cannabinoids JWH133, selective for non-psychoactive CB2 receptor ((6aR,10aR)-3-(1,1-Dimethylbutyl)-6a,7,10,10a-tetrahydro-6,6,9-trimethyl-6H-dibenzo[b,d] pyran), and WIN55,212-2, a CB1/CB2 receptor agonist ((R)-(+)-[2,3-Dihydro-5-methyl-3-(4-morpholinylmethyl) pyrrolo[1,2,3-de]-1,4-benzoxazin-6-yl]-1-naphthalenylmethanone)

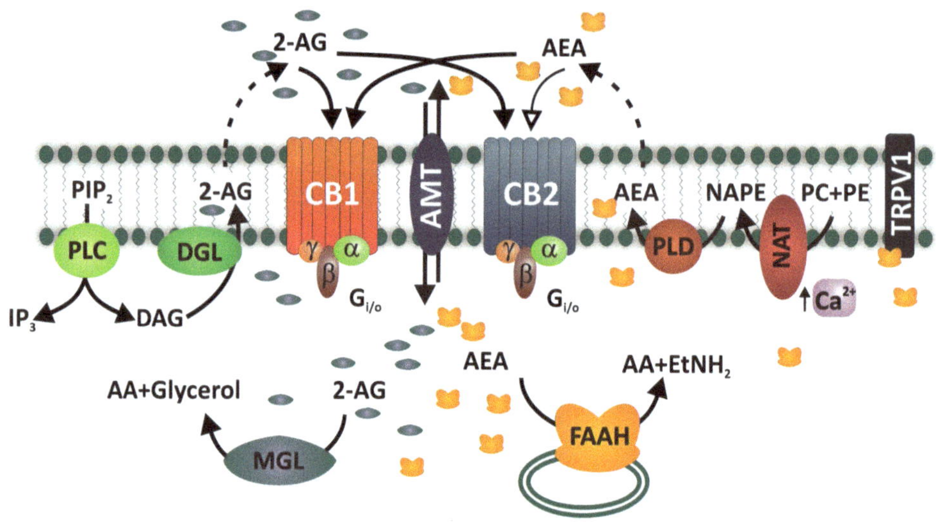

Fig. 11.2 The endocannabinoid system. Two arachidonic acid derivatives, anandamide (AEA) and 2-arachidonoyl glycerol (2-AG), are produced by cleavage of lipid precursors in response to increased calcium levels or receptor stimulation. AEA is generated from *N*-arachidonoylphosphatidylethanolamine (NAPE). NAPE originates from transfer of arachidonic acid from the *sn*-1 position of 1,2-*sn*-diarachidonoylphosphatidylcholine (PC) to phosphatidylethanolamine (PE), catalyzed by a Ca^{2+}-dependent *N*-acyltransferase (NAT). NAPE is cleaved by phospholipase D (PLD) to release AEA and phosphatidic acid. 2-AG is synthesized in two steps via generation of 1-acyl-2-arachidonoylglycerol (diacylglycerol, or DAG) form phospholipids (such as phosphatidylinositol-4,5-bisphosphate PIP_2) by phospholipase C (PLC) and subsequent hydrolysis of DAG by a diacylglycerol lipase (DGL). AEA and 2-AG are rapidly removed from the extracellular space possibly through a common purported high-affinity transporter (AMT), a carrier possibly working in both inward and outward directions. Once taken up by cells, AEA is a substrate for the fatty acid amide hydrolase (FAAH), which breaks the amide bond and releases arachidonic acid (AA) and ethanolamine ($EtNH_2$). 2-AG is primarily degraded by a specific monoacylglycerol lipase (MGL). Both AEA and 2-AG bind to and activate CB1 and CB2 receptors; however, AEA is a weaker agonist than 2-AG at CB1 and is only a partial agonist at CB2. The transient receptor potential cation channel subfamily V member 1 (TRPV1) is another key molecular target of AEA, but importantly not of 2-AG. The binding site of TRPV1 receptors for AEA is on an intracellular domain. All elements of the ECS are located in the plasma membrane except from FAAH, which is bound to intracellular membranes, and MGL, which is cytosolic

Endocannabinoid production and release by glial cells may underlie the neuroprotective properties of cannabinoids in experimental models, however its physiological significance is still unknown.

Although endocannabinoids, lipophilic in nature, can freely cross cell membranes, evidence suggests the existence of mechanisms facilitating endocannabinoid internalization. Endocannabinoids are transported into cells by a purported high-affinity membrane transporter to undergo enzymatic hydrolysis. Two degradative enzymes for endocannabinoids have been described so far: the fatty acid amide hydrolase (FAAH) and monoacylglycerol lipase (MGL), responsible mainly for AEA and 2-AG degradation, respectively (Fig. 11.2). FAAH and MGL exhibit a wide distribution in the rat and human central nervous system, but are mostly confined to neurons. FAAH is primarily a postsynaptic enzyme, whereas MGL localizes to presynaptic terminals. There are conflicting data on the expression of endocannabinoid metabolizing enzymes in glial cells in normal brain. FAAH is increased in hypertrophic astrocytes surrounding AD plaques or inflammatory infiltrates but is not detected in other types of glial cells (Pazos et al. 2005). MGL can be found in oligodendrocytes at different developmental stages, however in cultured oligodendrocyte progenitor cells (OPCs) MGL expression was lower than during maturation stages, suggesting that the increased levels of 2-AG are required for OPCs differentiation (Gomez et al. 2010).

Cannabinoids elicit a wide range of central and peripheral effects, which are mediated mostly through cannabinoid receptors (Howlett et al. 2002). There are two types of specific seven-transmembrane, $G_{i/o}$-protein-coupled receptors cloned so far, called CB1 and CB2 (Figs. 11.2 and 11.3), although an existence of additional cannabinoid-binding receptors has been suggested (Howlett et al. 2002; Stella 2004). CB1 and CB2 differ in their predicted amino acid sequence, tissue distribution and expression pattern. Overall homology between CB1 and CB2 is remarkably low (e.g. 44 % for human and 68 % for murine receptors) with significant disparities in the domains, which interact with G proteins and effector proteins. Thus, quite expectedly two cannabinoid receptors were shown to play distinct physiological roles and share only some common signaling mechanisms.

Many of the effects of cannabinoids rely on the CB1 receptor activation. CB1 is particularly abundant in discrete areas of the brain, such as cortex, hippocampus, basal ganglia and cerebellum, as well as in peripheral nerve terminals, where it mediates inhibition of neurotransmitter release and is involved in the control of motor activity, memory, cognition appetite and sensory perception. CB1 is also present in some extra-neural sites, such as testis, uterus, vascular endothelium, eye, spleen, and tonsils, controlling processes such as vascular tone, intraocular pressure and immune response (Howlett et al. 2002). Glial cells have been shown to express CB1 receptors, although their precise function in astrocytes, oligodendrocytes and microglia has been only partly unveiled.

By contrast, the CB2 receptor is predominantly expressed in cells and organs of the immune system (Howlett et al. 2002). The role of peripheral CB2 receptor activation under physiologic conditions is not well defined. CB2 signaling is involved in B-cell differentiation and modulation of immune response. Increased levels of CB2 are reported in tissues during development, inflammation, injury and cancer, revealing a critical role for the CB2 receptor in regulating these processes (Howlett et al. 2002). The CB2 receptor was believed to be absent from healthy brain, however, its expression has been detected in microglia – brain macrophages (Gong et al. 2006; Stella 2004), as well as in a small subpopulation of neurons (Gong et al. 2006; Van Sickle et al. 2005). Animal experiments show that CB2-selective agonists do not induce widespread psychoactive effects attributed to activation of the CB1 receptor (Guzman 2003; Valenzano et al. 2005).

Extensive molecular studies have demonstrated that activation of the CB1 and CB2 cannabinoid receptors upon agonist binding is canonically linked to inhibition of adenylyl cyclase via the α-subunit of $G_{i/o}$-protein. The consequent decrease in cyclic AMP (cAMP) production leads to down-regulation of protein kinase A (PKA) and impedes PKA – dependent signaling. The CB1 receptor, acting as a guardian on presynaptic membranes, is coupled to ion channels, inducing for example inhibition of voltage-gated L, N- and P/Q Ca^{2+} channels and activation of G-protein activated inwardly rectifying K^+ channels (Howlett et al. 2002). Cannabionoids have been also reported to affect several pathways that are more directly involved in the control of cell proliferation, differentiation and survival. Depending on a cell type and treatment conditions, signaling via cannabinoid receptors is linked to activation or inhibition of phosphatidylinositol 3-kinase (PI3K)/Akt pathway. Stimulation of either CB1 or CB2 receptor results in activation of three classes of mitogen-activated protein kinases (MAPKs): extracellular signal-regulated kinase (ERK), c-Jun N-terminal kinase (JNK) and p38 MAPK. Cannabinoid receptors are also coupled to activation of phospholipase C and a subsequent release of Ca^{2+} from inositol-1,4,5-trisphosphate (IP_3)-sensitive stores, as well as to modulation of the sphingomyelin cycle (Guzman 2003; Howlett et al. 2002).

There is emerging evidence suggesting additional pharmacological targets for cannabinoids. Peroxisome proliferator-activated receptors (PPARs) and transient receptor potential cation channel subfamily V member 1 (TRPV1, also known as capsaicin or vanilloid receptor) have been shown to be activated by cannabinoids, including Δ^9-THC and endogenous AEA. However, the precise role of these receptors in cannabinoid signaling is still unclear (Mackie and Stella 2006).

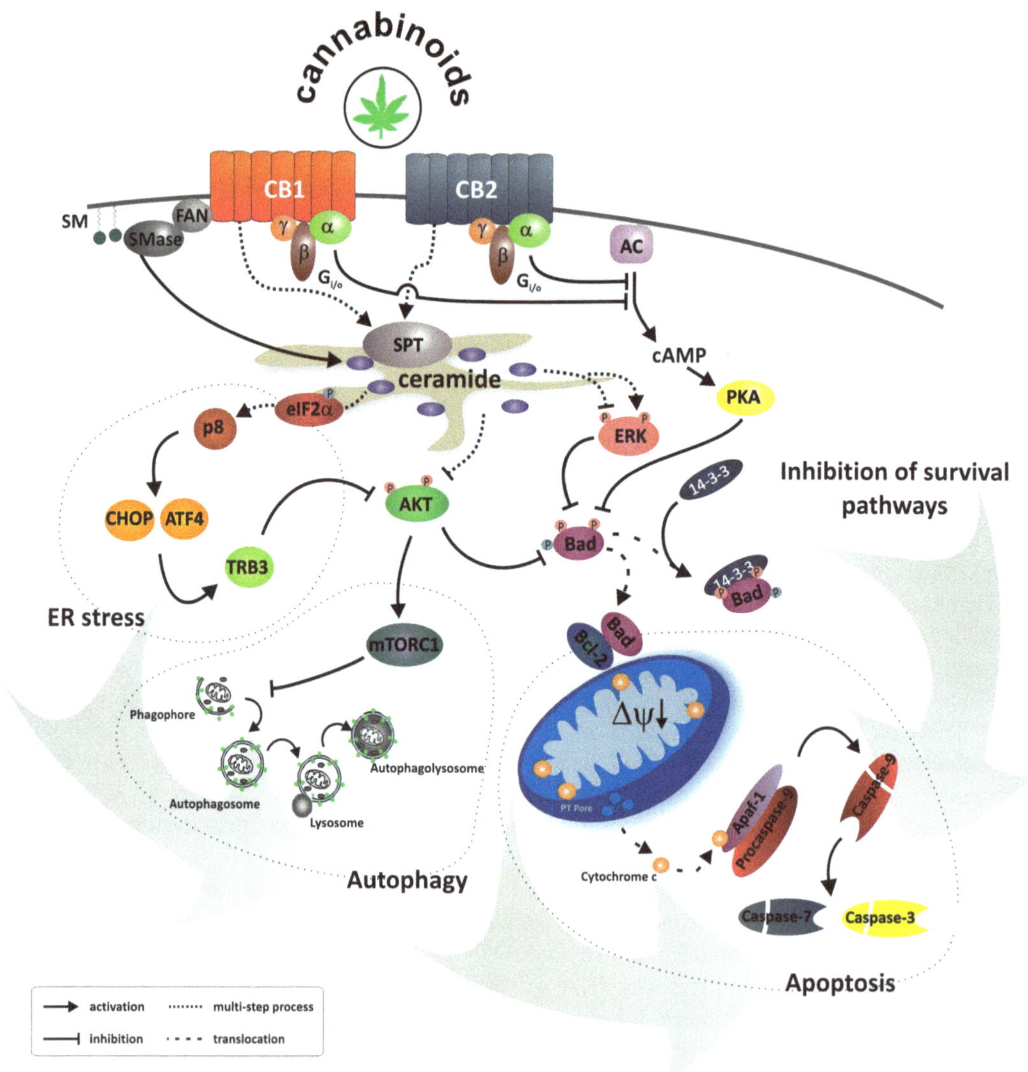

Fig. 11.3 Mechanisms of anti-tumoral action of cannabinoids. Increased ceramide synthesis *de novo* via induction of serine palmitoyltransferase (SPT) plays a central role in cannabinoid-induced cell death. Signals from CB1 via adaptor protein FAN (factor associated with neutral sphingomyelinase activation) trigger also the ceramide production from sphingomyelin breakdown, catalysed by neutral sphingomyelinase (SMase). Stimulation of cannabinoids receptors leads to inhibition of adenylyl cyclase (AC), reduction of cAMP levels and decreased protein kinase A (PKA) activity. Inhibition of PKA and pro-survival pathways (Akt and ERK signaling) stimulates translocation of Bad to the outer mitochondrial membrane and its pro-apoptotic function. Interaction between Bad and Bcl-2 triggers a decrease of the mitochondrial membrane potential (ΔΨ) and release of pro-apoptotic factors (such as cytochrome c) to the cytosol, where apoptosis is executed by caspase cascade. Alternatively, induction of apoptosis by cannabinoids can be mediated by ER (endoplasmic reticulum)-stress and autophagy. Cannabinoid-induced phosphorylation of eIF2α (eukaryotic translation initiation factor 2α) and subsequent up-regulation of the stress-regulated protein p8 and ER-stress-related downstream targets: ATF4 (activating transcription factor 4), CHOP (the C/EBP-homologous protein) and TRB3 (tribbles homologue 3) leads to inhibition of Akt, an upstream activator of mTORC1. Decreased activity of Akt/mTORC1 pathway contributes to initiation of autophagy, that precedes apoptosis of glioma cell

11.3 Cannabinoid System in Gliomas

Endocannabinoids, their receptors and specific machinery involved in biosynthesis, uptake and degradation, constitute the endocannabinoid system (ECS). A growing array of data suggests that alterations of a balance in the cannabinoid system between the levels of endogenous ligands and their receptors occur during malignant transformation in various types of cancer, including gliomas. While non-transformed astrocytes express only the CB1 cannabinoid receptor, both types of functional cannabinoid receptors have been found in several established human glioblastoma cell lines, as well as in primary cultures derived from the most malignant brain tumor, *glioblastoma multiforme* (GBM) (Galve-Roperh et al. 2000; Howlett et al. 2002; Sanchez et al. 2001). Immunohistochemical analysis of low and high grade human glioma surgical specimens revealed increased CB2 receptor expression in tumor cells, invading microglia/macrophages and endothelial cells of the tumor blood vessels, as compared to non-tumor brain samples (Ellert-Miklaszewska et al. 2007; Sanchez et al. 2001; Schley et al. 2009). We detected the presence of CB2 receptors in all analyzed biopsies of astrocytomas and glioblastomas. The proportion of malignant tumors expressing high levels of CB2 (10 out of 16, 62.5 %) was over 2-fold higher than that seen in the tumors of lower grade (7 out of 29, 24 %) (Ellert-Miklaszewska et al. 2007). Thus, the extent of CB2 expression correlated with the tumor malignancy grade. Interestingly, some benign pediatric astrocytic tumors, such as subependymal giant cell astrocytoma (SEGA), which may occasionally cause mortality owing to progressive growth in some patients, also displayed high CB2-imunoreactivity. Moreover, as observed by Sanchez et al., CB2 receptor immunoreactivity markedly prevailed over detected CB1 receptor levels in grade IV astrocytomas (Sanchez et al. 2001). The levels of CB1 receptor expression in tumor and tumor-associated endothelial cells were not significantly different from the control tissue and showed no dependence on tumor grade (Sanchez et al. 2001; Schley et al. 2009). However, in the recent study the decreased CB1 receptor expression was detected in high grade glioblastomas as compared to non-tumoral brain tissue (De Jesus et al. 2010).

A link between CB2 expression and malignancy grade of the tumor has been reported also in prostate, breast and pancreatic cancer, and the level of CB2 expression in transformed cells was higher than in the respective normal tissue (Caffarel et al. 2006; Carracedo et al. 2006a; Sarfaraz et al. 2005). In examined human tumors, CB2 receptors were usually located in the areas of intense tissue proliferation and invading cells. The enhancement of cannabinoid receptor expression in malignant versus healthy tissues might suggest a possible role of the endocannabinoid system in the tonic suppression of cell divisions and cancer growth. This hypothesis is partly supported by the finding of increased levels of anandamide and decreased levels of endocannabinoid metabolizing enzymes (i.e. FAAH) in human glioblastoma compared to human non-tumor brain tissue (Petersen et al. 2005).

11.4 Action of Cannabinoids in Glioma Cells

Cannabinoids induce significant inhibition of cell growth in tumor cells, due to modulation of proteins and nuclear factors involved in the control of cell survival, transformation and cell death. The programmed cell death of glioma cells after cannabinoid treatment was first described by Manuel Guzman and his co-workers (Sanchez et al. 1998). They showed that Δ^9-THC is able to inhibit growth of rat C6 glioma cells *in vitro* and induce cell death with features typical for apoptosis, a programmed cell death process (Sanchez et al. 1998). We reported an apoptotic death triggered by a mixed CB1/CB2 synthetic agonist WIN 55,212-2 in rat glioma cells (Ellert-Miklaszewska et al. 2005). Recent studies (Gomez del Pulgar et al. 2002; Salazar et al. 2009) and our own observations show effectiveness of Δ^9-THC, WIN55,212-2 and a CB2-selective synthetic cannabinoid JWH133 to induce apoptosis of cultured human glioblastoma cells and tumor-derived primary cultures (unpublished).

Cannabinoids exert anti-tumor effects *in vivo* leading to a significant regression of malignant gliomas in cannabinoid-treated animals (Duntsch et al. 2006; Galve-Roperh et al. 2000; Massi et al. 2004; Sanchez et al. 2001). Local administration of Δ^9-THC or the synthetic cannabinoid, WIN55,212-2, reduced the size of tumors generated by intracranial inoculation of C6 glioma cells in rats, leading to complete eradication of gliomas and increased survival in one third of the treated rats (Galve-Roperh et al. 2000). Studies performed in mouse xenograft models with intratumoral and intraperitoneal drug administration demonstrated that non-psychoactive phytocannabinoid cannabidiol (Massi et al. 2004), a CB2-selective agonist JWH133 (Sanchez et al. 2001) or a novel synthetic cannabinoid KM-233 (Duntsch et al. 2006) blocked the proliferation of human astrocytoma cells implanted subcutaneously in the flank of immune-deficient mice. Our preliminary data suggest that systemic cannabinoid administration can effectively hamper intracranial tumor growth in rats.

11.4.1 Mechanism of Cannabinoids Pro-Apoptotic Action – Inhibition of Pro-Survival Pathways

Several events and signal transduction pathways triggered by stimulation of the CB1 and CB2 receptors have already been described to participate in the cannabinoid-induced cell death in various tumor cells (Guzman 2003; Guzman et al. 2001). They include inhibition of PKA, activation of MAPK, superoxide generation, and a strong increase in intracellular calcium concentration (Howlett et al. 2002). However, the mechanisms underlying the antitumor effects of cannabinoid receptor activation are still not well understood, and experimental data suggest that these effects may be cell-type specific.

Sanchez and coworkers showed that Δ^9-THC was able to partially antagonize the forskolin-induced elevation of intracellular cAMP concentration, but did not affect basal cAMP levels in C6 glioma cells (Sanchez et al. 1997). As described in details in Sect. 3.5, glioma C6 cells are characterized by a very low constitutive level of cAMP, and rather increased, than decreased concentration of intracellular cAMP may be responsible for inhibition of cell proliferation. It suggest, that cannabinoid receptors may be coupled to inhibition of adenylyl cyclase in glioma cells, but down-regulation of cAMP levels is unlikely to play a role in the induction of apoptosis triggered by cannabinoids in these cells. Involvement of other signaling molecules or adaptor proteins (of still mostly unknown identity) in cannabinoid receptor signaling is a subject of ongoing studies.

The best characterized mechanism of cannabinoid-induced cell death of glioma cells involves sustained accumulation of pro-apoptotic sphingolipid ceramide (Fig. 11.3), which modulates signaling pathways crucial in the control of tumor cell growth and survival (Galve-Roperh et al. 2000; Sanchez et al. 2001). Activation of the CB1 receptor triggers two peaks of ceramide generation in glioma cells (Galve-Roperh et al. 2000; Gomez del Pulgar et al. 2002; Sanchez et al. 2001). Treatment with Δ^9-THC or another CB1 receptor agonist produces a rapid release of ceramide *via* enzymatic hydrolysis of sphingomyelin from the cell membrane, catalyzed by neutral sphingomyelinase (Fig. 11.3). This effect is G-protein independent and involves the adaptor protein FAN (factor associated with neutral sphingomyelinase activation). The second ceramide peak is generated within hours or days after receptor activation and depends on increase of ceramide synthesis *de novo* via induction of serine palmitoyltransferase, a regulatory enzyme of sphingolipid biosynthesis (Gomez del Pulgar et al. 2002). Selective CB2 receptor agonists, such as JWH133, are supposed to stimulate only the ceramide synthesis process, which is sufficient to turn on the cell death program (Gomez del Pulgar et al. 2002; Sanchez et al. 2001). Thus, enhanced production of ceramide *de novo* is considered as an important event in cannabinoids-induced apoptosis. However, still little is known about the signaling pathways underlying the promotion of ceramide synthesis through cannabinoid receptor activation.

Galve-Roperh and co-workers postulated that the increased ceramide levels reported upon cannabinoid challenge led to prolonged activation of Raf-1/MEK/ERK signaling cascade and thus mediated glioma cell cycle arrest and cell death (Galve-Roperh et al. 2000). The same authors showed also that pharmacological inhibition of ceramide synthesis *de novo* prevented the inhibition of protein kinase B/Akt triggered by cannabinoids (Gomez del Pulgar et al. 2002). Our studies revealed that rather down-regulation of ERK activity, together with inhibition of PI3K/Akt pathway, contributed to rat C6 glioma cell death induced by WIN55,212-2 (Ellert-Miklaszewska et al. 2005). The serine/threonine protein kinase Akt, activated downstream of PI3K, as well as the Ras-activated Raf1/MEK/ERK pathway are widely recognized as key mediators of growth factor-promoted cell survival in gliomas (Kapoor and O'Rourke 2003). Both survival pathways converge on a small pro-apoptotic member of a Bcl-2 family of proteins, Bad. Bad is also a substrate for PKA, which is negatively linked to both cannabinoid receptors via canonical $G_{i/o}\alpha$-mediated inhibition of adenylyl cyclase and subsequent decrease of cAMP levels. Phosphorylation of Bad by Akt, ERK and PKA retains the protein in the cytosol, where it is recognized by 14-3-3 regulatory proteins and sequestered (Zha et al. 1996). Otherwise, Bad translocates to mitochondria, and formation of heterodimers between non-phosphorylated Bad and anti-apoptotic proteins, such as Bcl-X_L or Bcl-2, may result in a loss of integrity of the outer mitochondrial membrane (Zha et al. 1996). The release of cytochrome c and other pro-apoptotic proteins from mitochondria triggers the executive phase of programmed cell death. We proposed a mechanism, in which the decrease of mitogenic/pro-survival signaling evoked by the synthetic cannabinoid WIN55,212-2 promotes the pro-apoptotic function of Bad (Fig. 11.3). Accordingly, we demonstrated changes in Bad phosphorylation level followed by collapse of the mitochondrial membrane potential in C6 glioma cells treated with WIN55,212-2. These events preceded activation of caspase 9 by factors released from disrupted mitochondria, subsequent processing of effector caspases and finally oligonucleosomal DNA fragmentation.

Our further studies, as well as some published data suggest that human glioma cells treated with cannabinoids enter the suicide cell death using the same mitochondria-dependent pathway (Carracedo et al. 2006b). This mechanism contributes to the induction of apoptosis by cannabinoids also in other types of tumor cells (Velasco et al. 2007). However, involvement of an alternative, death receptor–dependent pathway in the apoptotic process triggered by these compounds cannot be ruled out.

11.4.2 The Role of ER Stress and Autophagy in Cannabinoid-Induced Cell Death

Different experimental approaches showed that the pro-apoptotic and tumor growth-inhibiting activity of cannabinoids relies on the accumulation of *de novo*-synthesized ceramide, an event that occurs in the endoplasmic reticulum (ER) and eventually leads to execution of cell death via mitochondrial pathway. A recent study by Carracedo et al. suggested a new link between the two events (Carracedo et al. 2006b). They showed that Δ^9-THC treatment of glioma cells leads to up-regulation of the transcription co-activator p8 and its ER stress-related downstream targets: ATF4 (activating transcription factor 4), CHOP (the C/EBP-homologous protein) and TRB3 (pseudo-kinase tribbles homologue 3). Selective knockdown of ATF4 and TRB3 blocked cannabinoid-induced apoptosis in glioma cells. Inhibition of ceramide synthesis *de novo* prevented Δ^9-THC-induced p8, ATF4, CHOP and TRB3 up-regulation as well as ER dilation, indicating that ceramide accumulation is an early event in the cannabinoid-triggered ER stress and apoptosis in glioma cells (Carracedo et al. 2006b). Further studies by the same group indicated that Δ^9-THC induced Ser51-phosphorylation of eukaryotic translation initiation factor 2α (eIF2α) (Salazar et al. 2009). Activated eIF2α (after phosphorylation of Ser51 by a protein kinase-like endoplasmic reticulum kinase, PERK) is known to attenuate general protein translation, while enhancing the expression of several genes related to the ER stress response (Schroder and Kaufman 2005). Experiments performed with eIF2α S51A mutant cells have shown that phosphorylation of this factor is required for the up-regulation of the stress protein p8, as well as ATF4, CHOP and TRB3, implicated in cannabinoid-evoked ER stress and glioma cell death (Fig. 11.3).

Irradiation and chemotherapeutic drugs kill cancer cells typically through induction of apoptosis. However, most cancers (including gliomas) are resistant to therapies that induce apoptosis. Autophagy is an evolutionarily conserved catabolic process, where a cell self-digests its cytoplasmic contents. Autophagy is accompanied by the progressive development of vesicle structures from autophagosomes to autolysosomes, and could be activated in response to multiple stress stimuli during cancer progression, such as nutrient starvation, the unfolded protein response (the major ER stress pathway) and hypoxia. A detailed analysis of human astrocytoma cell lines and a primary culture of human glioma cells indicated that Δ^9-THC treatment led to formation of autophagosomes in tumor cells (Salazar et al. 2009). Autophagy was preceded by ER stress and followed by apoptosis in cannabinoid-treated human and mouse cancer cells. Autophagy-mediated cell death was associated with inhibition of the Akt/mTORC1 pathway. Mammalian target of rapamycin, complex 1 (mTORC1) is the best characterized cellular stress sensor controlling some of core autophagy pathway components and considered a key step in the early triggering of autophagy (Klionsky and Emr 2000). Treatment with Δ^9-THC induced TRB3 expression and promoted its interaction with Akt. This led to decreased phosphorylation of Akt, as well as of its direct substrates TSC2 (tuberous sclerosis protein 2, tuberin) and PRAS40 (the proline-rich Akt substrate of 40 kDa), which in turn resulted in mTORC1 inhibition (Salazar et al. 2009). Δ^9-THC-treatment decreased phosphorylation of p70S6 kinase (a well-established mTORC1 downstream target) and its substrate phospho-S6 ribosomal protein. Altogether, Δ^9-THC treatment triggered the following cascade of intracellular events: up-regulation of ER-stress-related TRB3, mTORC1 inhibition and induction of autophagy, which led to glioma cell death (Fig. 11.3). Moreover, this pathway was shown to be essential for cannabinoid anti-tumoral action *in vivo*. Administration of Δ^9-THC to mice bearing tumors derived from human astrocytoma cells resulted in increased TRB3 expression, inhibition of mTOR signaling pathway, appearance of autophagy markers and caspase-3 activation. These findings indicate that cannabinoid promotes the autophagy-mediated cell death through stimulation of ER stress in human glioma cells (Salazar et al. 2009).

11.5 Therapeutic Potential of Targeting Cannabinoid Signaling in Gliomas

Standard chemotherapeutics are a double edged sword; they eliminate cancer cells but affect severely healthy cells in the body. Based on evidence from *in vitro* and *in vivo* preclinical studies, as well as from a pilot phase I clinical trial in patients with recurrent *glioblastoma multiforme*, cannabinoids appear to have a favorable safety profile and do not produce the generalized toxic effects as most conventional chemotherapeutic drugs. Δ^9-THC induced apoptosis in several cancer cell lines but showed less efficacy in nontransformed cell counterparts (Galve-Roperh et al. 2000; Guzman 2003; McAllister et al. 2005). Cannabinoids promote survival of glial cells and neurons in different models of injury, suggesting that the anti-proliferative effect of cannabinoids is selective for brain tumor cells, while viability of normal brain cells remains unaffected or even favored by cannabinoid challenge (Guzman 2003; Molina-Holgado et al. 2002). Several mechanisms could be responsible for cannabinoid compounds targeting only the cancer cells, including differences in cannabinoid signaling in glioma and normal neural cells and selective over-expression of the CB2 receptor in tumor cells. In contrast to pro-apoptotic action of Δ^9-THC and WIN55,12-2 on transformed glial cells, treatment of primary cultured astrocytes with these CB1/CB2-activating cannabinoids does not trigger ceramide generation *de novo* or induction of ER stress-related genes. In our studies administration of the CB2-selective agonist JWH133 was effective toward tumor cells, and it did not affect survival or morphology of normal astrocytes (unpublished). Negligible CB2 receptor expression in the normal brain and its abundance in high grade gliomas confer a relative safety of CB2-selective agonists for targeted glioma therapy. Moreover, the CB2 selective compounds are devoid of undesirable psychodysleptic side-effects, attributed to marijuana abuse, which are mediated by the CB1 receptor.

Due to genetic and epigenetic alterations malignant glioblastomas are highly resistant to radiation and chemotherapy. Mainstream therapeutic strategies for the management of all primary brain tumors are still mostly palliative, known to leave survivors with devastating neurological deficits and frequently with a high risk of the disease recurrence. The potency of synthetic cannabinoids to induce apoptosis in glioblastoma cells has been tested by us and others on several cell lines and primary cell cultures derived from biopsies of human tumors, which to some extent may reflect the heterogeneity of glioma molecular characteristics (Duntsch et al. 2006; Ellert-Miklaszewska et al. 2005; Galve-Roperh et al. 2000; Guzman 2003; Massi et al. 2004; McAllister et al. 2005; Sanchez et al. 2001). Thus, cannabinoids were able to override alterations of growth regulatory and apoptotic pathways, caused by common mutations reported in primary and secondary glioblastomas. Pro-apoptotic action of cannabinoids relies on the generation of ceramide and disruption of signaling pathways crucial for regulation of cellular proliferation, differentiation or apoptosis (Ellert-Miklaszewska et al. 2005; Galve-Roperh et al. 2000; Gomez del Pulgar et al. 2002; Salazar et al. 2009). Their unique mechanism of action among standard oncology remedies justifies further research on their anti-tumoral properties. A promising field to explore is to prolong the effectiveness of endocannabinoids by interference with their biosynthesis, uptake and breakdown. Therefore future studies on cannabinoid signaling system in gliomas are clearly needed.

References

Caffarel MM, Sarrio D, Palacios J, Guzman M, Sanchez C (2006) Delta9-tetrahydrocannabinol inhibits cell cycle progression in human breast cancer cells through Cdc2 regulation. Cancer Res 66:6615–6621

Carracedo A, Gironella M, Lorente M, Garcia S, Guzman M, Velasco G, Iovanna JL (2006a) Cannabinoids induce apoptosis of pancreatic tumor cells via endoplasmic reticulum stress-related genes. Cancer Res 66:6748–6755

Carracedo A, Lorente M, Egia A, Blazquez C, Garcia S, Giroux V, Malicet C, Villuendas R, Gironella M, Gonzalez-Feria L, Piris MA, Iovanna JL, Guzman M, Velasco G (2006b) The stress-regulated protein p8 mediates cannabinoid-induced apoptosis of tumor cells. Cancer Cell 9:301–312

De Jesus ML, Hostalot C, Garibi JM, Salles J, Meana JJ, Callado LF (2010) Opposite changes in cannabinoid CB1 and CB2 receptor expression in human gliomas. Neurochem Int 56:829–833

Duntsch C, Divi MK, Jones T, Zhou Q, Krishnamurthy M, Boehm P, Wood G, Sills A, Ii BM (2006) Safety and efficacy of a novel cannabinoid chemotherapeutic, KM-233, for the treatment of high-grade glioma. J Neurooncol 77(2):143–152

Ellert-Miklaszewska A, Kaminska B, Konarska L (2005) Cannabinoids down-regulate PI3K/Akt and Erk signalling pathways and activate proapoptotic function of Bad protein. Cell Signal 17:25–37

Ellert-Miklaszewska A, Grajkowska W, Gabrusiewicz K, Kaminska B, Konarska L (2007) Distinctive pattern of cannabinoid receptor type II (CB2) expression in adult and pediatric brain tumors. Brain Res 1137:161–169

Galve-Roperh I, Sanchez C, Cortes ML, del Pulgar TG, Izquierdo M, Guzman M (2000) Anti-tumoral action of cannabinoids: involvement of sustained ceramide accumulation and extracellular signal-regulated kinase activation. Nat Med 6:313–319

Gomez del Pulgar T, Velasco G, Sanchez C, Haro A, Guzman M (2002) De novo-synthesized ceramide is involved in cannabinoid-induced apoptosis. Biochem J 363:183–188

Gomez O, Arevalo-Martin A, Garcia-Ovejero D, Ortega-Gutierrez S, Cisneros JA, Almazan G, Sanchez-Rodriguez MA, Molina-Holgado F, Molina-Holgado E (2010) The constitutive production of the endocannabinoid 2-arachidonoylglycerol participates in oligodendrocyte differentiation. Glia 58:1913–1927

Gong JP, Onaivi ES, Ishiguro H, Liu QR, Tagliaferro PA, Brusco A, Uhl GR (2006) Cannabinoid CB2 receptors: immunohistochemical localization in rat brain. Brain Res 1071:10–23

Guzman M (2003) Cannabinoids: potential anticancer agents. Nat Rev Cancer 3:745–755

Guzman M, Sanchez C, Galve-Roperh I (2001) Control of the cell survival/death decision by cannabinoids. J Mol Med 78:613–625

Howlett AC, Barth F, Bonner TI, Cabral G, Casellas P, Devane WA, Felder CC, Herkenham M, Mackie K, Martin BR, Mechoulam R, Pertwee RG (2002) International Union of Pharmacology. XXVII. Classification of cannabinoid receptors. Pharmacol Rev 54:161–202

Kapoor GS, O'Rourke DM (2003) Receptor tyrosine kinase signaling in gliomagenesis: pathobiology and therapeutic approaches. Cancer Biol Ther 2:330–342

Klionsky DJ, Emr SD (2000) Autophagy as a regulated pathway of cellular degradation. Science 290:1717–1721
Mackie K (2006) Cannabinoid receptors as therapeutic targets. Annu Rev Pharmacol Toxicol 46:101–122
Mackie K, Stella N (2006) Cannabinoid receptors and endocannabinoids: evidence for new players. AAPS J 8:E298–306
Massi P, Vaccani A, Ceruti S, Colombo A, Abbracchio MP, Parolaro D (2004) Antitumor effects of cannabidiol, a nonpsychoactive cannabinoid, on human glioma cell lines. J Pharmacol Exp Ther 308:838–845
McAllister SD, Chan C, Taft RJ, Luu T, Abood ME, Moore DH, Aldape K, Yount G (2005) Cannabinoids selectively inhibit proliferation and induce death of cultured human glioblastoma multiforme cells. J Neurooncol 74:31–40
Molina-Holgado E, Vela JM, Arevalo-Martin A, Almazan G, Molina-Holgado F, Borrell J, Guaza C (2002) Cannabinoids promote oligodendrocyte progenitor survival: involvement of cannabinoid receptors and phosphatidylinositol-3 kinase/Akt signaling. J Neurosci 22:9742–9753
Pazos MR, Nunez E, Benito C, Tolon RM, Romero J (2005) Functional neuroanatomy of the endocannabinoid system. Pharmacol Biochem Behav 81:239–247
Petersen G, Moesgaard B, Schmid PC, Schmid HH, Broholm H, Kosteljanetz M, Hansen HS (2005) Endocannabinoid metabolism in human glioblastomas and meningiomas compared to human non-tumour brain tissue. J Neurochem 93:299–309
Salazar M, Carracedo A, Salanueva IJ, Hernandez-Tiedra S, Lorente M, Egia A, Vazquez P, Blazquez C, Torres S, Garcia S, Nowak J, Fimia GM, Piacentini M, Cecconi F, Pandolfi PP, Gonzalez-Feria L, Iovanna JL, Guzman M, Boya P, Velasco G (2009) Cannabinoid action induces autophagy-mediated cell death through stimulation of ER stress in human glioma cells. J Clin Invest 119:1359–1372
Sanchez C, Velasco G, Guzman M (1997) Delta9-tetrahydrocannabinol stimulates glucose utilization in C6 glioma cells. Brain Res 767:64–71
Sanchez C, Galve-Roperh I, Canova C, Brachet P, Guzman M (1998) Delta9-tetrahydrocannabinol induces apoptosis in C6 glioma cells. FEBS Lett 436:6–10
Sanchez C, de Ceballos ML, del Pulgar TG, Rueda D, Corbacho C, Velasco G, Galve-Roperh I, Huffman JW, Ramon Y, Cajal S, Guzman M (2001) Inhibition of glioma growth in vivo by selective activation of the CB(2) cannabinoid receptor. Cancer Res 61:5784–5789
Sarfaraz S, Afaq F, Adhami VM, Mukhtar H (2005) Cannabinoid receptor as a novel target for the treatment of prostate cancer. Cancer Res 65:1635–1641
Schley M, Stander S, Kerner J, Vajkoczy P, Schupfer G, Dusch M, Schmelz M, Konrad C (2009) Predominant CB2 receptor expression in endothelial cells of glioblastoma in humans. Brain Res Bull 79:333–337
Schroder M, Kaufman RJ (2005) The mammalian unfolded protein response. Annu Rev Biochem 74:739–789
Stella N (2004) Cannabinoid signaling in glial cells. Glia 48:267–277
Valenzano KJ, Tafesse L, Lee G, Harrison JE, Boulet JM, Gottshall SL, Mark L, Pearson MS, Miller W, Shan S, Rabadi L, Rotshteyn Y, Chaffer SM, Turchin PI, Elsemore DA, Toth M, Koetzner L, Whiteside GT (2005) Pharmacological and pharmacokinetic characterization of the cannabinoid receptor 2 agonist, GW405833, utilizing rodent models of acute and chronic pain, anxiety, ataxia and catalepsy. Neuropharmacology 48:658–672
Van Sickle MD, Duncan M, Kingsley PJ, Mouihate A, Urbani P, Mackie K, Stella N, Makriyannis A, Piomelli D, Davison JS, Marnett LJ, Di Marzo V, Pittman QJ, Patel KD, Sharkey KA (2005) Identification and functional characterization of brainstem cannabinoid CB2 receptors. Science 310:329–332
Velasco G, Carracedo A, Blazquez C, Lorente M, Aguado T, Haro A, Sanchez C, Galve-Roperh I, Guzman M (2007) Cannabinoids and gliomas. Mol Neurobiol 36:60–67
Zha J, Harada H, Yang E, Jockel J, Korsmeyer SJ (1996) Serine phosphorylation of death agonist BAD in response to survival factor results in binding to 14-3-3 not BCL-X(L). Cell 87:619–628

Index

DOI 10.1007/978-94-007-4719-7,

FSC
www.fsc.org
MIX
Papier | Fördert
gute Waldnutzung
FSC® C083411

Kolibri 360 GmbH
Ferdinand-Jühlke-Straße 7
99095 Erfurt, Deutschland
produktsicherheit@kolibri360.de